T0144408

Introduction to Bioinformatics and Clinical Scientific Computing

This textbook provides an introduction to computer science theory, informatics best practice and the standards and legislation that apply to computing in a healthcare environment.

It delivers an accessible discussion of databases (construction, interrogation and maintenance); networking (design and low-level application); programming (best practice rather than the specifics of any one language – design, maintenance and safety).

It can be used to accompany the NHS Modernising Scientific Careers syllabus. It is also targeted towards those creating software rather than those using it, particularly computer scientists working in healthcare, specifically those in or close to the Physical Sciences, including radiotherapy, nuclear medicine, and equipment management and those working with genomics and health informatics.

There is an accompanying website (https://www.routledge.com/9781032324135) from which one can download some of the code referred to in the text, as well as colour versions of some of the images in colour.

Paul S. Ganney has had a long career in the world of Medical Physics, always with Scientific Computing at its heart. He has been Head of Scientific Computing at two NHS Trusts (Hull and East Yorkshire and University College London) and has lectured at Liverpool University, Hull University and Kings College London. Now mostly retired, he still writes articles and software as required.

Introduction to Bioinformatics and Clinical Scientific Computing

Paul S. Ganney

CRC Press
Taylor & Francis Group
Boca Raton London New York

CRC Press is an imprint of the
Taylor & Francis Group, an **informa** business

Cover image: spainter_vfx/Shutterstock

First edition published 2023
by CRC Press
6000 Broken Sound Parkway NW, Suite 300, Boca Raton, FL 33487-2742

and by CRC Press
4 Park Square, Milton Park, Abingdon, Oxon, OX14 4RN

ISBN: 978-1-032-32413-5 (hbk)
ISBN: 978-1-032-32693-1 (pbk)
ISBN: 978-1-003-31624-4 (ebk)

DOI: 10.1201/9781003316244

Typeset in Times
by SPi Technologies India Pvt Ltd (Straive)

Access the companion website: https://www.routledge.com/9781032324135

Contents

Acknowledgements

This book which you are holding has taken a lot of time to write. There is no way I'd have been able to accomplish any of this without a lot of help and encouragement.

The first acknowledgement must therefore be to Rachel Ganney who foolishly offered to turn a set of lecture notes into a book (which we self-published) and then even more foolishly offered to do a second (longer) set, which are pretty much the ones you have before you. Without her eye for detail, this book would still be on the "good idea – maybe one day" pile. She also drew some of the figures, re-wrote some of the text and generally put up with me when it wasn't going well. I'm sure there ought to be a paragraph in the marriage vows about this sort of thing.

I must also thank those with whom I have co-authored articles and book chapters, which I have adapted for inclusion here. Specifically, Sandhya Pisharody, Phil Cosgriff, Allan Green, Richard Trouncer, David Willis, Patrick Maw and Mark White. James Moggridge deserves specific thanks for inspiring and proof-reading the "NoSQL" chapter. I should also thank those who invited me to write for or with them, which encouraged me greatly. Usman Lula and Azzam Taktak join Richard from the previous list. Finally, I should thank my students, who have asked difficult questions (which made me work out the answers, some of which you have here) and laughed at the jokes. The ones they didn't laugh at have been retired.

1 Data Structures

1.1 INTRODUCTION

At some point in the construction and development of programs, the question as to what to do with the data[1] is going to arise. Issues such as the order in which data is required, whether it will change, whether new items will appear and old ones will need to be removed will determine the shape of the data and as such, the structure in which it is best stored.

A data structure defines how data is to be stored and how the individual items of data are to be related or located, if at all, in a way that is convenient for the problem at hand and easy to use. To return to an oft-quoted[2] equation:

$$Information = Data + Structure$$

There are many ways of arranging data, some of which we will examine here. The most important part is to select the most appropriate model for the data's use: e.g. a dictionary arranged in order of the number of letters in each word isn't as useful as one with the words in alphabetic order, even though it is still a valid sort (and, indeed, this is how a Scrabble dictionary is arranged).

1.2 ARRAYS

An array is the simplest form of data structure and exists as a part of most (if not all) high-level programming languages. An array is simply a collection of similar data items, one after another, referred to by the same identifier. Different items within the array are referenced by means of a number, called the index, which is the item's position in the array. There is no relationship between the value of the data item and its position in the array, unless one is imposed by the programmer.

A 1-dimensional array is often referred to as a vector and can be thought of as a simple list, such as in Figure 1.1.

An array's logical structure (which item follows which) mirrors its physical structure, which aids its simplicity but does reduce its flexibility, as we shall see.

Most high-level languages start array indices at 0, which also points towards the memory storage model: a 1D integer array will reserve 4 bytes per integer; thus this example has reserved 40 bytes. Element 7 is thus stored at the base address of the array plus 7 * 4 = 28 bytes. For this reason, languages that don't check for programmers exceeding array bounds (e.g. C) will still produce working programs, they'll just overwrite (or read from) memory that they weren't supposed to, with unpredictable results.

DOI: 10.1201/9781003316244-1

Index	Value
0	7
1	57
2	654
3	2
4	44
5	68
6	15
7	222
8	62
9	44

FIGURE 1.1   A 1D array, or vector.

FIGURE 1.2   A tiny picture of a cat in greyscale, from an original in colour.

An obvious extension to the 1D array is a multiply-dimensioned one. This has a similar structure to the vector above, but with additional columns, forming a table. This enables data to be kept together, e.g. a set of prices and discounts, or high and low values for physiological measurements. Another common usage is in holding pixel values for an image. For a greyscale image, this would be a 2D array as shown in Figure 1.3, from Figure 1.2, and for a colour image a 3D array is required as shown in Figure 1.4 (from the original colour image shown in Figure 1.2).

Taking our earlier example from Figure 1.1, we can extend the vector into an array as shown in Figure 1.5.

Note that in this example we have stored different data types in the columns. Most programming languages won't allow this, so the data will normally have to be converted (e.g. to strings) which in this case (a set of addresses) would be appropriate anyway. Our simple storage calculation no longer holds though, as the strings are of variable length. This is usually overcome by storing pointers to the data in the array, with the actual data being stored elsewhere.

2	2	1	1	1	2	2	2	2	2	2	2	2	2	2	2	2	2	2	2	2	2	2	2	2	2	2	1	7	7	1	2
5	5	9	8	8	2	5	5	5	5	5	5	5	5	5	5	5	5	5	5	5	5	5	5	5	5	4	3	3	3	8	5
5	2	4	3	5	6	5	5	5	5	5	5	5	5	5	5	5	5	5	5	5	5	5	5	5	5	4	1			8	5

2	2	1	1	1	1	2	2	2	2	2	2	2	2	2	2	2	2	2	2	2	2	2	2	2	2	1	7	7	7	1	2
5	3	8	8	8	8	2	5	5	5	5	5	5	5	5	5	5	5	5	5	5	5	5	5	5	5	2	2	2	2	2	5
5	9	3	3	3	4	8	4	5	5	5	5	5	5	5	5	5	5	5	5	5	5	5	5	5	5	4	9			3	5

2	2	1	1	1	1	1	2	2	2	2	2	2	2	2	2	2	2	2	2	2	2	2	2	1	7	7	1	7	8	2
5	2	8	8	9	8	8	4	5	5	5	5	5	5	5	5	5	5	5	5	5	5	5	5	6	2	9	3	7	1	4
5	2	3	7	6	3	5	0	5	5	5	5	5	5	5	5	5	5	5	5	5	5	5	5	7			8			0

FIGURE 1.3 Rows 1–3 of the greyscale array, the tips of the ears.

Note that 0=black and 255=white.

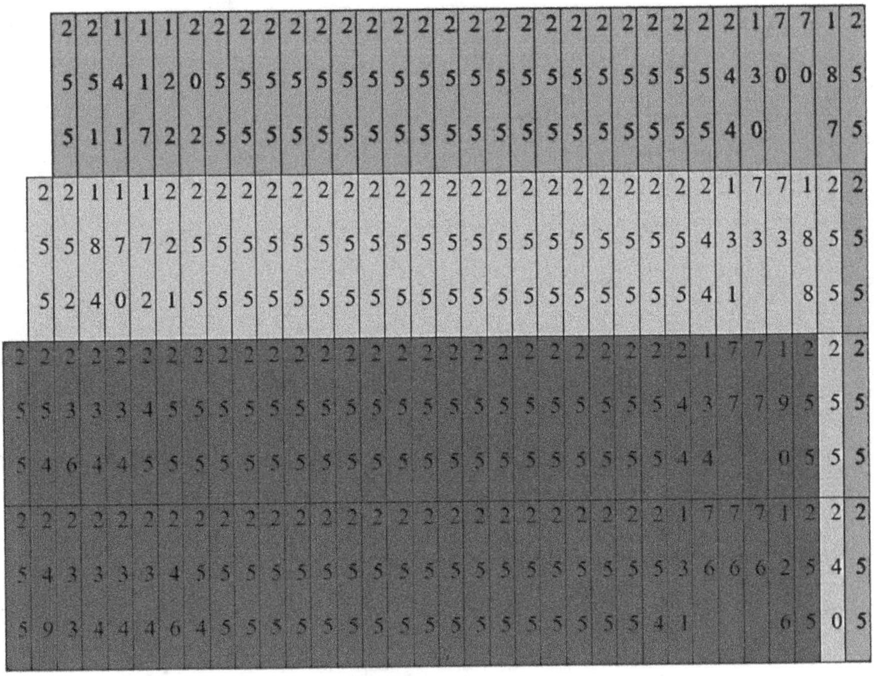

FIGURE 1.4 Rows 1–3 of the colour picture (the tips of the ears), as R, G, B, showing how a 3D array is now required. The bottom two rows with a dark grey background are red values, the light grey background shows the green values and the mid-grey background (top row and right-hand side) show the blue values.

Note that 0 = no colour and 255 = maximum colour.

Index	Value 1	Value 2	Value 3	Value 4	Value 5
0	7	Straight St	Hull	01482	541684
1	57	Long Ave	Hull	01482	564815
2	654	Long Ave	Bridlington	01262	541684
3	2	Round Close	Hull	01482	525150
4	44	Hamster Rd	Hull	01482	554382
5	68	Hamster Rd	Hull	01482	551943
6	15	Straight St	Hull	01482	541865
7	222	Long Ave	Bridlington	01262	564815
8	62	Hamster Rd	Hull	01482	554466
9	44	Straight St	Hull	01482	541745

FIGURE 1.5   A 2D array.

So far we have data but little information, as the structure we have only really allows data to be "thrown in" as it appears. Extracting data from such a structure is a simple task but not an efficient one. Therefore arrays like this are usually kept in order, which works well when the data changes little (if at all).[3] However, adding and inserting data into such a structure necessitates a lot of data moving, in order to create new rows (and hence move everything below it down) or remove old ones (moving all data "up one").

In order to extract information from our example array, we could sort it each time; e.g. to find all the addresses in Bridlington, we might sort on column "Value 3", giving rise to Figure 1.6.

Index	Value 1	Value 2	Value 3	Value 4	Value 5
0	654	Long Ave	Bridlington	01262	541684
1	222	Long Ave	Bridlington	01262	564815
2	7	Straight St	Hull	01482	541684
3	57	Long Ave	Hull	01482	564815
4	2	Round Close	Hull	01482	525150
5	44	Hamster Rd	Hull	01482	554382
6	68	Hamster Rd	Hull	01482	551943
7	15	Straight St	Hull	01482	541865
8	62	Hamster Rd	Hull	01482	554466
9	44	Straight St	Hull	01482	541745

FIGURE 1.6   The previous array sorted on the column "Value 3".

Index	Value 1	Value 2	Value 3	Value 4	Value 5	I1:V3
0	7	Straight St	Hull	01482	541684	2
1	57	Long Ave	Hull	01482	564815	3
2	654	Long Ave	Bridlington	01262	541684	0
3	2	Round Close	Hull	01482	525150	4
4	44	Hamster Rd	Hull	01482	554382	5
5	68	Hamster Rd	Hull	01482	551943	6
6	15	Straight St	Hull	01482	541865	7
7	222	Long Ave	Bridlington	01262	564815	1
8	62	Hamster Rd	Hull	01482	554466	8
9	44	Straight St	Hull	01482	541745	9

FIGURE 1.7 The previous array with an additional index on the "Value 3" column.

However, it is clear that for large amounts of information, this is hardly practical. Instead, we construct an index, which is an additional column[4] and gives rise to the array in Figure 1.7.

Note that there are no "draws" – equal values are still placed in order, in this case, by original order within the table. It is intuitively obvious that this technique is faster than sorting: in order to re-sort the array we need only re-write one column rather

Index	Value 1	Value 2	Value 3	Value 4	Value 5	I1:V3	I2:V5
0	7	Straight St	Hull	01482	541684	2	1
1	57	Long Ave	Hull	01482	564815	3	8
2	654	Long Ave	Bridlington	01262	541684	0	2
3	2	Round Close	Hull	01482	525150	4	0
4	44	Hamster Rd	Hull	01482	554382	5	6
5	68	Hamster Rd	Hull	01482	551943	6	5
6	15	Straight St	Hull	01482	541865	7	4
7	222	Long Ave	Bridlington	01262	564815	1	9
8	62	Hamster Rd	Hull	01482	554466	8	7
9	44	Straight St	Hull	01482	541745	9	3

FIGURE 1.8 The previous array with an additional index column on "Value 5".

than all columns. We can thus extend this idea to multiple indexes on this table, such as that in Figure 1.8.

In the worst case, adding a record becomes a matter of writing the new information and re-writing two columns, as opposed to writing the new information and then re-writing the whole table. For example, to add this row,

10	77	Long Ave	Bridlington	01482	561745	2	8

the index rows are re-written as shown in Figure 1.9.

I1:V3	I2:V5
~~2~~ 3	1
~~3~~ 4	~~8~~ 9
0	2
4 5	0
~~5~~ 6	6
~~6~~ 7	5
~~7~~ 8	4
1	~~9~~ 10
~~8~~ 9	7
~~9~~ 10	3

FIGURE 1.9   The alterations required to the index columns.

1.3   STACK OR HEAP

This data structure is a **"last in first out"** (LIFO) one (hence the name "stack" or "heap", which are synonymous in this context). It is useful for storing intermediate results, such as in maze solving. At each decision, one path is pursued and the other placed upon the stack for later analysis, should the current path prove unsuccessful.

For example, consider the maze in Figure 1.10.

We first mark the points at which a decision must be made, as shown in Figure 1.11.

At each decision point, we select a direction (according to a heuristic[5]) and place the other option(s) onto the stack. Should we reach a dead end, then we need to retrace. We do this by taking the top element from the stack and then follow this. We repeat until the maze is either solved, or all options are exhausted. For example, if we use a heuristic of "turn right first" (as all junctions only have two options, "left" and "right" are easy to determine, regardless of which way they actually lie) the solution is as shown in Figure 1.12.

Programmatically, a stack is easily implemented using a 1D array and a single pointer, to indicate where the top of the stack is. For example, at step 16, the stack would be as shown in Figure 1.13. Note that old values are not removed, only over-written, as memory always has a value so there is no advantage to writing an "empty" marker. This is why the values at indices 2 and 3 are not null even though they are not relevant to step 16, this doesn't matter because it is only ever the value at the index

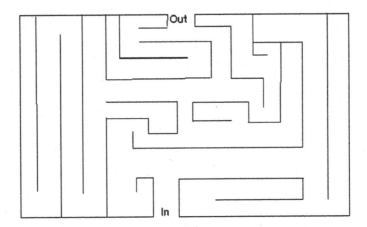

FIGURE 1.10   A simple maze.

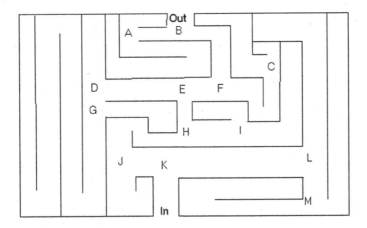

FIGURE 1.11   The maze with the decision points marked and labelled.

pointed to by the free space pointer (the top of the stack being one less than this) that is considered by the program.

A major use of the stack or heap in computing terms is in keeping track of variables that are in scope – especially useful when a local variable has the same name as a global one. Thus new variables are added to the stack at declaration (e.g. entering a new subroutine) and when referenced the topmost (i.e. first found) reference to a variable of this name is used – the local one. When the subroutine ends and the variable goes out of scope, the variable is removed from the stack (by adjusting the pointer) so that the topmost one with this name is now the global one and that is the one used.

Step	Action	Stack
1	Turn right @ K	left@K
2	Turn right @ L	left@K left@L
3	Turn right @ M	left@K left@L left@M
4	Dead end – retrace. Take top element from stack and follow this (turn left @ M)	left@K left@L
5	Dead end – retrace. Take top element from stack and follow this (turn left @ L)	left@K
6	Dead end – retrace. Take top element from stack and follow this (turn left @ K)	
7	Turn right @ J	left@J
8	Turn right @ H	left@J left@H
9	Turn right @ I	left@J left@H left@I

Step	Action	Stack
10	Dead end – retrace. Turn left @ I	left@J left@H
11	Dead end – retrace. Turn left @ H	left@J
12	Turn right @ E	left@J left@E
13	Turn right @ F	left@J left@E left@F
14	Turn right @ C	left@J left@E left@F left@C
15	Dead end – retrace. Turn left @ C	left@J left@E left@F
16	Dead end – retrace. Turn left @ F	left@J left@E
17	Turn right @ B	left@J left@E left@B
18	Success	

FIGURE 1.12 The solution to the maze using a "turn right first" heuristic.

Index	Value
5	
4	
3	left@C
2	left@F
1	left@E
0	left@J

FIGURE 1.13 The stack at step 16. The pointer value (to the next free space) is 2.

Memory is often depicted with the index 0 at the top of the page but is rendered here at the bottom in order to reinforce the concept of this data structure.

1.4 QUEUE

A queue is a similar structure to a stack, yet operates a **"first in, first out"** (FIFO) methodology. Like a stack, it can be easily implemented using an array but requires two pointers in order to indicate the start and the end of the queue. For example, after inserting 5 values into a queue, it may look like the one shown in Figure 1.14.

Index	Value
0	A
1	B
2	F
3	D
4	R
5	
6	
7	

Start pointer: 0

End pointer: 4

FIGURE 1.14 A queue implemented as an array plus two pointers.

Memory convention restored (see Figure 1.13).

If three values are removed from this queue, the queue would read as shown in Figure 1.15.

Note that the queue storage is circular, so the fourth addition appears at index 0. As with the stack, values are not overwritten until the storage is re-used. Both stacks

Index	Value
0	A
1	B
2	F
3	D
4	R
5	
6	
7	

Start pointer: 3

End pointer: 4

FIGURE 1.15 The queue with three removals; and if four more are added after this, the queue would read as shown in Figure 1.16.

Index	Value
0	R
1	B
2	F
3	D
4	R
5	Q
6	W
7	E

Start pointer: 3

End pointer: 0

FIGURE 1.16 The queue with four additions.

and queues are used internally within a computer – stacks record what an interrupt-driven operating system was doing before it was interrupted (so it can go back in order) and queues record the messages awaiting an event-driven operating system.

1.5   LINKED LIST

There are two ways to solve the problem posed by arrays (that of needing to keep re-writing memory every time something changes). We solved it first by using an index, but an alternative method (and for a third, see the next section) is to use a linked list.

A linked list is what it sounds like: a list of items linked together, as shown in Figure 1.17.

A linked list is an abstract data structure in which it is extremely unlikely that the physical and logical structures will ever be the same (the major exception being when the list is empty).

A linked list introduces us to the idea of a data element: in the simplest case, this is a pair of values – a data value and a pointer value, as above. The pointer value holds the index of the next element in the list. We also require three pointers: one to the first element in the list (the start pointer), another to the next free storage space (the free space pointer) and one to nowhere, used as a terminator. Conventionally, a value of –1 indicates the end of the list. An empty list is indicated by a terminator in the start pointer, and a full list by a terminator in the free space pointer.

The greatest strength of a linked list is its flexibility: insertion and deletion are very quick and easy to perform. Consequently, they are often used for applications where data changes often or where there are huge null elements making a conventional array inefficient (i.e. a sparse array).

In order to implement a linked list, we therefore require two arrays, a start pointer and a "next free element" pointer. For example, after inserting the values Q, W, E, R, T, Y into a linked list (using alphabetic order for ordering), a linked list implemented this way would appear as shown in Figure 1.18.

Finding a value in a linked list is a matter of tracing through the list by following pointers, until either the required value is found, a higher value is found (indicating "not found") or the terminator is reached. In the example above, we would begin at 2 (E) (as indicated by the start pointer), progress to 0 (Q), then to 3 (R), then to 4 (T), 1 (W) and finally to 5 (Y), performing comparisons as we go until the outcome of the search is known.

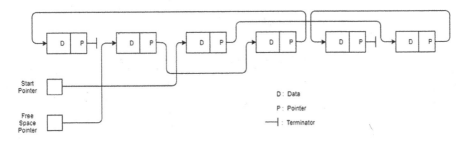

FIGURE 1.17   A linked list.

Index	Value	Pointer
0	Q	3
1	W	5
2	E	0
3	R	4
4	T	1
5	Y	-1
6		
7		
8		

Start pointer: 2

Free element: 6

FIGURE 1.18 A linked list implementation.

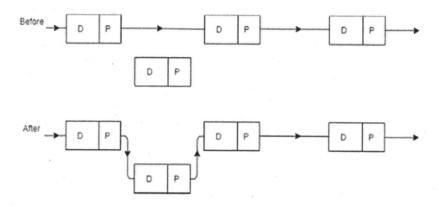

FIGURE 1.19 Linked list insertion.

Insertion and deletion are simply a matter of adjusting pointers. To insert a new element, we need to find its place in the list. This is done by tracing through the list until an element with a higher value is found, or we reach the end of the list. We then copy this index into our new element and re-direct the link to point to the new element. This can be illustrated graphically, as shown in Figure 1.19.

Or, if we wish to add the element "U" to our example, we have the list as shown in Figure 1.20.

Index	Value	Pointer
0	Q	3
1	W	5
2	E	0
3	R	4
4	T	6 ←—
5	Y	-1
6	U ←—	1 ←—
7		
8		

Start pointer: 2

Free element: 7 ←—

FIGURE 1.20   The linked list from our example with the element "U" inserted. The items indicated by an arrow have been altered.

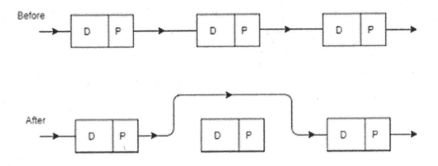

FIGURE 1.21   Linked list deletion.

This is clearly a very efficient method of data storage, in that very few write operations are required to add a new element in the correct place.

To delete an element from the list, we simply route around it by re-writing one link, as is shown graphically in Figure 1.21.

If there were a lot of additions and deletions, this would give a lot of wasted space, so the elements are re-used. This is achieved by holding the unused elements as a second linked list. So our list above should have read as shown in Figure 1.22.

Thus, deleting an element simply requires us to re-write two links and the free element pointer. For example, to remove "R" from this list the pointer values in positions 0 and 3 are re-written, along with the free element pointer, as Figure 1.23 shows.

Index	Value	Pointer
0	Q	3
1	W	5
2	E	0
3	R	4
4	T	6
5	Y	-1
6	U	1
7		8
8		-1

Start pointer: 2

Free element: 7

FIGURE 1.22 The linked list from our example with a "free element" list added.

Index	Value	Pointer
0	Q	4
1	W	5
2	E	0
3	R	7
4	T	6
5	Y	-1
6	U	1
7		8
8		-1

Start pointer: 2

Free element: 3

FIGURE 1.23 The linked list with the element "R" removed.

In our addresses example earlier (Figure 1.6–Figure 1.8), we needed two indices as we required two different ordered views of the data. With linked lists this is achieved by multiple pointers for each data element thus giving us a multiply linked list. Equally, the data could have several parts to it. We could therefore have the

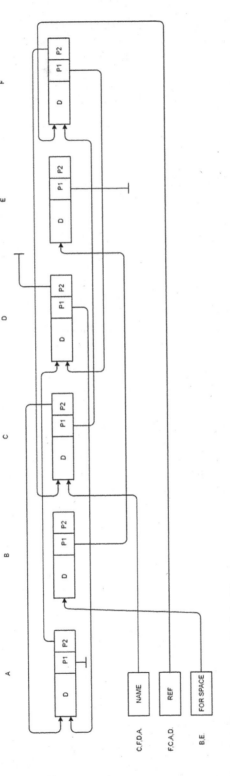

FIGURE 1.24 A doubly linked list.

scenario where each data part is kept in order via a pointer link. An example of a list ordered by both name and reference is shown in Figure 1.24.

A common form of doubly linked list contains pointers both forwards and backwards, which aids searching.

1.6   BINARY TREE

Another linked data structure is the binary tree. The main advantage of the binary tree is that the data structure intrinsically reflects the ordering of data stored within it. The binary tree has two pointers per element (or node), known respectively as the left and right pointers. The basic principle is that there exists a central node, called the root, which is (of course) at the top of the binary tree.

From the root, the binary tree spreads in two directions, left and right. Those nodes to the left of the root are ordered before it and those to the right are ordered after it. Moving down the tree to another node presents the same picture of two pointers and the same ordering rules. It is this feature that makes binary trees so useful: at any node (position in the tree) the structure below this point is also a binary tree (a sub-tree). A tree with no sub-trees is a terminal node and the absence of a sub-tree is recorded by the use of a terminator value in the pointer, in the same way as for linked lists.

For example, to insert the number 57 into the numeric tree shown in Figure 1.25, the process is as follows:

1. 57 is greater than 32, so branch right.
2. 57 is less than 85 so branch left.
3. 57 is less than 62 so branch left.
4. A terminator is reached, so place the data item 57 there, to the left of 62.

Deleting an item is slightly more complex, in that the pointer to the deleted node must be made to point to the left sub-tree of the deleted node. The right sub-tree of the deleted node becomes the rightmost element of the left sub-tree of the deleted

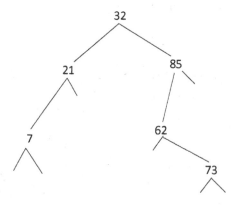

FIGURE 1.25   A binary tree holding numbers.

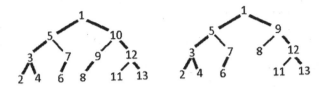

FIGURE 1.26   Deletion from a binary tree.

node. This is best illustrated graphically, as shown in Figure 1.26 where item 10 is deleted.

Listing the data in a binary tree (often called traversing the binary tree) is a simple recursive algorithm:

1. Traverse left sub-tree, if there is one.
2. Print node.
3. Traverse right sub-tree, if there is one.

This algorithm therefore requires a stack or heap to implement, as it will be last in first out, and is thus similar to the maze tracing done earlier.

Searching for an item in a binary tree is simply an application of the insertion principles, testing as we go for the required item, ceasing either when the item is found or a terminator is reached.

Binary trees are mainly used where data is inserted but not removed (as removal isn't simple, despite the above) and are often used for taxonomy or classification data, as well as expression parsers and solvers. Database indices are often stored as binary trees.

A key requirement for an efficient tree is that is it well balanced, for which the selection of an appropriate root node is critical. Algorithms therefore exist for re-balancing trees.

As with linked lists, n-way trees are also possible, simply by adding more pointers. These are often used for maintaining keys for an index of items

NOTES

1   Where and how to store it, not whether to bother keeping any.
2   By the author, at least.
3   As we noted earlier, the physical and logical structures are the same.
4   And it is not to be confused with the Index column, which denotes the physical position in the array. These indices indicate more logical positioning.
5   A heuristic is often called a "rule of thumb" and is a simple way of reaching a decision (as opposed to an algorithm). It has been defined as "a way of directing your attention fruitfully". The root is the Greek for "find and discover".

2 Databases[1]

2.1 INTRODUCTION

Whilst data can be stored in many ways and in many formats, in order to be useful, the storage must contain structure. It has been noted earlier (Section 1.1) that "Data + Structure = Information" and therefore we need structure to turn data into information.

The simplest data structure is the list: a single column of items. Arranging this list alphabetically, for example, brings further structure to the data and enables it to be used for rapid look-ups, using simple algorithms such as binary chop.[2]

It is a simple step to expand this list structure into multiple columns, allowing the overall structure to then be sorted by any one of the columns in order to yield different information. A basic spreadsheet[3] forms one such database. This information can equally be represented as a set of text with one line per row in the spreadsheet. This however introduces two new problems: how to separate the data in each row; and how to uniquely identify each row. Both of these are addressed by the spreadsheet structure, but not by the extracted data, or "flat file".

A flat file has no internal hierarchy, containing records that have no structured interrelationship. A typical flat file is a text file, containing no (or having had removed) word processing or other structure characters or markup (such as paragraph mark, tabs etc.). For example:

> Snail Brian Magic Roundabout brian@roundabout.com Flowerpot Bill Garden bill@weed.co.uk Cat Bagpuss Shop Window bagpuss@catworld.net Miller Windy The Windmill, Trumpton wmiller@chigley.ac.uk

A flat file can be given some structure by adding three things: a line separator; a unique identifier on each line; and a unique separator between each item in the lines.[4] Taking the third solution first, the most common separator is the comma (hence "comma separated values", or CSV, a common file format used for data exchange[5]); although, this can produce problems of its own when recording data containing commas, e.g. addresses. There are two common solutions to this: encapsulating the data within double inverted commas; and using an "escape character" (see the next paragraph). The first of these brings problems where double inverted commas are also present (e.g. addresses). The second, although more robust, requires slightly more processing.

DOI: 10.1201/9781003316244-2

An "escape character" is a character that changes the interpretation of the one that follows it. A common escape character is "\". Hence, a comma that separates data would stand alone: , whereas a comma that should be taken to be part of the data would be preceded: \, . This notation does not suffer from the others mentioned above, as a backslash simply becomes \\.

Returning to the problem of adding structure to a flat file we examine the unique identifier. A unique identifier may not need to be added, of course, if a part of the data is already unique. Such an item (whether added or already present) is called the "primary key" and its importance will become clearer as more structure and abstraction are added to our data. A spreadsheet already contains this identifier, in that each item of data can be uniquely referenced using its row and column identifiers.

This leads us onto the relational model of data, first proposed by Codd in 1969 and subsequently revised. The concept was introduced in the paper "A Relational Model of Data for Large Shared Data Banks" (Codd 1970). Modern relational databases simply implement a model that approximates the mathematical model defined by Codd. The relational model

> views information in a database as a collection of distinctly named tables. Each table has a specified set of named columns, each column name (also called an attribute) being distinct within a particular table, but not necessarily between tables. The entries within a particular column of a table must be atomic (that is, single data items) and all of the same type.
>
> (Pyle and Illingworth, 1996)

2.2　TERMINOLOGY

Relational database: A data structure through which data is stored in tables that are related to one another in some way. The way the tables are related is described through a relationship (see later definitions).

Data: Values stored in a database.

Entity: A person, place, thing, or event about which we want to record information; an object we're interested in.

Field: The smallest structure in a relational database, used to store the individual pieces of data about the object; represents an attribute.

Record: A single "row" in a table; the collection of information for a single occurrence of the entity that the table represents.

Table: The chief structure in a relational database, composed of fields and records, whose physical order is unimportant.[6] A single table collects together all of the information we are tracking for a single entity and represents that entity.

Figures 2.1 and 2.2 demonstrate some of this terminology.

Surname	Forename	Address	e-mail
Snail	Brian	Magic Roundabout	brian@roundabout.com
Flowerpot	Bill	Garden	bill@weed.co.uk
Cat	Bagpuss	Shop Window	bagpuss@catworld.net
Miller	Windy	The Windmill, Trumpton	wmiller@chigley.ac.uk

FIGURE 2.1   A set of data arranged in a table. The items in the first row are field names (but may also be thought of as column headings).

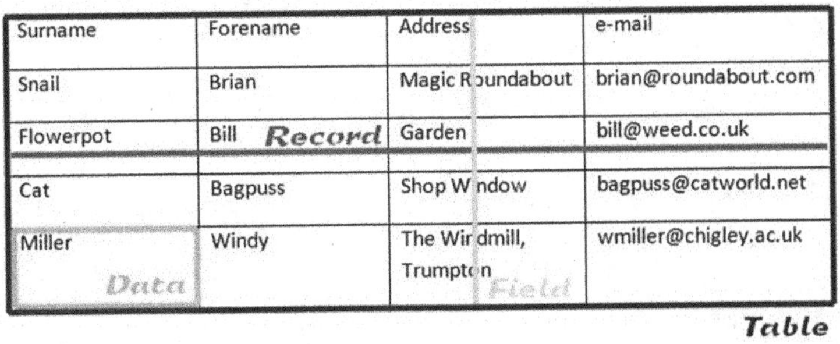

Surname	Forename	Address	e-mail
Snail	Brian	Magic Roundabout	brian@roundabout.com
Flowerpot	Bill *Record*	Garden	bill@weed.co.uk
Cat	Bagpuss	Shop Window	bagpuss@catworld.net
Miller *Data*	Windy	The Windmill, Trumpton *Field*	wmiller@chigley.ac.uk

Table

FIGURE 2.2   The table from Figure 2.1 labelled to show the terminology described above. An entity in this case is equivalent to a record.

2.3   THE GOALS OF DATABASE DESIGN

The goals are to:

- understand your data and why you're tracking it
- eliminate duplication of data
- eliminate redundant data
- eliminate meaningless data or data that is irrelevant
- promote accuracy of data
- promote consistency of data
- make sure we can retrieve the information we need from the database
- support the business functions that use the database
- build a database that lends itself to future growth

TABLE 2.1
The Field Names for the Single-Table Database

Media	CD/DVD/LP/MD/Tape/MP3 etc.
Artist	
Title	
Track1	
Track2	
(etc.)	
Price	In pence
Copies	
Total value	In pence

TABLE 2.2
A Sample Row

CD	Example	Playing in the Shadows	Skies Don't Lie	Stay Awake	Changed the Way You Kiss Me	999	2	1998

2.4 EXAMPLE

Let us now examine a specific example. Sam and Dakota decide to merge their music collections. Being computer scientists, they decide to create a database to catalogue the full collection. Their first attempt is a single-table database, such as described above and shown in Table 2.1. A sample row (with just 3 tracks for simplicity) is shown in Table 2.2.

It is clear that this design is inefficient and an inefficient design is difficult to maintain.

2.5 MORE TERMINOLOGY

Entity-Relationship Diagram (ERD): Identifies the data/information required by displaying the relevant entities and the relationships between them (see example in Figure 15.12).

Key: a field in the database (or an attribute in an ERD) that is used to uniquely identify records and establish relationships between tables or entities; used for the retrieval of data in the table.

Primary Key: uniquely identifies each record in a table, the primary key is part of the table for which it operates. (Note that this is normally a single field but may be a combination of fields – a **composite key**).

Foreign Key: A key from another table that is used to define a relationship to another record in another table. It has the same name and properties as the primary key from which it is copied.

Rules for foreign keys:
  1-1: The primary key from the main table is inserted into the second table
  1-Many: The primary key from the "1" table gets inserted into the "many" table
  Many-many: The primary key from each side gets placed into a third intermediate linking table that (usually) includes nothing but both keys.
Non-key: a "regular" field; it describes a characteristic of the table's subject.
Relationship: Establishes a connection or correspondence or link between a pair of tables in a database, or between a pair of entities in an ERD.
One-to-One Relationship[7]: A single record in table A is related to only one record in table B, and vice versa.
One-to-Many Relationship: A single record in table A can be related to one or more records in table B, but a single record in table B can be related to only one record in table A. One-to-many may be one-to-many (mandatory) where each record in the one table must have at least one entry in the many table; and one-to-many (optional) where zero related records in the many table are allowed (e.g. patient, clinic and admission tables where the patient is only recorded when they have a clinic appointment so has records in that table, but not been admitted so may not have any in that one).
Many-to-Many Relationship: A single record in table A can be related to one or more records in table B, and vice versa. There are problems with many-to-many relationships in that one of the tables will contain a large amount of redundant data; both tables will contain some duplicate data; it will be difficult to add/update/delete records because of the duplication of fields between tables.
Notation: There are many notations used to describe one-to-many relationships in an ERD. In the diagram in Figure 15.12, a simple "1" or "M" at opposite ends of the line indicating a relationship is used.
Others include those shown in Figure 2.3.

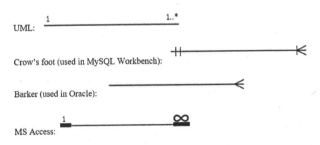

FIGURE 2.3   Some notations for one-to-many relationships.

2.6 EXAMPLE – MAKING THE DESIGN MORE EFFICIENT

First, we eliminate columns from the same table, creating a new column to hold this data.

Change Track1, Track2 etc. into Tracks, as shown in Table 2.3.

We then create separate tables for each group of related data (there is only one in this example) and identify each row with a unique column (the primary key – called "Item_ID" here), as shown in Table 2.4.

This is known as **first normal form** or **1NF**. Note that field names do not contain spaces or punctuation. Whilst some systems allow this, delimiters are then required.

To move to **second normal form** or **2NF**, we must remove subsets of data that apply to multiple rows of a table and place them in separate rows and create relationships between these new tables and their predecessors through the use of foreign keys. Here, the subsets are the Artist, the Title and the Track.[8] These new tables also require primary keys, called Artist_ID, Title_ID and Track_ID, respectively. These primary keys are then stored in the original table as foreign keys, creating the relationship between the tables, as shown in Tables 2.5–2.8.

TABLE 2.3
1NF Step 1

Media	Artist	Title	Tracks	Price	Copies	Value
CD	Example	Playing in the Shadows	Skies Don't Lie	999	2	1998
CD	Example	Playing in the Shadows	Stay Awake	999	2	1998
CD	Example	Playing in the Shadows	Changed the Way You Kiss Me	999	2	1998

TABLE 2.4
1NF Step 2

Item_ID	Media	Artist	Title	Tracks	Track_No	Price	Copies	Value
1	CD	Example	Playing in the Shadows	Skies Don't Lie	1	999	2	1998
2	CD	Example	Playing in the Shadows	Stay Awake	2	999	2	1998
3	CD	Example	Playing in the Shadows	Changed the Way You Kiss Me	3	999	2	1998

TABLE 2.5
2NF Step 1a

Item_ID	Media	Artist_ID	Title_ID	Track_ID	Track_No	Price	Copies	Value
1	CD	1	1	1	1	999	2	1998
2	CD	1	1	2	2	999	2	1998
3	CD	1	1	3	3	999	2	1998

TABLE 2.6
2NF Step 1b

Artist_ID	Artist
1	Example

TABLE 2.7
2NF Step 1c

Title_ID	Title
1	Playing in the Shadows

TABLE 2.8
2NF Step 1d

Track_ID	Track
1	Skies Don't Lie
2	Stay Awake
3	Changed the Way You Kiss Me

Finally, to move to **third normal form** or 3NF, we must remove columns that are not fully dependent upon the primary key. In this example, the fields Media, Price and Copies move to the Title table (where they are dependent on the Primary Key) and the field Value is removed, as it can be computed from the fields Price and Copies. Tables 2.9–2.12 show this final step.

We can see that this final structure has the following relationships (naming the tables Item, Artist, Title and Track, respectively):

- Artist to Item is one-to-many.
- Title to Item is one-to-many.
- Track to Item is one-to-one.

TABLE 2.9
3NF Step 1a

Item_ID	Artist_ID	Title_ID	Track_ID	Track_No
1	1	1	1	1
2	1	1	2	2
3	1	1	3	3

TABLE 2.10
3NF Step 1b

Artist_ID	Artist
1	Example

TABLE 2.11
3NF Step 1c

Title_ID	Title	Media	Price	Copies
1	Playing in the Shadows	CD	999	2

TABLE 2.12
3NF Step 1d

Track_ID	Track
1	Skies Don't Lie
2	Stay Awake
3	Changed the Way You Kiss Me

In this simple example, there is little efficiency gained from this abstraction, but by the addition of just one extra field (track length), it can be seen how this structure is more adaptable than the one we started with.

The table Item has a primary key (Item_ID) and many foreign keys (Artist_ID etc.) which are the primary keys in their own tables. An analogy from programming would be the use of pointers: not the data itself, but a link to where the data may be found (an analogy with the World Wide Web is similar).

A structure like this is dependent on and representative of the relationships between the data and is known as a **Relational Database**.

The three normal forms may be summed up as follows:

- Duplicate columns within the same table are eliminated. (1NF)
- Each group of related data is in its own table with a primary key to identify each row. (1NF)
- Subsets applying to multiple rows of a table are in their own table. (2NF)
- Relationships are maintained by creating foreign keys. (2NF)
- Any columns not dependent upon the primary key are removed from the table. (3NF)

An amendment to third normal form was proposed by Codd and Boyce in 1974 to address certain possible anomalies that may arise in 3NF. BCNF (Boyce–Codd Normal Form, sometimes rendered 3.5NF) is thus a slight variant on 3NF, and the terms are often used interchangeably as most data structures will not fall foul of the subtlety.[9]

One issue does arise through the use of normalising a database, however. The relationships are achieved through the use of indexes. These are rapidly changing files which therefore have a risk of corruption. If a corrupt index is used to retrieve records matching a certain key, then the records returned may not all match that key. It is therefore imperative (depending on the criticality of the data retrieved) that this data is checked prior to use. This can be as simple as ensuring that each returned record does contain the key searched for. This will ensure that no erroneous results are used, but does not ensure that all results have been returned. To achieve this, a redundant data item, such as a child record counter,[10] must be used. In most cases, this is not an issue but it does need to be considered when the criticality of the results is high (e.g. a pharmacy system).

2.7   FOURTH AND FIFTH NORMAL FORM

For **Fourth Normal Form** (4NF), introduced by Fagin in 1974, all **multi-valued dependencies** (MVDs) must be eliminated.[11] An MVD is as follows: in a relational database table with three columns X, Y and Z, then, in the context of a particular row, we can refer to the data in each column as x, y and z, respectively. An MVD $X \rightarrow \rightarrow Y,Z$ signifies that if we choose any x actually occurring in the table (call this choice x_c) and compile a list of all the $x_c yz$ combinations that occur in the table, we find that x_c is associated with the same y entries regardless of z. Thus, the presence of z provides no useful information to constrain the possible values of y.

For example, consider Table 2.13.

We can see that "Help!" appeared on two different labels and that U2 have two albums (in our collection). The key to this table uses all three columns in order to uniquely identify a row, and the MVD is Artist$\rightarrow \rightarrow$Title, Label as "Beatles" will always associate to "Help!" regardless of Label and "U2" will always associate to "Island" regardless of Title.

4NF is achieved by decoupling to separate tables for the MVDs and creating a new table, Artist, as shown in Tables 2.14–2.16[12].

TABLE 2.13
4NF Example Source

Artist	Title	Label
Beatles	Help!	Parlophone
Beatles	Help!	United Artists
U2	The Joshua Tree	Island
U2	Boy	Island
Example	Playing in the Shadows	Ministry of Sound
Slade	Slayed?	Polydor
Deep Purple	Shades of Deep Purple	Harvest

TABLE 2.14
4NF Example Solution Part 1

Artist
Beatles
U2
Example
Slade
Deep Purple

TABLE 2.15
4NF Example Solution Part 2

Artist	Album
Beatles	Help!
U2	The Joshua Tree
U2	Boy
Example	Playing in the Shadows
Slade	Slayed?
Deep Purple	Shades of Deep Purple

TABLE 2.16
4NF Example Solution Part 3

Artist	Label
Beatles	Parlophone
Beatles	United Artists
U2	Island
Example	Ministry of Sound
Slade	Polydor
Deep Purple	Harvest

For a truly efficient design, the artist name would be replaced by ID numbers (as in our 3NF example) but are left textual here (and in the next example) for clarity. A further worked example can be found at: https://www.slideshare.net/kosalgeek/database-normalization-1nf-2nf-3nf-bcnf-4nf-5nf

To move to 5NF, the attributes of MVDs are related. As an example, we add the owner of the label to the database as well. As a subset, we now have the data as shown in Table 2.17.

This table is not in 5NF as the MVD is Title →→ Owner, Label and Label is related to Owner. We move to 5NF by replacing this table with the six tables shown in Tables 2.18–2.23.

TABLE 2.17
5NF Example Source

Title	Owner	Label
Playing in the Shadows	Sony	Ministry of Sound
Shades of Deep Purple	Warner Brothers	Parlophone
Shades of Deep Purple	Warner Brothers	Harvest
Shades of Deep Purple	Warner Brothers	Warner Brothers
Help!	MGM	United Artists
Help!	Apple Corps	Apple

TABLE 2.18
5NF Example Solution Part 1

Title
Playing in the Shadows
Shades of Deep Purple
Help!

TABLE 2.19
5NF Example Solution Part 2

Title	Owner
Playing in the Shadows	Sony
Shades of Deep Purple	Warner Brothers
Help!	MGM
Help!	Apple Corps

TABLE 2.20
5NF Example Solution Part 3

Owner

Sony
Warner Brothers
MGM
Apple Corps

TABLE 2.21
5NF Example Solution Part 4

Title	Label
Playing in the Shadows	Ministry of Sound
Shades of Deep Purple	Parlophone
Shades of Deep Purple	Harvest
Shades of Deep Purple	Warner Brothers
Help!	United Artists
Help!	Apple

TABLE 2.22
5NF Example Solution Part 5

Label

Ministry of Sound
Parlophone
Harvest
Warner Brothers
United Artists
Apple

TABLE 2.23
5NF Example Solution Part 6

Owner	Label
Sony	Ministry of Sound
Warner Brothers	Parlophone
Warner Brothers	Harvest
Warner Brothers	Warner Brothers
MGM	United Artists
Apple Corps	Apple

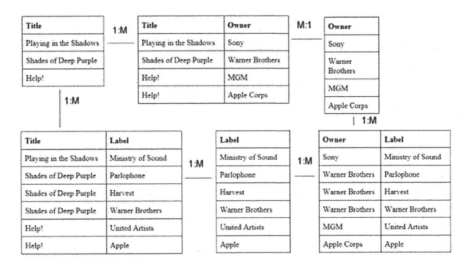

FIGURE 2.4   A 5NF schema.

These six tables are related as shown in Figure 2.4.

Whilst most database designs will not need to progress beyond 3NF, 4NF can be of use in organising the data and eliminating anomalies and redundancy that 3NF has failed to do. 5NF is of little practical use to database designers but is of interest from a theoretical point of view and is included here for completeness.

2.8   MANY-TO-MANY RELATIONSHIPS

A Many-to-Many relationship is one where an entity in table A can be related to many entities in table B and vice versa. For example, consider songs and songwriters. A song can have multiple writers and a writer can compose multiple songs. In relational databases, such relationships are inefficient[13] and so are normally implemented via an intermediate table, called a cross-reference or associative table, which has a one-to-many relationship with each table. An example is shown in Table 2.24, which becomes Table 2.25, and then as shown in Figure 2.5. The addition of the associative table makes it as shown in Figure 2.6, thus replacing the M:M relationship with a 1:M and a M:1.

2.9   DISTRIBUTED RELATIONAL SYSTEMS AND DATA REPLICATION

Replication of a database is what the name implies: a copy exists of the data enabling it to be switched to in the event of a failure in the primary system. The simplest replication to achieve is a copy, taken at a specified time, to the secondary system. The two systems are therefore only ever briefly synchronised and in the event of a system failure bringing the secondary system online means that the data will be at worst out-of-date by the time interval between copying. The most common use of this form of

TABLE 2.24

Many-to-Many Example Source Part 1

Song	Songwriters
Smoke on the Water	Blackmore, Gillan, Glover, Lord, Paice
Perfect Strangers	Blackmore, Gillan, Glover
Stargazer	Blackmore, Dio
Concerto for Group and Orchestra	Lord

TABLE 2.25

Many-to-Many Example Source Part 2

Song	Songwriter
Smoke on the Water	Blackmore
Smoke on the Water	Gillan
Smoke on the Water	Glover
Smoke on the Water	Lord
Smoke on the Water	Paice
Perfect Strangers	Blackmore
Perfect Strangers	Gillan
Perfect Strangers	Glover
Stargazer	Blackmore
Stargazer	Dio
Concerto for Group and Orchestra	Lord

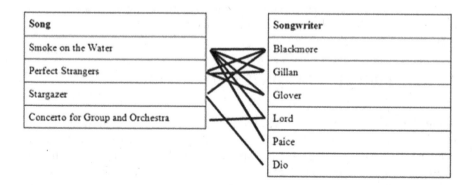

FIGURE 2.5   Many-to-many example source part 3.

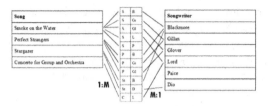

FIGURE 2.6   Many-to-many example solution with associative table.

replication is in data repositories, thereby removing some of the workload from the primary system and improving its reliability (and response time).

The more complex method of replication is synchronised: i.e. both copies are exactly the same at all times. Usually achieved by an interface engine, messages to the primary server are also sent to the secondary (updates only – SELECT queries are irrelevant and therefore should be filtered out by the interface engine).

Additionally, a database can be split across hardware and indeed across software. The simplest method of hardware distribution is splitting the database onto multiple physical hard drives. Some even go a step further and allow splitting the contents of a table over multiple disks. When using multiple physical disks, I/O operations speed up significantly since the multiple heads fetch data in parallel.

The next level is to split the database onto separate server installations. Whilst this has been possible in SQL Server clusters for a while, the term NewSQL is being used to describe this technology. NewSQL is a class of relational database management systems that seek to provide the same scalable performance of NoSQL systems for online transaction processing read-write workloads while still maintaining the guarantees of a traditional database system. There are (loosely) three categories:

- Completely new database platforms. These are designed to operate in a distributed cluster of shared-nothing nodes,[14] in which each node owns a subset of the data.[15] These databases are often written from scratch with a distributed architecture in mind and include components such as distributed concurrency control, flow control and distributed query processing. Example systems in this category are CockroachDB, Clustrix and NuoDB.
- Highly optimised storage engines for SQL. These systems provide the same programming interface as SQL, but scale better than built-in engines. Examples of these new storage engines include MySQL Cluster, Infobright, TokuDB, MyRocks and the now-defunct InfiniDB.
- Transparent sharding.[16] These systems provide a sharding middleware layer to automatically split databases across multiple nodes. Examples include ScaleBase and Vitess.

2.10   COLUMNSTORE AND DATA WAREHOUSING

A columnstore index is a technology for storing, retrieving and managing data by using a columnar data format, called a columnstore. Columnstore indexes are the standard for storing and querying large data warehousing fact tables (which we will meet in the next section). They provide improved performance for data warehousing and reporting workloads and, in combination with clustered columnstore indexes, there is a huge compression benefit over rowstore indexes.[17] A columnstore index uses column-based data storage and query processing to achieve up to 10 times query performance gains in the data warehouse over traditional row-oriented storage and up to 10 times data compression over the uncompressed data size.

They are created via the SQL command

CREATE CLUSTERED COLUMNSTORE INDEX <name> ON <table>

The key terms and concepts are:

- A *rowstore* is data that is logically organised as a table with rows and columns, and then physically stored in a row-wise data format. This has been the traditional way to store relational table data. In MS SQL Server, rowstore refers to tables where the underlying data storage format is a heap, a clustered index or a memory-optimised table.
- A *columnstore* is data that is logically organised as a table with rows and columns and physically stored in a column-wise data format.
- A *rowgroup* is a group of rows that are compressed into columnstore format at the same time.[18] A rowgroup usually contains the maximum number of rows per rowgroup which is 1,048,576 rows in SQL Server. For high performance and high compression rates, the columnstore index slices the table into groups of rows, called rowgroups, and then compresses each rowgroup in a column-wise manner. The number of rows in the rowgroup must be large enough to improve compression rates and small enough to benefit from in-memory operations – if the required data will fit into memory then the speed of querying far exceeds that of when the data must be queried from disc.
- A *column segment* is a column of data from within the rowgroup. Each rowgroup contains one column segment for every column in the table and each column segment is compressed together and stored on physical media, as illustrated in Figure 2.7.[19]

The set of rows is divided into rowgroups that are converted to column segments and dictionaries (labelled "dict" in Figure 2.7) that are then stored using SQL Server blob storage.

A *clustered columnstore index* is the physical storage for the entire table. To reduce fragmentation of the column segments and improve performance, the

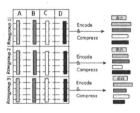

FIGURE 2.7   Creation and storage of a column store index.

columnstore index might store some data temporarily into a clustered index (which is called a deltastore – see the next item) and a btree[20] list of IDs for deleted rows. The deltastore operations are handled by the SQL engine and require no user intervention. To return the correct query results, the clustered columnstore index combines query results from both the columnstore and the deltastore.

- Used with clustered column store indexes only, a *deltastore* is a clustered index that improves columnstore compression and performance by storing rows until the number of rows reaches a threshold. They are then moved into the columnstore.
- A *nonclustered columnstore index* and a clustered columnstore index function in the same way. The difference is that a nonclustered index is a secondary index created on a rowstore table, whereas a clustered columnstore index is the primary storage for the entire table. The nonclustered index contains a copy of part or all of the rows and columns in the underlying table. The index is defined as one or more columns of the table and may have a condition (it is optional) that filters the rows. A nonclustered columnstore index enables real-time operational analytics in which the **Online Transaction Processing** (OLTP) workload uses the underlying clustered index, while analytics run concurrently on the columnstore index.
- *Batch execution* is a query processing method in which queries process multiple rows together. Queries on columnstore indexes use batch mode execution which improves query performance typically by 2–4 times. Batch execution is closely integrated with, and optimised around, the columnstore storage format. Batch-mode execution is sometimes known as vector-based or vectorised execution.

A columnstore index can provide a very high level of data compression which reduces the data warehouse storage cost significantly. Additionally, for analytics they offer a performance which is an order of magnitude better than a btree index. They are the preferred data storage format for data warehousing and analytics workloads.

The reasons why columnstore indexes are so fast are:

- Columns store values from the same domain and commonly have similar values, which results in high compression rates. This minimises or eliminates I/O bottleneck in the system while significantly reducing the memory footprint.
- High compression rates improve query performance by using a smaller in-memory footprint, enabling SQL Server to perform more query and data operations in-memory.
- Batch execution improves query performance, typically 2–4 times, by processing multiple rows together.
- Queries often select only a few columns from a table, which reduces total I/O from the physical media – unrequired data need not be retrieved.

A clustered columnstore index is best used to store fact tables and large dimension tables for data warehousing workloads and to perform analysis in real-time on an OLTP workload.

In summary, the differences between rowstore and columnstore are the following:

- Rowstore indexes perform best on queries that seek into the data, searching for a particular value, or for queries on a small range of values. Rowstore indexes are best used with transactional workloads since they tend to require mostly table seeks instead of table scans.
- Columnstore indexes give high performance gains for analytic queries that scan large amounts of data, especially on large tables. Columnstore indexes are best used on data warehousing and analytics workloads, especially on fact tables, since they tend to require full table scans rather than table seeks.

2.11   OLAP CUBES

Modification of data (add, edit, delete) is generally known as OLTP, whereas data retrieval is referred to as **Online Analytical Processing** (OLAP). For OLTP, the relational model is very efficient as it reduces data redundancy. For OLAP, the most important requirements are query performance and schema simplicity. Columnstore, which we have just examined, is one of the main ways of implementing OLAP.

An OLAP cube is a multi-dimensional dataset, sometimes called a hypercube when the dimension exceeds 3.[21] It is a data structure that overcomes the query limitations of relational databases by providing rapid analysis of data. In database theory, an OLAP cube is an abstract representation of a projection of a **Relational Database Management System** (RDBMS) relation. Conceiving data as a cube with hierarchical dimensions can lead to conceptually straightforward operations in order to facilitate analysis.[22] OLAP cubes can display and sum large amounts of data while also providing users with searchable access to any data points. The data can thus be rolled up,[23] drilled down, sliced and diced as needed to provide the required information.[24]

A roll-up operation summarises data along a dimension. This summarisation could be an aggregate function, such as computing totals along a hierarchy or applying a set of formulas such as "episode = end date – start date". General aggregation functions may be costly to compute when rolling up: if they cannot be determined from the cells of the cube, they must be computed from the base data, either computing them online (which is slow) or precomputing them for possible later rollouts (which may require large storage). Aggregation functions that can be determined from the cells are known as decomposable aggregation functions and allow efficient computation. For example, COUNT, MAX, MIN and SUM are simple to support in OLAP, as these can be computed for each cell of the OLAP cube and then rolled up, since an overall sum (or count etc.) is the sum of sub-sums, but it is difficult to support MEDIAN, as the median of a set is not the median of medians of subsets so must be computed for every view separately. Figure 2.8 shows a drill-down operation, moving from the summary category "Pulse oximeters" to see the installed units for the individual manufacturers. Roll-up is moving in the opposite direction, grouping the equipment into a summary category.

A slice operation is the selection of a rectangular subset of a cube by choosing a single value for one of its dimensions, creating a new cube with one fewer

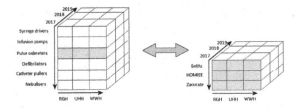

FIGURE 2.8   Roll-up and drill-down on an OLAP cube.

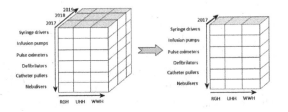

FIGURE 2.9   A slice of an OLAP cube.

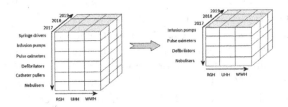

FIGURE 2.10   A dice of an OLAP cube.

dimension. A slice of a 3D cube is thus any two-dimensional section of the data cube. A slice selects values corresponding to one value on one or more dimensions in order to filter information. Figure 2.9 shows a slicing operation: the equipment in all hospitals in all equipment categories in the years 2018 and 2019 are "sliced" out of the data cube.

A dice operation is the "rotation" of the cube to reveal another, different slice of data. A data cube can be diced by exchanging one dimension for other dimensions. A dice selects values corresponding to one or more consecutive slices on more than two dimensions of the cube, producing a subcube by allowing the selection of specific values of multiple dimensions. Figure 2.10 shows a dicing operation: the new cube shows the installed units for a limited number of equipment categories, the time and hospital ranges remaining the same.

The query language used to interact and perform tasks with OLAP cubes is **multidimensional expressions** (MDX). The MDX language was originally developed by Microsoft in the late 1990s and has been adopted by many other vendors of multidimensional databases.

Although it stores data like a traditional database does, an OLAP cube is structured very differently. OLTP Databases are designed according to the requirements of the IT systems that use them. OLAP cubes, being used for advanced analytics, are designed using business logic and understanding. They are optimised for analytical purposes, so that they can report on millions of records at a time.

OLAP data is typically stored in a star or snowflake schema in a relational database warehouse or special-purpose data management system. Measures are derived from the records in the fact table and dimensions are derived from the dimension tables.

2.12 STAR SCHEMA

A star schema is the simplest and therefore most widely used OLAP schema. It consists of one or more fact tables referencing any number of dimension tables. Figure 2.11 shows a sample star schema.

The fact table (at the centre) usually contains measurements and is the only type of table with multiple joins to other tables. Surrounding it are dimension tables, which are related to the fact table by a single join. Dimension tables contain data that describe the different characteristics, or dimensions, of the data. Facts change regularly and dimensions either do not change or do so very slowly.

The fact table contains the columns for the measurements of the database subject (in this case, equipment maintenance). It also includes a set of columns that form a concatenated or composite key. Each column of the concatenated key is a foreign key drawn from a dimension table primary key. Fact tables usually have few columns and many rows.

Dimension tables store descriptions of the characteristics of the data subject. A dimension is usually descriptive information that qualifies a fact. For example, in Figure 2.11, the Item, Staff, Location and Contract dimensions describe the measurements in the fact table. These dimensions may also be computed fields, such as sums and averages.

Dimension tables contain few records and many columns and each dimension table in a star schema database has a single-part primary key joined to the fact table.

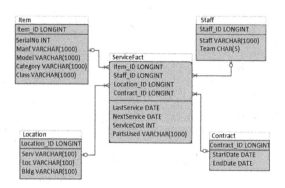

FIGURE 2.11   A star schema for an equipment maintenance database.

An important design characteristic of a star schema database is the ability to quickly browse a single dimension table. This is possible because dimension tables are flat and denormalised. The required SQL is considerably simpler, being of the form

SELECT field1,field2 FROM table

As opposed to

SELECT table1.field1,table2.field2 FROM table1 JOIN table2 ON table1. foreignkeyfield=table2.primarykeyfield

Or even

SELECT   table1.field1,SUM(table2.field2),AVG(table3.field3)   FROM table1 JOIN table2 ON table1.foreignkeyfield1=table2.primarykeyfield JOIN table3 ON table1.foreignkeyfield2=table3.primarykeyfield

In designing a star schema (i.e. moving from OLTP to OLAP) the dimension tables are effectively lookups, so the fact table contains the fields that are being searched on/for. After determining the questions to be answered, a schema can be created that answers those questions with as few joins as possible, by placing fields into the fact table.

It is important to remember that this schema doesn't have to be able to answer all questions. The star schema in Figure 2.11 will answer servicing queries quickly and efficiently, but not so for queries concerning equipment under contract in a specific location. If there is something that the star schema is unable to answer then a new OLAP structure can be devised to do this as all the data comes from the OLTP and creating/destroying OLAP structures doesn't damage the data. It is therefore common to write scripts to perform this translation which can be modified and adapted as required.

2.13 DATABASE STANDARDS AND STANDARDS FOR INTEROPERABILITY AND INTEGRATION

Interoperability is the ability of a system to use and share information or functionality of another system by adhering to common standards. There are many standards for interoperability, especially in web programming (which are covered in Chapter 11). These include XML and SOAP. Figure 2.12 shows the exchange of data between the systems Obstetrix and MedRex and between Obstetrix and HRrex, demonstrating interoperability between the two systems. There is no interoperability between MedRex and HRrex.

Integration can thus be understood as the process of joining distinct systems in such a way that they appear to be a single system. In this case, patient data from MedRex and human resource data from HRrex are joined and integrated in the

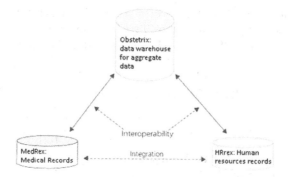

FIGURE 2.12   An example of interoperability.

Obstetrix data warehouse. MedRex and HRrex are integrated, but there is no interoperability between them. Obstetrix is interoperable with both MedRex and HRrex.

If Obstetrix were to be calculating the statistic "deliveries per midwife per maternity unit for July 2019" then it would issue two requests: one to MedRex (deliveries by maternity unit for July 2019) and one to HRrex (midwives by maternity unit for July 2019). Each system would aggregate and return the requested data for Obstetrix to collate.

It is common for the interoperability request to include both the standard for data exchange and the data which is to be exchanged. These two components and their internal relationship are therefore similar to the messenger and the message or the container and the content.

Interoperability is thus a sub-set of integration, being more directly operational than what is often understood by the more open term of integration.

> Studies have shown that companies with high levels of standardization reduce the cost of supporting end users by as much as 35 percent or more as compared to companies with low levels of standardization.
>
>   (Mullins 2012)

The following standards should be considered in database administration:

2.13.1   Database Naming Conventions

Without standard naming conventions (including for all database objects), database objects are difficult to correctly identify and administration tasks become more difficult. The naming standard should align with all other IT naming standards in the organisation (hospital or department, as NHS-wide naming standards do not exist). This, though, should not be at the expense of impairing the database environment. For example, many organisations have conventions for naming files, but this may not be appropriate for database tables.

The database naming standard should be designed to minimise name changes across environments. For example, pre- or post-fixing a "T" or "P" into the name (for "test" and "production", respectively) is a bad idea, as user-visible database objects such as columns, tables and views will require renaming in queries once testing is over. Minimising name changes simplifies the migration of databases from one environment to another. A similar (and better) method of identifying the objects is to make all the names the same but assign each environment to a different instance or subsystem. The instance or subsystem name, rather than the database object names, will differentiate the environments.

There is, of course, a difference between database objects that may be visible to end users (e.g. in report generation) and those that are only visible to developers. For objects not accessed by end users, a way to differentiate types of database objects can be useful (e.g. starting indexes with I or X and databases with D).[25]

The database naming convention needs to be developed logically as imposing unnecessary restrictions on the names of objects accessed by end users (such as fixed length) reduces the user friendliness of the system – field names that are difficult to remember increase query development time and may thus reduce the use of the system.

Table names should be as descriptive as possible as should the names of all "table-like" objects (such as views, synonyms and aliases). As each of these objects appears similar to the end user their names should follow the same convention. One approach would be to have table names prefixed by a 2- or 3-byte application identification,[26] followed by an underscore and then a clear, user-friendly name.[27] For example, a good name for the table containing employee information in a human resources system would be HR_Employee.

It is worth noting that some database object names will, in some cases, be presented to end users. For instance, most **Database Management Systems** (DBMSs) externalise constraint names when the constraint is violated and an error message is produced. Keeping the names consistent across environments allows the error messages to be consistent. If the DBMS delivers the same error message in the development, test, integration and production environments, debugging and error correction will be easier.

If abbreviations are to be used (e.g. "HR" as above), they should be used only when the full text is too long to be supported as an object name or when it renders the object name unwieldy or difficult to remember. A list of standard abbreviations should be created, documented and distributed and the use of nonstandard abbreviations should be forbidden.[28] For example, if "ORG" is the standard abbreviation for "organisation," variants such as "ORGS" should not be used. Using standard abbreviations minimises mistyping and makes it easier for users to remember database object names. This practice makes it easier to understand the database objects within the environment, leading to faster query development.

2.13.2   DATA ADMINISTRATION STANDARDS

While taking note of (and complying with) the organisation's policy on data,[29] including its importance to the organisation, data administration standards should address:

- Responsibility for creating and maintaining logical data models.
- Data creation, data ownership, data storage and data stewardship.

- Conceptual and logical data modelling.
- Metadata management.
- Software tool usage.
- The documentation of when physical databases deviate from the logical data model.
- Communication between data administration and database administration.

2.13.3   DATABASE ADMINISTRATION STANDARDS

Database administration standards ensure the success of the **Database Administrator** (DBA) role. They serve as a guide to the DBA services offered and to specific approaches to supporting the database environment. Database administration standards should address:

- How requests are made to create a new database or make changes to existing databases.
- The types of database objects and DBMS features that are permitted (and which are not).
- Backup and recovery procedures (including disaster recovery plans).
- Methods used to transform a logical data model into a physical database implementation.
- Database performance monitoring and tuning.

These DBA standards should be shared with application development staff as the more the application programmers understand the DBMS and the role of the DBA, the better the working relationship between DBA and development will be – resulting in a more efficient database environment.

2.13.4   SYSTEM ADMINISTRATION STANDARDS

System administration standards should address[30]:

- DBMS installation and testing.
- Upgrades.
- Maintenance, including bug fixes.
- Notification of impending changes (who more than why).
- Interface considerations.
- DBMS storage, usage and monitoring.

2.13.5   DATABASE APPLICATION DEVELOPMENT STANDARDS

As the development of database applications differs from typical program development, the special development considerations required when writing programs that access databases require documenting. These database application development standards are an adjunct to the standard application development procedures within the organisation and should address:

- The difference between database access and flat file access.
- SQL coding (including performance tips and techniques).
- Program preparation and how to embed SQL in an application program.
- Interpretations of SQLSTATEs and error codes.

2.13.6 DATABASE SECURITY STANDARDS

Database security standards and procedures should address:

- The level of authority to grant for specific types of situations. For example, if a program is being migrated to production status, the DBMS authorisation required to be granted so that the program will operate in production.
- Documentation of any special procedures.
- Documentation for governance- and compliance-related requests.
- Who can approve database authorisation requests.[31]
- Interfaces used to connect DBMS security with operating system security products.
- The use of the WITH GRANT OPTION clause of the SQL GRANT statement.
- The handling of cascading REVOKEs.
- Notifying a requester that database security has been granted.
- Removing security from retiring, relocating and terminated[32] employees.

2.13.7 APPLICATION MIGRATION AND TURNOVER PROCEDURES

The minimum number of environments for supporting database applications is two: test and production. For complex systems, multiple environments to support different phases of the development life cycle may be appropriate, including:

- Unit testing – for developing and testing individual programs.
- Integration testing – for testing how individual programs interoperate.
- User acceptance testing – for end user testing prior to production status.
- Quality assurance – for debugging programs.
- Education – for training end users in the application system.

Procedures are required for migrating database objects and programs between environments. Specific guidelines are therefore required to accomplish this migration in a manner conducive to the usage of each environment. Issues include (which the standards should address):

- The data volume required for each environment.
- Assuring data integrity when testing.
- Whether the data is migrated, or just the database structures.
- How existing data in the target environment is to be treated (it may be kept, or overlaid with new data)

2.13.8 OPERATIONAL SUPPORT STANDARDS

Operational support is that part of the organisation that oversees the database environment and assures that applications are run according to schedule. Operational support standards ensure that the operational support staff understand the special requirements of database applications. Whenever possible, operational support personnel should be trained to resolve simple database-related problems without involving the DBA because the DBA is a more expensive resource. It is also more user-friendly for their support call to be handled immediately, rather than awaiting the DBA's availability.

NOTES

1 The first part of this chapter (up to and including Section 2.6) is reproduced from Ganney et al. (2022).
2 An algorithm where the ordered data is divided into two parts and the mid-point is examined. If the sought data is before it, then the latter half can be ignored. If the sought data is after this mid-point, then the first half can be ignored. The search then repeats on the half of interest until the sought item is found.
3 i.e. one without any formulae or computed cells.
4 There is some debate over this as many common flat file formats include line separators. Strictly, they shouldn't, but in practice they often do.
5 A CSV file is one in which table data is placed in lines of ASCII text with the value from each table cell separated by a comma and each row represented by a new line. A CSV file is one of the most common flat files, as it represents relational data in a text file.
6 As the order is placed on the table externally.
7 This and the following two relationships are intentionally similar to the rules for foreign keys.
8 It could be argued that the Track is not a subset – however, a compilation album by the same artist will contain the same track name as the original album. Likewise, an album owned on multiple formats will contain the same tracks.
9 See https://en.wikipedia.org/wiki/Boyce%E2%80%93Codd_normal_form for an example of a database that does.
10 A field that stores the number of records linked to this one.
11 Assuming we already have a 3NF or 3.5NF database.
12 Giving us three tables, one for each MVD and the new Artist table.
13 See https://stackoverflow.com/questions/7339143/why-no-many-to-many-relationships for a discussion.
14 A node is a processor/memory/storage unit.
15 A key point being that the data does not appear in more than one node, as opposed to shared-everything nodes. Shared-nothing avoids processing contention but may lack resilience. The distribution of the data across the nodes is key to its efficiency.
16 A database shard is a horizontal partition of data in a database or search engine. Each individual partition is referred to as a shard or database shard. Each shard is held on a separate database server instance, to spread the load. Some data within a database remains present in all shards, but some appears only in a single shard (in the case of shared-nothing nodes all data will only appear in a single shard). Each shard (or server) acts as the single source for this subset of data.
17 The type we have examined so far.
18 i.e. compressed together.

19 It may be simpler just to think of it as a column of data, but it has been stored together making access in a columnar fashion quicker.
20 Binary tree – see Section 1.6.
21 As it is easier to visualise with 3 dimensions, that is what we will use here.
22 Certainly true for 3 dimensions, debatable for more.
23 Also known as "drill-up".
24 Other operations such as pivot, drill-across and drill-through also exist.
25 This is similar to Hungarian Notation, see Section 16.1.4.
26 Although this may not be applicable for tables used by multiple applications.
27 Also similar to Hungarian Notation.
28 Remember that this is a standard so it can be ignored – thus "forbidden".
29 Which will also include the policy on data sharing.
30 These standards should include both policy and procedures.
31 The type may be different for each person.
32 The employment is terminated, not the employee.

REFERENCES

Codd EF, 1970, *A Relational Model of Data for Large Shared Data Banks*, *Association for Computing Machinery (ACM)*, Vol. 13, No. 6, June 1970, pp. 377–387.

Ganney P, Maw P, White M, ed. Ganney R, 2022, *Modernising Scientific Careers The ICT Competencies*, 7th edition, Tenerife: ESL.

Mullins CS, 2012, *Database Administration: The Complete Guide to DBA Practices and Procedures*, 2nd Edition. Pearson [online]. Available: http://www.informit.com/articles/article.aspx?p=1963781&seqNum=4 [Accessed 29/03/22].

Pyle IC and Illingworth V, 1996, *Dictionary of Computing*, Fourth Edition. Oxford: Oxford University Press.

3 SQL[1]

3.1 INTRODUCTION

Operations on relational databases are often carried out using **Structured Query Language** (SQL). As with any language, there are various dialects, but the underlying principles are the same. The SQL syntax described here is based upon MySQL version 5.7 (MySQL 5.7 Reference Manual 2022 [online] (1)), unless otherwise referenced.[2] SQL commands need not be in upper case but are rendered so here in order to highlight them.

Some of the clauses we will come across will have been deprecated, meaning that their use is discouraged because they have been superseded or are no longer used in some versions of SQL. They are included here because the reader may come across them in older (or less compliant) programs.

3.2 COMMON COMMANDS: SELECT, INSERT, UPDATE AND DELETE

The four most common commands are SELECT, INSERT, UPDATE and DELETE. The basic structure of a SELECT command is:

SELECT {fields} FROM {table} WHERE {condition}

For example, in the given example (and using the database we began with in Chapter 2), "SELECT Track FROM Track WHERE Track_ID = 2" will return one result: "Stay Awake". The WHERE clause is a logical statement (using Boolean logic and can therefore include Boolean operators) which will evaluate to either TRUE or FALSE. The statement thus finds all rows for which the WHERE clause is TRUE and then returns the fields listed in the SELECT clause.

In order to make use of the relational structure, however, we need to return data from more than one table. There are two main ways of doing this:

SELECT {fields} FROM {tables} WHERE {condition, including the relationship}

And

SELECT {fields} FROM {table} JOIN {table} ON {relationship} WHERE {condition}

DOI: 10.1201/9781003316244-3

TABLE 3.1
The Data Returned

Artist	Track
Example	Skies Don't Lie
Example	Stay Awake
Example	Changed the Way You Kiss Me

Examples of these might be:

SELECT Artist.Artist,[3] Track.Track FROM Artist, Track, Item WHERE
   Artist.Artist_ID = Item.Artist_ID AND Track.Track_ID = Item.
   Track_ID AND Item.Title_ID=1

which returns the data given in Table 3.1.

The notation "Table.Field" can become unwieldy and so an alias[4] can be used to make the statement easier to read:

SELECT a.Artist, t.Track FROM Artist a, Track t, Item i WHERE a.Artist_ID
   = i.Artist_ID AND t.Track_ID = i.Track_ID and i.Title_ID=1

The notation "Artist a" (in the FROM clause) provides an alias for the table name which can be used throughout the command.

The second form, in order to achieve the same result, might be written as:

SELECT a.Artist, t.Track FROM Item i JOIN Artist a ON a.Artist_ID
   = i.Artist_ID JOIN Track t ON t.Track_ID = i.Track_ID WHERE
   i.Title_ID=1

A wildcard (*) will list all fields, e.g.

SELECT * FROM Artist

will list all artist names and ID numbers in the database.

An INSERT statement adds a record to a single table. To add the track "The Way" to the example above would require the following statements:

INSERT INTO Item (Item_ID, Artist_ID, Title_ID, Track_ID, Track_No)
   VALUES (4, 1, 1, 4, 3)
INSERT INTO Track (Track_ID, Track) VALUES (4, "The Way")

This gives the results shown in Tables 3.2 and 3.3.

TABLE 3.2
Table Item

Item_ID	Artist_ID	Title_ID	Track_ID	Track_No
1	1	1	1	1
2	1	1	2	2
3	1	1	3	3
4	1	1	4	3^5

TABLE 3.3
Table Track

Track_ID	Track
1	Skies Don't Lie
2	Stay Awake
3	Changed the Way You Kiss Me
4	The Way

The field list is not always necessary: if the list is omitted, it is assumed that the values are in the same order as the fields. If the primary key is an Autonumber field (i.e. one that the system increments and assigns) then this cannot be specified, meaning that the field list is required (If the primary key is an Autonumber field, then adding data may require several steps, for the value assigned will have to be retrieved so it can be provided to the other INSERT statements).

An UPDATE statement has the form:

UPDATE {table} SET {field1=value1, field2=value2, ...} WHERE
    {condition}

Two examples of statements to correct the error introduced by the INSERT example which has produced 2 track number 3s might be:

UPDATE Item SET Track_No=Track_no+1 WHERE Title_ID=1 AND
    Track_ID>3
UPDATE Item SET Track_No=4 WHERE Item_ID=4

Note that the first form may update multiple records, whereas the second will update only one as it uses the primary key to uniquely identify a single record.

Both forms give the same resultant table, shown in Table 3.4.

TABLE 3.4
Table Item

Item_ID	Artist_ID	Title_ID	Track_ID	Track_No
1	1	1	1	1
2	1	1	2	2
3	1	1	3	3
4	1	1	4	4

Finally, the DELETE statement has the form:

DELETE FROM {table} WHERE {condition}

If the condition is omitted, then all records from the specified table will be deleted. An example might be:

DELETE FROM Track WHERE Track_ID=4

This statement alone would create a referential integrity error, in that Item now refers to a record in Track that no longer exists. In order to correct this, either a new Track with Track_ID of 4 must be created, or the following must be done:

DELETE FROM Item WHERE Track_ID=4

This gives the tables shown in Table 3.5 and Table 3.6.

TABLE 3.5
Table Item

Item_ID	Artist_ID	Title_ID	Track_ID	Track_No
1	1	1	1	1
2	1	1	2	2
3	1	1	3	3

TABLE 3.6
Table Track

Track_ID	Track
1	Skies Don't Lie
2	Stay Awake
3	Changed the Way You Kiss Me

Relational databases are very common in hospital informatics and clinical computing. Examples include oncology management systems, equipment management systems, the electronic patient record and cardiology patient monitoring systems.

3.3 SOME USEFUL COMMANDS/FUNCTIONS

MAX() and MIN() work as you might expect, e.g.

> SELECT MAX(Track_No) FROM Item

will return the largest track number (in this example, 3). It can then be used as part of a more complex SELECT, e.g.

> SELECT Title_ID FROM Item WHERE (Track_No)=(SELECT
>   MAX(Track_No) FROM Item)

will return the IDs of the highest numbered tracks (in this example, 3).

COUNT() returns the number of records matching the WHERE clause, rather than the records themselves, e.g.

> SELECT COUNT(Title) from Title WHERE Artist_ID=1

will return the number of titles for the artist with ID of 1 (in this example, 1).

DAY(), MONTH() and YEAR() will return the required part of a date.

3.4 SELECT MODIFIERS

The full form of the SELECT query in MySQL is:

```
SELECT
    [ALL | DISTINCT | DISTINCTROW]
        [HIGH_PRIORITY]
        [STRAIGHT_JOIN]
        [SQL_SMALL_RESULT] [SQL_BIG_RESULT]
        [SQL_BUFFER_RESULT]
        [SQL_CACHE | SQL_NO_CACHE] [SQL_CALC_FOUND_ROWS]
    select_expr [, select_expr ...]
    [FROM table_references
        [PARTITION partition_list]
    [WHERE where_condition]
    [GROUP BY {col_name | expr | position}
        [ASC | DESC], ... [WITH ROLLUP]]
    [HAVING where_condition]
    [ORDER BY {col_name | expr | position}
        [ASC | DESC], ...]
    [LIMIT {[offset,] row_count | row_count OFFSET offset}]
```

```
[PROCEDURE procedure_name(argument_list)]
[INTO OUTFILE 'file_name'
  [CHARACTER SET charset_name]
  export_options
  | INTO DUMPFILE 'file_name'
  | INTO var_name [, var_name]]
[FOR UPDATE | LOCK IN SHARE MODE]]
```

And we will examine each of these parameters here.

The ALL and DISTINCT[6] modifiers specify whether duplicate rows should be returned. ALL (the default, so if nothing is specified this is assumed) specifies that all matching rows should be returned, including duplicates. DISTINCT specifies the removal of duplicate rows from the result set. It is an error to specify both modifiers.

For example, using the database described above,

SELECT Artist.Artist FROM Artist, Item WHERE Artist.Artist_ID = Item.
Artist_ID

returns Table 3.7, whereas

SELECT DISTINCT Artist.Artist FROM Artist, Item WHERE Artist.Artist_
ID = Item.Artist_ID

returns Table 3.8.

It is possible to use DISTINCT across multiple columns.[7]

HIGH_PRIORITY, STRAIGHT_JOIN and modifiers beginning with SQL_ are MySQL extensions to standard SQL, the use of which is quite common across SQL engines. Microsoft, originally one of the worst offenders for proprietary extensions to languages, had an Indexing Service which used a SQL syntax containing

TABLE 3.7
The Data Returned

Artist

Example
Example
Example

TABLE 3.8
The Data Returned

Artist
Example

extensions to the subset of SQL-92 and SQL3 queries for relational database systems. This is no longer the case as of Windows 8, although the documentation says it works up to XP and not from 8, giving no mention as to what happens if you try it in Windows 7.

HIGH_PRIORITY gives the SELECT higher priority than a statement that updates a table. This should therefore only be used for queries that are very fast and must be done at once. A SELECT HIGH_PRIORITY query that is issued while the table is locked for reading runs even if there is an update statement waiting for the table to be free. HIGH_PRIORITY cannot be used with SELECT statements that are part of a UNION.

STRAIGHT_JOIN forces the query optimiser to join the tables in the order in which they are listed in the FROM clause. This modifier can therefore be used to speed up a query if the optimiser otherwise joins the tables in nonoptimal order.[8]

SQL_BIG_RESULT or SQL_SMALL_RESULT can be used with GROUP BY or DISTINCT to tell the optimiser that the result set has many rows or is small, respectively. For SQL_BIG_RESULT, MySQL uses disk-based temporary tables if needed and prefers sorting to using a temporary table with a key on the GROUP BY elements. For SQL_SMALL_RESULT, MySQL uses in-memory (which is fast) temporary tables to store the resulting table instead of using sorting. This parameter should not normally be needed.

SQL_BUFFER_RESULT forces the result to be put into a temporary table. This enables MySQL to free the table locks early and helps in cases where it will take a long time to send the result set to the client. This modifier can be used only for top-level SELECT statements, not for subqueries or following UNION.

SQL_CALC_FOUND_ROWS tells MySQL to calculate how many rows there would be in the result set, disregarding any LIMIT clause (but does not actually fetch them).[9] The number of rows can then be retrieved with SELECT FOUND_ROWS().

The SQL_CACHE and SQL_NO_CACHE modifiers affect the caching of query results in the query cache.[10] SQL_CACHE tells MySQL to store the result in the query cache if it is cacheable and the value of the system variable query_cache_type is 2 or DEMAND. With SQL_NO_CACHE, the server does not use the query cache. It neither checks the query cache to see whether the result is already cached, nor does it cache the query result.

These two modifiers are mutually exclusive and an error occurs if they are both specified. Also, these modifiers are not permitted in subqueries (including subqueries in the FROM clause) or in SELECT statements in unions other than the first SELECT.

For views, SQL_NO_CACHE applies if it appears in any SELECT in the query. For a cacheable query, SQL_CACHE applies if it appears in the first SELECT of a view referred to by the query.

PARTITION can only be used on tables that have been partitioned, either at creation or later. For example:

```
CREATE TABLE[11] employees (
    employee_id INT NOT NULL AUTO_INCREMENT PRIMARY KEY,
    forename VARCHAR(25) NOT NULL,
```

```
    surname VARCHAR(25) NOT NULL,
    hospital_id INT NOT NULL,
    department_id INT NOT NULL,
    salary INT NOT NULL,
    band_ceiling INT NOT NULL
)
    PARTITION BY RANGE(employee_id) (
    PARTITION p0 VALUES LESS THAN (5),
    PARTITION p1 VALUES LESS THAN (10),
    PARTITION p2 VALUES LESS THAN (15),
    PARTITION p3 VALUES LESS THAN MAXVALUE
)
INSERT INTO employees VALUES
    (', 'Harry', 'Potter', 3, 2, 30000, 35000), (', 'Sirius', 'Black', 1, 2, 33000,
        35000),
    (', 'Hermione', 'Granger', 3, 4, 37000, 40000), (', 'Luna', 'Lovegood', 2, 4,
        16500, 35000),
    (', 'Albus', 'Dumbledore', 1, 1, 28750, 35000), (', 'Peter', 'Pettigrew', 2, 3,
        35000, 35000),
    (', 'Severus', 'Snape', 2, 1, 52000, 60000), (', 'Dolores', 'Umbridge', 3, 1,
        35000, 35000),
    (', 'Ron', 'Weasley', 1, 3, 27000, 35000), (', 'Cho', 'Chang', 2, 4, 31000,
        35000),
    (', 'Draco', 'Malfoy', 1, 4, 28750, 35000), (', 'Pansy', 'Parkinson', 3, 2,
        35000, 35000),
    (', 'Ginny', 'Weasley', 1, 2, 32000, 35000), (', 'Padma', 'Patil', 3, 3, 42000,
        45000),
    (', 'Rubeus', 'Hagrid', 2, 3, 29000, 35000), (', 'Elphias', 'Doge', 2, 2, 31000,
        35000),
    (', 'Gellert', 'Grindelwald', 3, 3, 28000, 35000), (', 'Lucius', 'Malfoy', 3, 2,
        32000, 40000)
```

We can check that the partitions have been created using

```
SELECT PARTITION_NAME,TABLE_ROWS FROM INFORMATION_
    SCHEMA.PARTITIONS WHERE TABLE_NAME = 'employees'
```

which returns the data shown in Table 3.9.
Then if we use the query

```
SELECT * FROM employees PARTITION (p1)[12]
```

We retrieve only the rows that fall into this partition, as shown in Table 3.10.
This is, of course, the same result as

```
SELECT * FROM employees WHERE employee_id BETWEEN[13] 5 AND 9.
```

TABLE 3.9
The Data Returned

PARTITION_NAME	TABLE_ROWS
p0	4
p1	5
p2	5
p3	4

TABLE 3.10
The Data Returned

Employee_id	Forename	Surname	Hospital_id	Department_id	Salary	Band_ceiling
5	Albus	Dumbledore	1	1	28750	35000
6	Peter	Pettigrew	2	3	35000	35000
7	Severus	Snape	2	1	52000	60000
8	Dolores	Umbridge	3	1	35000	35000
9	Ron	Weasley	1	3	27000	35000

Rows from multiple partitions can be returned using a comma-delimited list. For example,

SELECT * FROM employees PARTITION (p1, p3)

returns all rows from partitions p1 and p3 while excluding rows from the remaining partitions. This is just one example of the types of partition available (RANGE) – others are by LIST, COLUMNS, HASH and KEY. It is also possible to partition by more than one column.

A table can only have one set of partitions so re-partitioning drops the existing one (although HASH and KEY are harder to drop than RANGE, LIST or COLUMNS). For example:

ALTER TABLE employees PARTITION BY KEY(employee_id)
    PARTITIONS 4

Also creates 4 partitions on employee_id but they are different, as the engine uses modulo arithmetic to determine the boundaries, as shown in Table 3.11.

Partitions are dynamically updated. If we insert all the employees again (possible as the primary key is AUTO_INCREMENT so they have different unique identifiers) we get the data in Table 3.12.

TABLE 3.11
The Partitions Returned

Partition_Name	Table_Rows
p0	5
p1	4
p2	4
p3	5

TABLE 3.12
The Partitions Returned After a Second INSERT

Partition_Name	Table_Rows
p0	9
p1	9
p2	9
p3	9

TABLE 3.13
The Data Returned

COUNT(hospital_id)	department_id
3	1
6	2
5	3
4	4

GROUP_BY is often used with aggregate functions, such as COUNT() to group the result-set by one or more columns, e.g.

    SELECT COUNT(hospital_id), department_id FROM employees GROUP
        BY department_id

will return the data in Table 3.13.

The WITH ROLLUP modifier adds an extra row to the returned data, being the sum of the non-grouped column, e.g.

    SELECT COUNT(hospital_id), department_id FROM employees GROUP
        BY department_id WITH ROLLUP

returns the data in Table 3.14.

TABLE 3.14
The Data Returned

COUNT(hospital_id)	department_id
3	1
6	2
5	3
4	4
18	NULL

TABLE 3.15
The Data Returned

Surname	Count(department_id)
Weasley	2
Malfoy	2

A HAVING clause must come after any GROUP BY clause and before any ORDER BY clause. It is applied nearly last, just before items are sent to the client, with no optimisation (LIMIT is applied after HAVING). It is similar to the WHERE clause, except that it allows aggregate functions, whereas WHERE does not.

Therefore, it is inefficient to use HAVING for items that should be in the WHERE clause, because it is evaluated almost last so more data is returned than need be.

For example, this:

    SELECT col_name FROM tbl_name HAVING col_name > 0

Should instead be:

    SELECT col_name FROM tbl_name WHERE col_name > 0

As mentioned, HAVING can refer to aggregate functions, which the WHERE clause cannot, e.g.

    SELECT surname,count(department_id) FROM employees GROUP BY
        surname HAVING count(department_id)>1

Which cannot be written as a WHERE clause and returns the data shown in Table 3.15.

The SQL standard requires that HAVING must reference only columns in the GROUP BY clause or columns used in aggregate functions. However, MySQL supports an extension to this behaviour and permits HAVING to refer to columns in the

TABLE 3.16
The Data Returned

Surname	Band_Ceiling	Salary	(Sand_Ceiling-salary)
Snape	60000	52000	8000
Patil	45000	42000	3000
Granger	40000	37000	3000

SELECT list and columns in outer subqueries as well.[14] Preference is given to standard SQL behaviour, though, so if a HAVING column name is used both in GROUP BY and as an aliased column in the output column list, preference is given to the column in the GROUP BY column.

ORDER_BY simply sorts the returned values according to the specified criteria. The default is ascending, which can be explicitly required via ASC. To sort in reverse order, use DESC, e.g.

SELECT surname, band_ceiling, salary, (band_ceiling-salary) FROM employees WHERE salary>35000 ORDER BY salary DESC

returns the data shown in Table 3.16.

As noted earlier, LIMIT is applied after HAVING and thus operates on the data returned by the rest of the SELECT statement. The LIMIT clause can be used to constrain the number of rows returned by the SELECT statement. LIMIT takes one or two numeric arguments, which must both be nonnegative integer constants, with these exceptions:

- Within prepared statements, LIMIT parameters can be specified using ? Placeholder markers.
- Within stored programs, LIMIT parameters can be specified using integer-valued routine parameters or local variables.

With two arguments, the first argument specifies the offset of the first row to return, and the second specifies the maximum number of rows to return. The offset of the initial row is 0 (not 1), e.g.

SELECT employee_id FROM employees LIMIT 5,10

Retrieves rows 6–15, as shown in Table 3.17.

To retrieve all rows from, say, the 96th to the last a very large number is used, e.g.

SELECT * FROM tbl LIMIT 95,18446744073709551615

TABLE 3.17
The Data Returned

Employee_Id

6
7
8
9
10
11
12
13
14
15

TABLE 3.18
The Data Returned

Employee_Id

1
2
3
4
5

With one argument, the LIMIT value specifies the number of rows to return from the beginning of the result set, e.g.

SELECT employee_id FROM employees LIMIT 5

Retrieves the first 5 rows, as shown in Table 3.18.

LIMIT row_count is thus equivalent to LIMIT 0, row_count.

For compatibility with PostgreSQL, MySQL also supports the LIMIT row_count OFFSET offset syntax.

If LIMIT occurs within a subquery and also is applied in the outer query, the outermost LIMIT takes precedence. Therefore, this query

(SELECT employee_id FROM employees LIMIT 1) LIMIT 2

will return two rows, not one.

The PROCEDURE clause names a procedure that should process the data in the result set. It is not permitted in a UNION statement.[15]

INTO OUTFILE writes the result set into the specified file, by default in /var/lib/mysql/<database_name> so it is essential to make sure the current user has write permission in that directory.[16] SELECT ... INTO OUTFILE writes the resulting rows to a file and allows the use of column and row terminators to specify a particular output format. The file must not exist as it cannot be overwritten.

The CHARACTER SET clause specifies the character set in which the results are to be written. Without the clause, no conversion takes place. In this case, if there are multiple character sets, the output will contain these too and may not easily be reloaded, e.g. to produce a file in CSV format:

```
SELECT employee_id, surname, forename
INTO OUTFILE '/var/lib/mysql-files/employees.txt'
FIELDS TERMINATED BY ','
OPTIONALLY ENCLOSED BY '"'
LINES TERMINATED BY '\n'
FROM employees
```

This produces the file employees.txt, containing:

```
1,"Harry","Potter"
2,"Sirius","Black"
3,"Hermione","Granger"
4,"Luna","Lovegood"
5,"Albus","Dumbledore"
6,"Peter","Pettigrew"
7,"Severus","Snape"
8,"Dolores","Umbridge"
9,"Ron","Weasley"
10,"Cho","Chang"
11,"Draco","Malfoy"
12,"Pansy","Parkinson"
13,"Ginny","Weasley"
14,"Padma","Patil"
15,"Rubeus","Hagrid"
16,"Elphias","Doge"
17,"Gellert","Grindelwald"
18,"Lucius","Malfoy"
```

When used with a storage engine that uses page or row locks, the FOR UPDATE modifier causes rows examined by the query to be write-locked until the end of the current transaction. Using LOCK IN SHARE MODE sets a shared lock that permits other transactions to read the examined rows but not to update or delete them.

Finally in this section, we examine two useful select_expr items.

Firstly, LIKE(pattern) is true if the pattern is matched, e.g. LIKE("P%")[17] will be true for anything beginning with "P", e.g.

```
SELECT surname FROM employees WHERE surname LIKE("P%")
```

Returns the data in Table 3.19.

TABLE 3.19
The Data Returned

Surname

Potter
Pettigrew
Parkinson
Patil

TABLE 3.20
The Data Returned

Surname

Snape
Parkinson

Where as % matches any number of characters, _ matches only one, e.g.

SELECT surname FROM employees WHERE surname LIKE("P___l")

returns just the last row above,

SELECT surname FROM employees WHERE surname LIKE("P__l")

returns no rows
    and

SELECT surname FROM employees WHERE surname LIKE("%s%n%")[18]

Returns the data shown in Table 3.20.

To search for a literal % or _, the character needs to be escaped using a \ (MySQL) or [] (SQL Server)

Note that in many SQL engines, MySQL included, "like" is not case sensitive. To make it so, use the BINARY modifier.

e.g. SELECT 'abc' LIKE 'ABC' returns 1

whereas SELECT 'abc' LIKE BINARY 'ABC' returns 0
Secondly, CONCAT() concatenates fields, e.g.

SELECT CONCAT(surname,', ',forename) AS full_name FROM employees
        ORDER BY full_name LIMIT 3

Returns the data shown in Table 3.21.

TABLE 3.21
The Data Returned

full_name

Potter, Harry
Black, Sirius
Granger, Hermione

Earlier we looked at joining tables together thus creating (or invoking) their relationship. We did this in two ways:

> SELECT {fields} FROM {tables} WHERE {condition, including the relationship}

And

> SELECT {fields} FROM {table} JOIN {table} ON {relationship} WHERE {condition}

There are four different types of join in SQL[19]:

- (INNER) JOIN: Returns records that have matching values in both tables. This is sometimes called a simple join and is the one created by the use of the WHERE clause to create the join.
- LEFT (OUTER) JOIN: Return all records from the left table and the matched records from the right table (i.e. only those rows from the right table where the joined fields are equal (the join condition is met)).
- RIGHT (OUTER) JOIN: Return all records from the right table and the matched records from the left table (i.e. only those rows from the left table where the joined fields are equal (the join condition is met)).
- FULL (OUTER) JOIN: Return all records when there is a match in either left or right table.

The syntax is <left table – table1> <type of join> <right table – table2> ON <criteria> and is best illustrated as shown in Figure 3.1.

For example, for the tables shown in Tables 3.22 and 3.23.

> SELECT Tests.TestID, Patients.PatientName
> FROM Tests
> INNER JOIN Patients ON Tests.PatientID = Patients.PatientID;

Returns the data in Table 3.24.

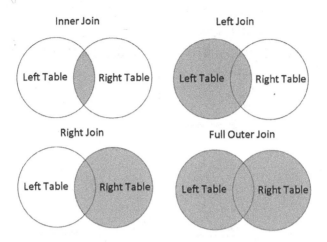

FIGURE 3.1   The four different JOIN types.

TABLE 3.22
The Table Tests

TestID	PatientID	StaffID	OrderDate
10308	2	7	2006-09-18
10309	37	3	2006-09-19
10310	77	8	2006-09-20

TABLE 3.23
The Table Patients

PatientID	PatientName	Address	City	Postal Code	Country
1	Hans Eins	Obere Str. 57	Berlin	12209	Germany
2	Manos Dos	Avda. de la Constitución 2222	México D.F.	05021	Mexico
3	Manos Tres	Mataderos 2312	México D.F	05023	Mexico

TABLE 3.24
The Data Returned by an INNER JOIN

TestID	PatientName
10308	Manos Dos

TABLE 3.25

The Data Returned by a LEFT JOIN

TestID	PatientName
10308	Manos Dos
10309	
10310	

TABLE 3.26

The Data Returned by a RIGHT JOIN

TestID	PatientName
	Hans Eins
10308	Manos Dos
	Manos Tres

Whereas

    SELECT Tests.TestID, Patients.PatientName
    FROM Tests
    LEFT JOIN Patients ON Tests.PatientID = Patients.PatientID;

Returns the data in Table 3.25.
    And

    SELECT Tests.TestID, Patients.PatientName
    FROM Tests
    RIGHT JOIN Patients ON Tests.PatientID = Patients.PatientID;

Returns the data in Table 3.26.
    Finally,

    SELECT Tests.TestID, Patients.PatientName
    FROM Tests
    FULL JOIN Patients ON Tests.PatientID = Patients.PatientID;

Returns the data in Table 3.27.
    There is also a SELF JOIN, which is a regular join where the table is joined with itself.

TABLE 3.27
The Data Returned by a FULL JOIN

TestID	PatientName
10308	Manos Dos
10309	
10310	
	Hans Eins
	Manos Tres

TABLE 3.28
The Data Returned by a SELF JOIN

PatientName1	PatientName2	City
Manos Dos	Manos Tres	México D.F.
Manos Tres	Manos Dos	México D.F.

For example

```
SELECT A.PatientName AS PatientName1, B.PatientName AS
    PatientName2, A.City
FROM Patients A, Patients B
WHERE A.PatientID <> B.PatientID
AND A.City = B.City
ORDER BY A.City;
```

matches patients that are from the same city and produces the data in Table 3.28.

The final SELECT modifier we will examine is one that aids in optimising queries: EXPLAIN, sometimes rendered DESCRIBE. The DESCRIBE and EXPLAIN statements are synonyms but in practice, the DESCRIBE keyword is more often used to obtain information about table structure,[20] whereas EXPLAIN is used to obtain a query execution plan. It is this latter that we are interested in, so will use EXPLAIN.

EXPLAIN works with SELECT, DELETE, INSERT, REPLACE and UPDATE statements. When EXPLAIN is used with an explainable statement, MySQL explains how it would process the statement, including information about how tables are joined and in which order. It does this by getting information from the optimiser about the statement execution plan.

With the help of EXPLAIN, it can be seen where indexes should be added to tables so that the statement executes faster by using indexes to find rows (which we will cover later). EXPLAIN can also be used to check whether the optimiser joins the tables in an optimal order. SELECT STRAIGHT_JOIN rather than just SELECT gives a hint to the optimiser to use a join order corresponding to the order in which the tables are named in the SELECT statement. However, STRAIGHT_JOIN may prevent indexes from being used because it disables semi-join transformations.[21]

The output format from EXPLAIN is determined via the FORMAT option: TRADITIONAL (or nothing) presents the output in tabular format. JSON displays the information in JSON format (see Section 12.3 for a description of JSON).

> EXPLAIN returns a row of information for each table used in the SELECT statement. It lists the tables in the output in the order that MySQL would read them while processing the statement. MySQL resolves all joins using a nested-loop join method. This means that MySQL reads a row from the first table, and then finds a matching row in the second table, the third table, and so on. When all tables are processed, MySQL outputs the selected columns and backtracks through the table list until a table is found for which there are more matching rows. The next row is read from this table and the process continues with the next table.
>
> (MySQL 5.7 Reference Manual 2022 [online] (2))

For example,

EXPLAIN SELECT * FROM employees

Returns the data in Table 3.29.

The id is the SELECT ID, and the type (select_type) can be one of the values listed in Table 3.30[22]:

Table is the name of the table to which the row of output refers, and it can be:

- <unionM,N>: The row refers to the union of the rows with id values of M and N.
- <derivedN>: The row refers to the derived table result for the row with an id value of N. A derived table may result, for example, from a subquery in the FROM clause.
- <subqueryN>: The row refers to the result of a materialised subquery for the row with an id value of N.

Partitions comes between table and type and it lists the partitions from which records would be matched by the query. It is NULL for nonpartitioned tables, as in the above example (which is why it doesn't appear).

Type is the JOIN type and possible_keys indicates the indexes from which MySQL can choose to find the rows in this table. Note that this column is totally

---

TABLE 3.29

The Return Values from an EXPLAIN Statement

id	select_type	table	type	possible_keys	Key	key_len	ref	rows	Extra
1	SIMPLE	employees	ALL	NULL	NULL	NULL	NULL	18	

TABLE 3.30
Possible Values for the select_type Column

select_type Value	Meaning
SIMPLE	Simple SELECT (not using UNION or subqueries)
PRIMARY	Outermost SELECT
UNION	Second or later SELECT statement in a UNION
DEPENDENT UNION	Second or later SELECT statement in a UNION, dependent on outer query
UNION RESULT	Result of a UNION
SUBQUERY	First SELECT in subquery
DEPENDENT SUBQUERY	First SELECT in subquery, dependent on outer query
DERIVED	Derived table SELECT (subquery in FROM clause)
MATERIALIZED	Materialised subquery
UNCACHEABLE SUBQUERY	A subquery for which the result cannot be cached and must be re-evaluated for each row of the outer query
UNCACHEABLE UNION	The second or later select in a UNION that belongs to an uncacheable subquery (see UNCACHEABLE SUBQUERY)

independent of the order of the tables as displayed in the output from EXPLAIN. That means that some of the keys in possible_keys might not be usable in practice with the generated table order. NULL indicates that there are no relevant indexes. In this case, it may be possible to improve the performance of the query by examining the WHERE clause to check whether it refers to some column or columns that would be suitable for indexing. If so, the index should be created (see Section 3.6) and checked with EXPLAIN.

The join types are, ordered from the best type to the worst:

- system, where the table has only one row (= system table). This is a special case of the const join type.
- const, where the table has at most one matching row, which is read at the start of the query. Because there is only one row, values from the column in this row can be regarded as constants by the rest of the optimiser. const tables are very fast because they are read only once, e.g.

  SELECT * FROM tbl_name WHERE primary_key=1

  which will return only one row due to the primary key being in the query, with an explicit value to match.
- eq_ref, where one row is read from the table for each combination of rows from the previous tables. Other than the system and const types, this is the best possible join type. It is used when all parts of an index are used by the join and the index is a PRIMARY KEY or UNIQUE NOT NULL index. eq_ref can be used for indexed columns that are compared using the = operator.

The comparison value can be a constant or an expression that uses columns from tables that are read before this table, e.g.

> SELECT * FROM ref_table,other_table WHERE ref_table.key_
> column=other_table.column

Here, one row is read from the table in order to match the query as it is uniquely indexed.

- ref, where all rows with matching index values are read from this table for each combination of rows from the previous tables. ref is used if the join uses only a leftmost prefix of the key or if the key is not a PRIMARY KEY or UNIQUE index (in other words, if the join cannot select a single row based on the key value). If the key that is used matches only a few rows, this is a good join type. ref can be used for indexed columns that are compared using the = or <=>[23] operator, e.g.

> SELECT * FROM ref_table WHERE key_column=expr

Here several rows are read from the table in order to match the query as it is not uniquely indexed. However, it is still not having to read the whole table.

- fulltext, where the join is performed using a FULLTEXT index.
- ref_or_null, which is like ref, but with the addition that MySQL does an extra search for rows that contain NULL values. This join type optimisation is used most often in resolving subqueries, e.g.

> SELECT * FROM ref_table WHERE key_column=expr OR key_
> column IS NULL

Like ref, several rows are read from the table in order to match the query as it is not uniquely indexed. However, it then has to search for NULL values as well. It is still not having to read the whole table, though.

- index_merge, where the Index Merge optimisation is used. In this case, the key column in the output row contains a list of indexes used, and key_len contains a list of the longest key parts for the indexes used.
- unique_subquery is just an index lookup function that replaces the subquery completely for better efficiency, replacing eq_ref for some IN subqueries of the form:

> value IN (SELECT primary_key FROM single_table WHERE some_expr)

but it is essentially similar to eq_ref.

- index_subquery, which is similar to unique_subquery. It replaces IN sub-queries, but it works for nonunique indexes in subqueries of the form:

> value IN (SELECT key_column FROM single_table WHERE some_expr)

- range, where only rows that are in a given range are retrieved, using an index to select the rows. The key column in the output row indicates which index is used. The key_len contains the longest key part that was used. The ref column is NULL for this type. range can be used when a key column is compared to a constant using any of the =, <>, >, >=, <, <=, IS NULL, <=>, BETWEEN, or IN() operators, e.g.

  SELECT * FROM tbl_name WHERE key_column BETWEEN 10 and 20

- Index, which is the same as ALL, except that the index tree is scanned. This occurs in two ways:
  - If the index is a covering index for the queries and can be used to satisfy all data required from the table, only the index tree is scanned. In this case, the Extra column says "Using index". An index-only scan usually is faster than ALL because the size of the index usually is smaller than the table data.
  - A full table scan is performed using reads from the index to look up data rows in index order. "Uses index" does not appear in the Extra column.
- ALL, where a full table scan is done for each combination of rows from the previous tables. This is normally not good if this table is the first table not to be marked as const, and usually very bad in all other cases (Basically, it's not a good thing but you may get away with it if "ALL" is the first non-"const" in the list). Normally, ALL can be avoided by adding indexes that enable row retrieval from the table based on constant values or column values from earlier tables. This is the value in the above example and shows that optimisation is certainly possible.

The key column indicates the key (index) that MySQL actually used. If MySQL decides to use one of the possible_keys indexes to look up rows, that index is listed as the key value. It is possible that key will name an index that is not present in the possible_keys value. This can happen if none of the possible_keys indexes are suitable for looking up rows, but all the columns selected by the query are columns of some other index, i.e., the named index covers the selected columns, so although it is not used to determine which rows to retrieve, an index scan is more efficient than a data row scan. MySQL can be forced to use or ignore an index listed in the possible_keys column, by using FORCE INDEX, USE INDEX, or IGNORE INDEX in the query.

The key_len column indicates the length of the key that MySQL decided to use. The value of key_len enables determination of how many parts of a multiple-part key MySQL actually uses. If the key column is NULL then the key_len column is also NULL. Due to the key storage format, the key length is one greater for a column that can be NULL than for a NOT NULL column.

The ref column shows which columns or constants are compared to the index named in the key column to select rows from the table. If the value is "func", the value used is the result of some function, which can be displayed by using SHOW WARNINGS following EXPLAIN to see the extended EXPLAIN output. The function might actually be an operator such as an arithmetic operator.

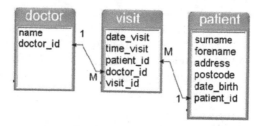

FIGURE 3.2　A sample database.

The rows column indicates the number of rows MySQL believes it must examine to execute the query.

The filtered column comes between rows and extra and indicates an estimated percentage of table rows that will be filtered by the table condition; i.e., rows shows the estimated number of rows examined and rows × filtered / 100 shows the number of rows that will be joined with previous tables.

Extra contains additional information about how MySQL resolves the query. A description of the different values and an example of how to use EXPLAIN output to optimise a query can be found at https://dev.mysql.com/doc/refman/5.7/en/explain-output.html.

As a further example of using EXPLAIN, using the database in Figure 3.2, let us examine this query:

> EXPLAIN SELECT surname,forename FROM patient,visit,doctor WHERE patient.patient_id=visit.patient_id AND visit.doctor_id=doctor.doctor_id AND name="Smith" AND date_visit='2019-01-03'

This returns values listed in Table 3.31.

A good indication of how good a join is can be found by taking the product of the values in the rows column of the EXPLAIN output. In this case, it yields 1028, which is not very efficient as we shall see later.

TABLE 3.31
Using EXPLAIN to Optimise a Query, Part 1

id	select_ type	table	type	possible_ keys	Key	key_ len	ref	rows	Extra
1	SIMPLE	visit	ALL	NULL	NULL	NULL	NULL	1028	Using where
1	SIMPLE	patient	eq_ref	PRIMARY	PRIMARY	5	appts.visit. patient_id	1	
1	SIMPLE	doctor	eq_ref	PRIMARY	PRIMARY	5	appts.visit. doctor_id	1	Using where

As another example,

> EXPLAIN SELECT Surname, Forename FROM patient WHERE (date_
> birth)=(SELECT MAX(date_birth) FROM patient)

returns the values listed in Table 3.32.

which may seem quite efficient, but the product (1369) is not, especially when compared to the two separate queries, the results of which are shown in Tables 3.33 and 3.34, respectively.

> EXPLAIN SELECT MAX(date_birth) FROM patient

TABLE 3.32
Using EXPLAIN to Optimise a Query, Part 2

id	select_type	table	type	possible_ keys	key	key_ len	ref	rows	Extra
1	PRIMARY	patient	ALL	NULL	NULL	NULL	NULL	37	Using where
2	SUBQUERY	patient	ALL	NULL	NULL	NULL	NULL	37	

TABLE 3.33
Using EXPLAIN to Optimise a Query, Part 3

id	select_ type	table	type	possible_ keys	key	key_len	ref	rows	Extra
1	SIMPLE	patient	ALL	NULL	NULL	NULL	NULL	37	

TABLE 3.34
Using EXPLAIN to Optimise a Query, Part 4

id	select_type	table	type	possible_ keys	Key	key_len	ref	rows	Extra
1	SIMPLE	patient	ALL	NULL	NULL	NULL	NULL	37	Using where

And

> EXPLAIN SELECT Surname, Forename FROM patient WHERE
> (date_birth)='1973-10-10'

which examines a total of 37+37=74 rows, demonstrating that elegance (the single query solution) is not necessarily the most efficient.

3.4.1 MySQL 8.0

We have considered the SELECT statement as implemented in MySQL 5.7, noting where statements have been deprecated in later versions. MySQL8.0 (the latest version) has these additional clauses:

- The clause "INTO OUTFILE" (complete with all modifiers) is a separate element (called "into_option") that may additionally appear after "select_ expr" and after "LIMIT".
- [WINDOW window_name AS (window_spec) [, window_name AS (window_spec)] ...] (appearing after HAVING)[24]
- The "FOR UPDATE" clause is expanded to

> [FOR {UPDATE I SHARE}
>   [OF tbl_name [, tbl_name] ...]
>   [NOWAIT I SKIP LOCKED]
>   I LOCK IN SHARE MODE]

3.5   CREATE/ALTER TABLE

A new table may be created using the syntax:

> CREATE TABLE table_name (
>   column1 column_definition,
>   column2 column_definition,
>   column3 column_definition,
>
>   ....
>)

There are lots of additional syntaxes which may be used (such as TEMPORARY which creates a table that is visible only within the current session and is dropped automatically when the session is closed[25]).

The example we used earlier (ignoring the partitioning) was

> CREATE TABLE employees (
> employee_id INT NOT NULL AUTO_INCREMENT PRIMARY KEY,
> forename VARCHAR(25) NOT NULL,

```
surname VARCHAR(25) NOT NULL,
hospital_id INT NOT NULL,
department_id INT NOT NULL,
salary INT NOT NULL,
band_ceiling INT NOT NULL
)
```

The key points of which are:

- The field names are employee_id, forename, surname, hospital_id, depart-ment_id, salary and band_ceiling.
- All of the fields must contain data (NOT NULL).
- VARCHAR(25) is a variable-width text field storing up to 25 characters – CHAR(25) would always store 25 and is thus less efficient (but may be important, e.g. for postcodes or telephone numbers).
- PRIMARY KEY declares the primary key for the table – this also creates an index for the field. This field must always have NOT NULL as a property.
- AUTO_INCREMENT means that values for the field should not be speci-fied when using an INSERT statement, but the value will be generated by the database engine. This is common practice when using ID fields.

Another important modifier is:

- [UNIQUE] INDEX. This creates an index on the named field. A UNIQUE index creates a constraint such that all values in the index must be distinct. An error occurs if a new row is added with a key value that matches an exist-ing row. For all engines, a UNIQUE index permits multiple NULL values for columns that can contain NULL.

Once a table has been created, it can be modified using very similar syntax via the ALTER TABLE command. Key clauses are:

- ADD COLUMN col_name column_definition, which adds the specified column.
- ALTER [COLUMN] col_name {SET DEFAULT literal l DROP DEFAULT}, which alters a column.
- CHANGE [COLUMN] old_col_name new_col_name column_definition [FIRSTlAFTER col_name], which alters a column.
- MODIFY [COLUMN] col_name column_definition [FIRST l AFTER col_name] , which alters a column.
- DROP [COLUMN] col_name, which removes a column.
- ADD {INDEXlKEY} [index_name] [index_type] (index_col_name,...) [index_option] ..., which adds an index.
- DROP INDEX index_name, which removes an index.
- DROP PRIMARY KEY, which removes a primary key (and its index).

- RENAME {INDEX|KEY} old_index_name TO new_index_name, which renames an index.
- RENAME [TO|AS] new_tbl_name, which renames the table.
- ADD PARTITION (partition_definition), which adds a partition
- DROP PARTITION partition_names, which removes a partition.
- REMOVE PARTITIONING, which removes all partitioning.

ALTER, CHANGE and MODIFY[26] all modify columns but are subtly different:

- CHANGE can rename a column and change its definition, or both. It has more capability than MODIFY, but at the expense of convenience for some operations. CHANGE requires naming the column twice if not renaming it. With FIRST or AFTER, it can reorder columns.
- MODIFY can change a column definition but not its name. It is more convenient than CHANGE to change a column definition without renaming it. With FIRST or AFTER, it can reorder columns.
- ALTER is used only to change a column default value.

To alter a column to change both its name and definition, CHANGE is used, specifying the old and new names and the new definition. For example, to rename an INT NOT NULL column from a to b and change its definition to use the BIGINT data type while retaining the NOT NULL attribute[27]:

ALTER TABLE t1 CHANGE a b BIGINT NOT NULL

To change a column definition but not its name, CHANGE or MODIFY is used. With CHANGE, the syntax requires two column names, so the same name must be specified twice to leave the name unchanged. For example, to change the definition of column b:

ALTER TABLE t1 CHANGE b b INT NOT NULL

MODIFY is more convenient to change the definition without changing the name because it requires the column name only once:

ALTER TABLE t1 MODIFY b INT NOT NULL

To change a column name but not its definition, CHANGE is used. The syntax requires a column definition, so to leave the definition unchanged, the definition the column currently has must be respecified. For example, to rename an INT NOT NULL column from b to a:

ALTER TABLE t1 CHANGE b a INT NOT NULL

For column definition changes using CHANGE or MODIFY, the definition must include the data type and all attributes that should apply to the new column, other than index attributes such as PRIMARY KEY or UNIQUE. Attributes present in the original definition but not specified for the new definition are not carried forward. If a column col1 is defined as INT UNSIGNED DEFAULT 1 COMMENT 'my column' and the column is modified as follows, intending to change only INT to BIGINT:

> ALTER TABLE t1 MODIFY col1 BIGINT

That statement changes the data type from INT to BIGINT, but it also drops the UNSIGNED, DEFAULT and COMMENT attributes. To retain them, the statement must include them explicitly:

> ALTER TABLE t1 MODIFY col1 BIGINT UNSIGNED DEFAULT 1
> COMMENT 'my column'

For data type changes using CHANGE or MODIFY, MySQL tries to convert existing column values to the new type as well as possible. This conversion may result in alteration of data, however. For example, if a string column is shortened, values may be truncated. To prevent the operation from succeeding if conversions to the new data type would result in loss of data, strict SQL mode should be invoked before using ALTER TABLE.

If CHANGE or MODIFY are used to shorten a column for which an index exists on the column, and the resulting column length is less than the index length, MySQL shortens the index automatically.

A table is removed using the DROP TABLE tablename command.

Table structure can be checked by using DESCRIBE or SHOW COLUMNS FROM, e.g.

> DESCRIBE doctor

Returns the data in Table 3.35.

TABLE 3.35
The Return from a DESCRIBE Command

Field	Type	Null	Key	Default	Extra
name	varchar(50)	YES		NULL	
doctor_id	decimal(10,0)	NO	PRI	NULL	

3.6   INDEXES

We have seen how to create, modify and remove indexes and in the chapter on Data Structures (Figure 1.7) discussed the use of indexes in sorting and searching. We return to the query

> SELECT surname,forename FROM patient,visit,doctor WHERE patient.
> patient_id=visit.patient_id AND visit.doctor_id=doctor.doctor_id AND
> name="Smith" AND date_visit='2017-01-03'

which we saw (using EXPLAIN) as examining 1028 rows. The foreign keys in the visit table (patient_id and doctor_id) are clearly going to be frequently used. It therefore makes sense to create indexes on these fields even though this will mean an additional overhead during INSERT and DELETE operations.

If we add these:

> ALTER TABLE visit ADD INDEX (patient_id)
> ALTER TABLE visit ADD INDEX (doctor_id)

We can see what indexes a table has, using SHOW INDEX FROM visit, as shown in Table 3.36.

We can thus see that the indexes have been created. If we now examine the query again, we get the data as shown in Table 3.37.

We obtain a row score of $12*5*1 = 60$, which is considerably better than the 1028 previously. Indexes are the most common way to speed up a slow SQL query. Using ADD and DROP means that an index could be created purely to run a query and removed afterwards – this removes the maintenance overhead for all other database operations, but creating the index may be slower than the original query.

In the EXPLAIN output in Table 3.37, we can also see an example of multiple possible keys and the optimiser selecting the one to use. It is possible that the optimiser will not select the most optimal one and so a SELECT query can provide a hint. The syntax for this is:

> tbl_name [[AS] alias] [index_hint_list]

where index_hint_list is

> index_hint [index_hint] ...

index_hint is

> USE {INDEX|KEY}
> [FOR {JOIN|ORDER BY|GROUP BY}] ([index_list])
> | IGNORE {INDEX|KEY}
> [FOR {JOIN|ORDER BY|GROUP BY}] (index_list)
> | FORCE {INDEX|KEY}
> [FOR {JOIN|ORDER BY|GROUP BY}] (index_list)

TABLE 3.36
The Effect of Adding an INDEX, Part 1

Table	Non_unique	Key_name	Seq_in_index	Column_name	Collation	Cardinality	Sub_part	Packed	Null	Index_type	Comment	Index_comment
visit	0	PRIMARY	1	visit_id	A	825	NULL	NULL		BTREE		
visit	1	patient_id	1	patient_id	A	165	NULL	NULL	YES	BTREE		
visit	1	doctor_id	1	doctor_id	A	165	NULL	NULL	YES	BTREE		

TABLE 3.37

The Effect of Adding an INDEX, Part 2

id	select_ type	table	type	possible_keys	key	key_len	ref	rows	Extra
1	SIMPLE	doctor	ALL	PRIMARY	NULL	NULL	NULL	12	Using where
1	SIMPLE	visit	ref	patient_ id,doctor_id	doctor_id	6	appts. doctor. doctor_id	5	Using where
1	SIMPLE	Patient	eq_ref	PRIMARY	PRIMARY	5	appts.visit. patient_id	1	

And index_list is

> index_name [, index_name] ...

Examples of such usage might be:

> SELECT * FROM table1 USE INDEX (col1_index,col2_index) WHERE
>   col1=1 AND col2=2 AND col3=3
> SELECT * FROM table1 IGNORE INDEX (col3_index) WHERE col1=1
>   AND col2=2 AND col3=3

Determining which indexes to create can be difficult, but an order for investigation (using a particular SQL statement) might be:

1. Create indexes on primary keys in JOINs
2. Create indexes on foreign keys in JOINs
3. Create indexes on filtered columns (only really useful if the query returns a lot of data)
4. Create indexes on aggregate functions
5. Create indexes on columns in the ORDER BY statement

3.7 PRIVILEGES

Most DBMSs restrict access to certain parts of the system. This is achieved through the GRANT command, which grants privileges to user accounts and has the syntax:

> GRANT
>   priv_type [(column_list)]
>   [, priv_type [(column_list)]] ...
>   ON [object_type] priv_level
>   TO user [auth_option] [, user [auth_option]] ...
>   [REQUIRE {NONE I tls_option [[AND] tls_option] ...}]
>   [WITH {GRANT OPTION I resource_option} ...]

```
GRANT PROXY ON user
    TO user [, user] ...
    [WITH GRANT OPTION]
object_type: {
    TABLE
    | FUNCTION
    | PROCEDURE
}
priv_level: {
    *
    | *.*
    | db_name.*
    | db_name.tbl_name
    | tbl_name
    | db_name.routine_name
}
auth_option: {
    IDENTIFIED BY 'auth_string'
    | IDENTIFIED WITH auth_plugin
    | IDENTIFIED WITH auth_plugin BY 'auth_string'
    | IDENTIFIED WITH auth_plugin AS 'hash_string'
    | IDENTIFIED BY PASSWORD 'hash_string'
}
tls_option: {
    SSL
    | X509
    | CIPHER 'cipher'
    | ISSUER 'issuer'
    | SUBJECT 'subject'
}
resource_option: {
    | MAX_QUERIES_PER_HOUR count
    | MAX_UPDATES_PER_HOUR count
    | MAX_CONNECTIONS_PER_HOUR count
    | MAX_USER_CONNECTIONS count
}
```

To use GRANT, the DBA (as no-one else ought to be doing this) must have the GRANT OPTION privilege and must have the privileges that are being granted.
    An example, including creating the user in the first place, might be[28]:

```
CREATE USER 'paul'@'localhost' IDENTIFIED BY 'password'
GRANT ALL ON db1.* TO 'paul'@'localhost'
GRANT SELECT ON db2.invoice TO 'paul'@'localhost'
ALTER USER 'paul'@'localhost' WITH MAX_QUERIES_PER_HOUR 90
```

The opposite of GRANT is REVOKE, which removes the privileges. The permissible privileges for GRANT and REVOKE are shown in Table 3.38 (MySQL 5.7 Reference Manual 2022 [online] (3)).

TABLE 3.38
Privilege Definitions

Privilege	Meaning and Grantable Levels
ALL [PRIVILEGES]	Grant all privileges at specified access level except GRANT OPTION and PROXY.
ALTER	Enable use of ALTER TABLE. Levels: Global, database, table.
ALTER ROUTINE	Enable stored routines to be altered or dropped. Levels: Global, database, procedure.
CREATE	Enable database and table creation. Levels: Global, database, table.
CREATE ROUTINE	Enable stored routine creation. Levels: Global, database.
CREATE TABLESPACE	Enable tablespaces and log file groups to be created, altered, or dropped. Level: Global.
CREATE TEMPORARY TABLES	Enable use of CREATE TEMPORARY TABLE. Levels: Global, database.
CREATE USER	Enable use of CREATE USER, DROP USER, RENAME USER and REVOKE ALL PRIVILEGES. Level: Global.
CREATE VIEW	Enable views to be created or altered. Levels: Global, database, table.
DELETE	Enable use of DELETE. Level: Global, database, table.
DROP	Enable databases, tables and views to be dropped. Levels: Global, database, table.
EVENT	Enable use of events for the Event Scheduler. Levels: Global, database.
EXECUTE	Enable the user to execute stored routines. Levels: Global, database, table.
FILE	Enable the user to cause the server to read or write files. Level: Global.
GRANT OPTION	Enable privileges to be granted to or removed from other accounts. Levels: Global, database, table, procedure, proxy.
INDEX	Enable indexes to be created or dropped. Levels: Global, database, table.
INSERT	Enable use of INSERT. Levels: Global, database, table, column.
LOCK TABLES	Enable use of LOCK TABLES on tables for which the user has the SELECT privilege. Levels: Global, database.
PROCESS	Enable the user to see all processes with SHOW PROCESSLIST. Level: Global.
PROXY	Enable user proxying. Level: From user to user.
REFERENCES	Enable foreign key creation. Levels: Global, database, table, column.
RELOAD	Enable use of FLUSH operations. Level: Global.
REPLICATION CLIENT	Enable the user to ask where master or slave servers are. Level: Global.
REPLICATION SLAVE	Enable replication slaves to read binary log events from the master. Level: Global.
SELECT	Enable use of SELECT. Levels: Global, database, table, column.
SHOW DATABASES	Enable SHOW DATABASES to show all databases. Level: Global.
SHOW VIEW	Enable use of SHOW CREATE VIEW. Levels: Global, database, table.
SHUTDOWN	Enable use of mysqladmin shutdown. Level: Global.
SUPER	Enable use of other administrative operations such as CHANGE MASTER TO, KILL, PURGE BINARY LOGS, SET GLOBAL and mysqladmin debug command. Level: Global.
TRIGGER	Enable trigger operations. Levels: Global, database, table.
UPDATE	Enable use of UPDATE. Levels: Global, database, table, column.
USAGE	Synonym for "no privileges".

Global privileges are administrative or apply to all databases on a given server. They are assigned using the ON *.* syntax, e.g.:

    GRANT ALL ON *.* TO 'someuser'@'somehost'
    GRANT SELECT, INSERT ON *.* TO 'someuser'@'somehost'

Database privileges apply to all objects in a given database.[29] They are assigned using the ON db_name.* syntax, e.g.:

    GRANT ALL ON mydb.* TO 'someuser'@'somehost' \
    GRANT SELECT, INSERT ON mydb.* TO 'someuser'@'somehost'

Table privileges apply to all columns in a given table. They are assigned using the ON db_name.tbl_name syntax, e.g.:

    GRANT ALL ON mydb.mytbl TO 'someuser'@'somehost'
    GRANT SELECT, INSERT ON mydb.mytbl TO 'someuser'@'somehost'

Column privileges apply to single columns in a given table. Each privilege to be granted at the column level must be followed by the column or columns, enclosed within parentheses, e.g.:

    GRANT SELECT (col1), INSERT (col1,col2) ON mydb.mytbl TO
        'someuser'@'somehost'

Stored routine privileges apply to stored routines (procedures and functions). They can be granted at the global and database levels. Except for CREATE ROUTINE, these privileges can be granted at the routine level for individual routines. For example:

    GRANT CREATE ROUTINE ON mydb.* TO 'someuser'@'somehost'
    GRANT EXECUTE ON PROCEDURE mydb.myproc TO
        'someuser'@'somehost'

This is, of course, inefficient for large groups of users. In this case it is better to create roles, assign the roles to users and then GRANT (or REVOKE) privileges to that role, which then affects all users with that role. For example:

    CREATE ROLE 'mydb_ReadOnly'
    GRANT 'mydb_ReadOnly' TO 'someuser'@'somehost'
    GRANT SELECT ON mydb.* TO 'mydb_ReadOnly'

GRANT/REVOKE can be applied to the role at any point and other users can be added to the role, who will inherit the privileges (or otherwise) of that role.

3.8   LOADING LARGE DATA SETS

While the INSERT statement works well for inserting a small quantity of data, repeatedly typing it is both boring and inefficient for large data sets. Either of these can lead to simple errors being made as well. There are two main ways of loading a large data set.

The first utilises script files. A script file is simply a text file that consists of a set of SQL commands that is executed against the database engine. There are two main ways to do this in MySQL:

- At invocation of MySQL, e.g.
  - mysql < text_file

Note that this will require the USE statement and any other setup commands to be part of the script file.

- Within MySQL, once the database has been selected, e.g.
  - source file_name

It is often useful to display messages as the script file progresses and this can be done either by starting MySQL with the –verbose option, which then displays each command before it is run, or by embedding statements such as SELECT '<info_to_display>' AS ' ' within the script.

In order to create such a script file from, say a large Excel spreadsheet, code needs to be written to create each line as an INSERT statement.

The second form of data loading is the LOAD DATA INFILE statement. This complements the SELECT ... INTO OUTFILE statement and in its simplest form is:

LOAD DATA INFILE 'file_name' INTO TABLE tbl_name

There are several modifiers[30] of which those detailing how the data is to be interpreted are probably the most useful. This is accomplished using the FIELDS [TERMINATED BY 'string'] [[OPTIONALLY] ENCLOSED BY 'char'] [ESCAPED BY 'char'] and LINES [STARTING BY 'string'] [TERMINATED BY 'string'] clauses. If no FIELDS or LINES clauses are specified, the defaults are the same as this:

FIELDS TERMINATED BY '\t' ENCLOSED BY 'ESCAPED BY '\\'
LINES TERMINATED BY '\n' STARTING BY'[31]

If all the lines to read in have a common prefix that should be ignored, LINES STARTING BY 'prefix_string' can be used to skip over the prefix, and anything before it. If a line does not include the prefix, the entire line is skipped. For example, the statement:

LOAD DATA INFILE '/tmp/test.txt' INTO TABLE test FIELDS
TERMINATED BY ',' LINES STARTING BY 'xxx'

With a data file of:

    xxx"abc",1
    something xxx"def",2
    "ghi",3123

will load the rows ("abc",1) and ("def",2). The third row in the file is skipped because it does not contain the prefix.

The IGNORE number LINES option can be used to ignore lines at the start of the file. For example, IGNORE 1 LINES can be used to skip over an initial header line containing column names:

    LOAD DATA INFILE '/tmp/test.txt' INTO TABLE test IGNORE 1 LINES

As an example, this statement would load in a .csv[32] file:

    LOAD DATA INFILE '/var/lib/mysql-files/data.csv'[33] INTO TABLE
        tbl_name
    FIELDS TERMINATED BY ',' ENCLOSED BY '"'
    LINES TERMINATED BY '\r\n'
    IGNORE 1 LINES

By default, when no column list is provided at the end of the LOAD DATA INFILE statement, input lines are expected to contain a field for each table column. In order to load only some of a table's columns, a column list should be specified:

    LOAD DATA INFILE 'persondata.txt' INTO TABLE persondata
        (col1,col2,...)

A column list must also be specified if the order of the fields in the input file differs from the order of the columns in the table. Note that all of the data from the input file is expected to be loaded – the column list specifies which columns to load the data into and in which order.

3.9   STORED ROUTINES

A stored routine is a set of SQL statements that are stored in the server. Once this has been done, the individual statements are invoked by referring to the stored routine instead. Stored routines can be particularly useful in certain situations:

- When multiple client applications are written in different languages or work on different platforms but need to perform the same database operations.
- When security is paramount. Banks, for example, use stored procedures and functions for all common operations. This provides a consistent and

secure environment, and routines can ensure that each operation is properly logged. In such a setup, applications and users would have no access to the database tables directly and can only execute specific stored routines.

Stored routines can provide improved performance because less information needs to be sent between the server and the client. The trade-off is that this does increase the load on the database server because more of the work is done on the server side and less is done on the client (application) side.

Stored routines also enable libraries of functions to be present in the database server.

A stored routine is either a procedure or a function. They are created with the CREATE PROCEDURE and CREATE FUNCTION statements. A procedure is invoked using a CALL statement and can only pass back values using output variables. A function can be called from inside a statement just like any other function (that is, by invoking the function's name) and can return a scalar value. The body of a stored routine can use compound statements.

Stored routines can be dropped with the DROP PROCEDURE and DROP FUNCTION statements and altered with the ALTER PROCEDURE and ALTER FUNCTION statements.

These examples show the command line for clarity[34]:
Procedure:

```
mysql> delimiter //[35]
mysql> CREATE PROCEDURE simpleproc (OUT param1 INT)
-> BEGIN
->   SELECT COUNT(*) INTO param1 FROM t;
-> END//
Query OK, 0 rows affected (0.00 sec)
mysql> delimiter ;
mysql> CALL simpleproc(@a);
Query OK, 0 rows affected (0.00 sec)
mysql> SELECT @a; Table 3.39 shows the console return values.
```

Function:

```
mysql> CREATE FUNCTION hello (s CHAR(20))
mysql> RETURNS CHAR(50) DETERMINISTIC
-> RETURN CONCAT('Hello, ',s,'!');
Query OK, 0 rows affected (0.00 sec)
mysql> SELECT hello('world'); Table 3.40 shows the console return values.
```

TABLE 3.39
User Console Return Values

@a
3

TABLE 3.40
Return Values from the Stored Procedure

hello('world')
Hello, world!

Some SQL engines (but not MySQL) require that the procedure contains statements to indicate actions that the procedure/function may take:

- CONTAINS SQL indicates that the routine does not contain statements that read or write data. This is the default if none of these characteristics is given explicitly. Examples of such statements are SET @x = 1 or DO RELEASE_LOCK('abc'), which execute but neither read nor write data.
- NO SQL indicates that the routine contains no SQL statements.
- READS SQL DATA indicates that the routine contains statements that read data (for example, SELECT), but not statements that write data.
- MODIFIES SQL DATA indicates that the routine contains statements that may write data (for example, INSERT or DELETE).

Stored routines often benefit from control mechanisms such as CASE, IF/ELSE, IFNULL, NULLIF, LOOP and WHILE. These operate as you would expect, for example:

```
CREATE procedure SevenTimesTable()
    label:BEGIN
    DECLARE times INT ;
    DECLARE times_table VARCHAR(255);
    SET times =1;
    SET times_table = ';
    loop_label: LOOP
        IF times > 10 THEN
            LEAVE loop_label;
        END IF;
        SET times_table = CONCAT(times_table,'7*',times,'=',7*times,',');
        SET times = times + 1;
        ITERATE loop_label;
    END LOOP;
    SELECT times_table;
END//
```

which returns the value as shown in Table 3.41.

TABLE 3.41

The Return Value from the Stored Procedure

times_table

7*1=7,7*2=14,7*3=21,7*4=28,7*5=35,7*6=42,7*7=49,7*8=56,7*9=63,7*10=70,

3.10   TRIGGERS

Triggers are actions that are performed by the SQL engine under certain conditions (i.e. when the condition is triggered). An example might be, for a table "mainte-nance_cost" with two fields (equipment_id INT, amount DECIMAL(10,2))

> CREATE TRIGGER ins_sum BEFORE INSERT ON maintenance_cost FOR EACH ROW SET @sum = @sum + NEW.amount

This CREATE TRIGGER statement creates a trigger named ins_sum that is associated with the maintenance_cost table. It also includes clauses that specify the trigger action time, the triggering event and what to do when the trigger activates:

- The keyword BEFORE indicates the trigger action time. In this case, the trigger activates before each row is inserted into the table. The other permitted keyword here is AFTER.
- The keyword INSERT indicates the trigger event; that is, the type of operation that activates the trigger. In the example, INSERT operations cause the trigger activation. Triggers for DELETE and UPDATE operations can also be created.
- The statement following FOR EACH ROW defines the trigger body; that is, the statement to execute each time the trigger activates, which occurs once for each row affected by the triggering event. In the example, the trigger body is a simple SET that accumulates into a user variable the values inserted into the amount column. The statement refers to the column as NEW.amount which means "the value of the amount column to be inserted into the new row."

In this case, the code is used as follows (command line prompt again included for clarity – the results are shown in Table 3.42):

> mysql> SET @sum = 0;
> mysql> INSERT INTO maintenance_cost VALUES (137,14.98),
>   (141,1937.50),(97,-100.00);
> mysql> SELECT @sum AS 'Total amount inserted';

Triggers are destroyed using the DROP TRIGGER statement.

TABLE 3.42
TRIGGER Example Part 1

Total Amount Inserted

1852.48

As a fuller example:

```
CREATE TABLE psg1(num1 INT);
CREATE TABLE psg2(num2 INT);
CREATE TABLE psg3(num3 INT NOT NULL AUTO_INCREMENT
    PRIMARY KEY);
CREATE TABLE psg4(
    num4 INT NOT NULL AUTO_INCREMENT PRIMARY KEY,
    num5 INT DEFAULT 0
);
delimiter |
CREATE TRIGGER psg_update BEFORE INSERT ON psg1
    FOR EACH ROW
    BEGIN
        INSERT INTO psg2 SET num2 = NEW.num1;
        DELETE FROM psg3 WHERE num3 = NEW.num1;
        UPDATE psg4 SET num5 = num5 + 1 WHERE num4 = NEW.num1;
    END;
|
delimiter ;
INSERT INTO psg3 (num3) VALUES
    (NULL), (NULL), (NULL), (NULL), (NULL),
    (NULL), (NULL), (NULL), (NULL), (NULL);
INSERT INTO psg4 (num4) VALUES
    (0), (0), (0), (0), (0), (0), (0), (0), (0), (0);
```

We now have in the tables the values shown in Table 3.43.
If we then insert into psg1:

```
INSERT INTO psg1 VALUES (1), (3), (1), (7), (1), (8), (4), (4)
```

The tables now contain the values shown in Table 3.44.

It is possible to have an INSERT statement that triggers an INSERT on another table, which in turn triggers an INSERT in another (and so on). It is therefore vitally important to check that loops do not exist.

Triggers can slow database performance, so ought to be used for simple and quick tasks, unless part of a routine that takes a long time anyway (e.g. generating data for reporting).

TABLE 3.43
TRIGGER Example Part 2

psg1	psg2	psg3	psg4	
num1	num2	num3	num4	num5
		1	1	0
		2	2	0
		3	3	0
		4	4	0
		5	5	0
		6	6	0
		7	7	0
		8	8	0
		9	9	0
		10	10	0

TABLE 3.44
TRIGGER Example Part 3

psg1	psg2	psg3	psg4	
num1	num2	num3	num4	num5
1	1	2	1	3
3	3	5	2	0
1	1	6	3	1
7	7	9	4	2
1	1	10	5	0
8	8		6	0
4	4		7	1
4	4		8	1
			9	0
			10	0

3.11 COLUMNSTORE

(See Section 2.10 for a description of columnstore vs rowstore).

All relational tables, unless specified as a clustered columnstore index, use rowstore as the underlying data format. CREATE TABLE creates a rowstore table unless the WITH CLUSTERED COLUMNSTORE INDEX option is specified.

A rowstore table can be converted into a columnstore, using the CREATE COLUMNSTORE INDEX statement.

3.12   CONCURRENCY CONTROL AND TRANSACTION MANAGEMENT

Transaction management (TM) handles all transactions properly in DBMS.
• Database transactions are the events or activities such as series of data read/write operations on data object(s) stored in database system.

(Larson et al. 2011)

Concurrency control (CC) is a process to ensure that data is updated correctly and appropriately when multiple transactions are concurrently executed in DBMS.

(Connolly and Begg 2014)

In a multi-user database it is possible that two users will be modifying the same data at the same time. The larger the database, the higher the probability of it happening. TM and CC are methods to prevent such incidents corrupting the database. Consider a stored procedure for inserting records into two tables. The first table has an autoincrement primary key that must be retrieved (once generated) and then stored into the second table. For example,

> INSERT INTO Title (Title, Media, Price, Copies) VALUES ("Single Release","CD",2.00,1)
> INSERT INTO Item (Artist_ID, Title_ID, Track_ID, Track_No) VALUES (1,SELECT MAX(Title_ID) FROM Title,1,1)

This requires the two SQL statements to be run one after the other in order to keep the database accurate. However, if another "INSERT INTO Title" statement appears in between them, then the database will be corrupt.

Other examples might be a SELECT summing account balances while an UPDATE applies an interest rate and thus increases them.

In general, concurrency control is an essential part of TM. It is a mechanism for ensuring correctness when two or more database transactions that access the same data or data set are executed concurrently with a time overlap. Data interference is usually caused by a write operation among transactions on the same set of data in a DBMS.

A transaction is a logical unit of work that contains one or more SQL statements. A transaction is therefore an atomic unit.[36] The effects of all of the SQL statements in a transaction can be either all committed (applied to the database) or all rolled back (undone from the database).

There are two main kinds of concurrency control mechanisms: pessimistic and optimistic.

The pessimistic concurrency control delays the transactions if they conflict with other transactions at some time in the future by locking the data or by using a timestamping technique.

The optimistic concurrency control assumes that conflict is rare and thus allows concurrent transactions to proceed without imposing delays then checks for conflicts only at the end, when a transaction commits.

There is another mechanism, the semi-optimistic technique, which uses lock operations in some situations (if they may violate some rules) and does not lock in other circumstances.

Pessimistic and optimistic concurrency control mechanisms provide different performance metrics, e.g., different average transaction completion rates or throughput.

There is a tradeoff between the concurrency control techniques.

For pessimistic concurrency control, the strengths are:

- It guarantees that all transactions can be executed correctly.
- Data is properly consistent by either rolling back to the previous state (Abort operation) or by accepting new content (Commit operation) when the transaction conflict is cleared.
- Database is relatively stable and reliable.

Its weaknesses are:

- Transactions are slow due to the delay caused by locking or time-stamping event.
- Runtime is longer. Transaction latency increases significantly.
- Throughput (e.g. read/write, update, rollback operations, etc.) is reduced.

For optimistic concurrency control, the strengths are:

- Transactions are executed more efficiently.
- Data content is relatively safe.
- Throughput is much higher.

Its weaknesses are:

- There is a risk of data interference among concurrent transactions since conflict may occur during execution. In this case, data is no longer correct.
- Database may have some hidden errors with inconsistent data; even the conflict check is performed at the end of transactions.
- Transactions may be in deadlock that causes the system to hang.

Transaction control is achieved through these statements:

- START TRANSACTION or BEGIN start a new transaction.
- COMMIT commits the current transaction, making its changes permanent.
- ROLLBACK rolls back the current transaction, cancelling its changes.
- SET autocommit disables or enables the default autocommit mode for the current session. (Vallejo et al. 2011)

By default, MySQL runs with autocommit mode enabled. This means that as soon as a statement that updates a table is executed, MySQL makes it permanent. The change cannot be rolled back.

The START TRANSACTION statement thus disables autocommit mode implicitly for a single series of statements, e.g.

```
START TRANSACTION;
SELECT @A:=SUM(cost) FROM maintenance WHERE type=1;
UPDATE op_report SET summary=@A WHERE type=1;
COMMIT;
```

With START TRANSACTION, autocommit remains disabled until the transaction is ended with COMMIT or ROLLBACK. The autocommit mode then reverts to its previous state.

Some statements cannot be rolled back. In general, these include **Data Definition Language** (DDL) statements, such as those that create or drop databases, those that create, drop, or alter tables or stored routines. These statements implicitly end any transaction active in the current session, as if a COMMIT had taken place before executing the statement. Examples are ALTER DATABASE, ALTER PROCEDURE, ALTER TABLE, CREATE DATABASE, CREATE INDEX, CREATE PROCEDURE, CREATE TABLE, CREATE TRIGGER, DROP DATABASE, DROP INDEX, DROP PROCEDURE, DROP TABLE, DROP TRIGGER.

The concept of a transaction is simply illustrated using a banking database. When a bank customer transfers money from a savings account to a current account, the transaction can consist of three separate operations:

- Decrement the savings account
- Increment the current account
- Record the transaction in the transaction journal

The DBMS must allow for two situations. If all three SQL statements can be performed to maintain the accounts in proper balance, the effects of the transaction can be applied to the database. However, if a problem such as insufficient funds, invalid account number, or a hardware failure prevents one or two of the statements in the transaction from completing, the entire transaction must be rolled back so that the balance of all accounts remains correct. For example,

```
START TRANSACTION
UPDATE saving_account SET balance=balance-500 WHERE
      account_number=1048
UPDATE current_account SET balance=balance+500 WHERE
      account_number=576
INSERT INTO log_file VALUES ('','transfer',1048,576,500)
COMMIT
```

Furthermore, if at any time during the execution a SQL statement causes an error, all effects of the statement are rolled back. The effect of the rollback is as if that statement had never been run. This operation is a statement-level rollback.

Errors discovered during SQL statement execution cause statement-level rollbacks. An example of such an error is attempting to insert a duplicate value in a primary key. Single SQL statements involved in a deadlock (competition for the same data) can also cause a statement-level rollback. Errors discovered during SQL statement parsing, such as a syntax error, have not yet been run, so they do not cause a statement-level rollback.

A SQL statement that fails causes the loss only of any work it would have performed itself. It does not cause the loss of any work that preceded it in the current transaction. If the statement is a DDL statement, then the implicit commit that immediately preceded it is not undone.

Intermediate markers called savepoints can be declared within the context of a transaction. Savepoints divide a long transaction into smaller parts. Using savepoints provides the option of rolling back work performed before the current point in the transaction, but after a declared savepoint within the transaction. For example, savepoints can be used throughout a long complex series of updates, so an error does not require every statement to be resubmitted.

Savepoints are similarly useful in application programs. If a procedure contains several functions, then a savepoint can be created before each function begins. Then, if a function fails, it is easy to return the data to its state before the function began and re-run the function with revised parameters or perform a recovery action.

The syntax is:

SAVEPOINT identifier
ROLLBACK [WORK] TO [SAVEPOINT] identifier
RELEASE SAVEPOINT identifier

Another way to implement transaction management is with locks – either at record or table level.

The syntax is:

LOCK TABLES
    tbl_name [[AS] alias] lock_type
    [, tbl_name [[AS] alias] lock_type] ...
lock_type:
    READ [LOCAL]
    | [LOW_PRIORITY] WRITE
UNLOCK TABLES

A table lock protects only against inappropriate reads or writes by other sessions. A session holding a WRITE lock can perform table-level operations such as DROP TABLE or TRUNCATE TABLE. For sessions holding a READ lock, DROP TABLE and TRUNCATE TABLE operations are not permitted.

For the two types of lock:
READ [LOCAL] lock:

• The session that holds the lock can read the table (but not write it).
• Multiple sessions can acquire a READ lock for the table at the same time.

- Other sessions can read the table without explicitly acquiring a READ lock.
- The LOCAL modifier enables nonconflicting INSERT statements (concurrent inserts) by other sessions to execute while the lock is held. However, READ LOCAL cannot be used if the database will be manipulated using processes external to the server while the lock is held.

[LOW_PRIORITY] WRITE lock:

- The session that holds the lock can read and write the table.
- Only the session that holds the lock can access the table. No other session can access it until the lock is released.
- Lock requests for the table by other sessions block while the WRITE lock is held.

If the LOCK TABLES statement must wait due to locks held by other sessions on any of the tables, it blocks until all locks can be acquired.

A session that requires locks must acquire all the locks that it needs in a single LOCK TABLES statement. While the locks thus obtained are held, the session can access only the locked tables. For example, in the following sequence of statements, an error occurs for the attempt to access t2 because it was not locked in the LOCK TABLES statement (the console return is shown in Table 3.45):

```
mysql> LOCK TABLES t1 READ;
mysql> SELECT COUNT(*) FROM t1;
mysql> SELECT COUNT(*) FROM t2;
```

UNLOCK TABLES commits a transaction only if any tables currently have been locked with LOCK TABLES to acquire nontransactional table locks. A commit does not occur for UNLOCK TABLES following FLUSH TABLES WITH READ LOCK because the latter statement does not acquire table-level locks.

Record-level locking is a part of database management but is normally done by the DBMS and not the user. A record locked by another SQL statement will normally only cause a delay in another one accessing the same record. At user level, transaction processing is the best way to achieve the desired effect.

3.13   DATABASE PERFORMANCE TUNING

Database tuning describes a group of activities used to optimise and homogenise the performance of a database. It usually overlaps with query tuning, but refers to the

TABLE 3.45
User Console Return

COUNT(*)
3 ERROR 1100 (HY000): Table 't2' was not locked with LOCK TABLES

design of the database files, selection of the database management system application and configuration of the database's environment (operating system, CPU, etc.) – of these only the first is relevant to this book and has been covered earlier in the movement through normal forms (see Section 2.6).

Database tuning can be an incredibly difficult task, particularly when working with large-scale data (such as an **Electronic Health Record System** (EHRS)) where even the most minor change can have a dramatic (positive or negative) impact on performance.

Techniques and tips for tuning the queries (which is the major part of any database activity) are plentiful, but include:

- Use of EXPLAIN as described earlier.
- Use of STATISTICS. This gathers information from the database in order to assist the optimiser in, for example, selection of the most appropriate index. For example:
  - SELECT * FROM patient WHERE city = 'London' AND phone = '123-456-789'
  - If the table is indexed on both city and phone, then the optimiser will only use one of these. phone is likely to return less records (so that is the index to use), but it is the statistics that will enable the optimiser to assess this.
- Avoid functions on the right hand side of the operator, For example:
  - SELECT * FROM patient WHERE YEAR(appointment) == 2015 AND MONTH(appointment) = 6
  - If appointment has an index, this query changes the WHERE clause in such a way that this index cannot be used anymore. Rewriting the query as following will increase the performance by enabling the index to be used:
  - SELECT * FROM patient WHERE appointment BETWEEN '1/6/2015' AND '30/6/2015' [37]
- Predetermine expected growth. As discussed earlier, indexes can have a negative impact on **Data Manipulation Language** (DML)[38] queries. One way of minimising this negative affect is to specify an appropriate value for fill factor when creating indexes. When an index is created, the data for indexed columns is stored on the disk. When new rows of data are inserted into the table or the values in the indexed columns are changed, the database may have to reorganise the storage of the data to make room for the new rows. This reorganisation can take an additional toll on DML queries. However, if new rows are expected on a regular basis in any table, the expected growth for an index can be specified.
- Use index hints as described earlier.
- Select limited data. The fewer the data that is retrieved, the faster the query will run. Rather than filtering on the client, as much filtering as possible should take place on the server. This will result in less data being transmitted and results will appear quicker. Obvious or computed columns should be avoided. For example:
  - SELECT FirstName, LastName, City WHERE City = 'Barcelona'

- The "City" column can be eliminated from the query results as it will always be "Barcelona". For large result sets, especially if transmitted over a slow connection, this can be significant.
- SELECT * is therefore to be avoided as much as possible.
- Drop indexes before loading data. This makes the INSERT statement run faster. Once the inserts are completed, the index can be recreated. Additionally, on an insert of thousands of rows in an online system, a temporary table is more efficient to load data (provided it has no index). Moving data from one table to another is much faster than loading from an external source, so the indexes on the primary table can be dropped when the data moves from the temporary to the final table, and finally, the indexes are recreated.
- Avoid correlated subqueries. A correlated subquery is one which uses values from the parent query. This type of query tends to run row-by-row, once for each row returned by the outer query, and thus decreases SQL query performance. For example:
  - SELECT c.Name, c.City, (SELECT HospitalName FROM Hospital WHERE ID = c.HospitalID) AS HospitalName FROM Customer c
  - The problem here is that the inner query (SELECT HospitalName…) is run for each row returned by the outer query (SELECT c.Name…).
  - A more efficient query would be to rewrite the correlated subquery as a join:
  - SELECT c.Name, c.City, co.HospitalName FROM Customer c LEFT JOIN Hospital co ON c.HospitalID = co.HospitalID
  - This accesses the Hospital table just once and JOINs it with the Customer table.
- Use the most efficient command. For example, when checking whether a record exists, a common query would be:
  - IF (SELECT COUNT(1) FROM employees WHERE forename LIKE '%JOHN%') > 0
  - PRINT 'YES'
  - However, COUNT() scans the entire table, counting up all entries matching the condition. The command EXISTS() exits as soon as it finds the result. This not only improves performance but also provides clearer code:
  - IF EXISTS(SELECT forename FROM employees WHERE forename LIKE '%JOHN%')
  - PRINT 'YES'
- Use the built-in performance reports. SQL Server 2005 includes over 20 reports providing performance statistics for the server.

3.14   HINTS AND TIPS

In answer to the stackoverflow.com question "What are some of your most useful database standards?" The following (with some debate) were listed (Stack Overflow 2012 [online]):

3.14.1   NAMING STANDARDS

Schemas are named by functional area (Products, Orders, Shipping)
No Hungarian Notation: No type names in object names (no strFirstName)
Do not use registered keywords for object names
No spaces or any special characters in object names (Alphanumber + Underscore are the only characters allowed)
Name objects in a natural way (FirstName instead of NameFirst)
Table name should match Primary Key Name and Description field (SalesType – SalesTypeId, SalesTypeDescription)
Do not prefix with tbl_ or sp_
Name code by object name (CustomerSearch, CustomerGetBalance)
CamelCase database object names
Column names should be singular
Table names may be plural
Give business names to all constraints (MustEnterFirstName)

3.14.2   DATA TYPES

Use same variable type across tables (Zip code – numeric in one table and varchar in another is not a good idea)
Use nNVarChar for customer information (name, address(es)) etc. you never know when you may go multinational

3.14.3   IN CODE

Keywords always in UPPERCASE
Never use implied joins (Comma syntax) - always use explicit INNER JOIN OUTER JOIN
One JOIN per line
One WHERE clause per line
No loops – replace with set based logic
Use short forms of table names for aliases rather than A, B, C
Avoid triggers unless there is no recourse
Avoid cursors like the plague (read http://www.sqlservercentral.com/articles/T-SQL/66097/)

3.14.4   DOCUMENTATION

Create database diagrams
Create a data dictionary

3.14.5   NORMALISATION AND REFERENTIAL INTEGRITY

Use single column primary keys as much as possible. Use unique constraints where required.
Referential integrity will be always enforced
Avoid ON DELETE CASCADE
OLTP must be at least 4NF
Evaluate every one-to-many relationship as a potential many-to-many relationship
Non user generated Primary Keys
Build insert based models instead of update based
Primary key to foreign key must be same name (Employee.EmployeeId is the same field as EmployeeSalary.EmployeeId) except when there is a double join (Person.PersonId joins to PersonRelation.PersonId_Parent and PersonRelation.PersonId_Child)

3.14.6   MAINTENANCE: RUN PERIODIC SCRIPTS TO FIND

Schema without table
Orphaned records
Tables without primary keys
Tables without indexes
Non-deterministic user defined functions
Backup, Backup, Backup

3.14.7   BE GOOD

Be Consistent
Fix errors now

NOTES

1 The first part of this chapter (up to and including "Some useful commands/functions") is reproduced from Ganney et al. (2022).
2 The latest version is 8.0, but 5.7 is very widespread. 8.0 generally extends the syntax rather than altering it.
3 The syntax is Table.Field and is used so that each field is uniquely identified. The importance of this can be seen from the table structures: several field names appear in multiple tables (as can be seen from the WHERE clause) – SQL requires that no ambiguity exists in the statement.
4 An abbreviation.
5 The error is intentional and will be corrected later.
6 DISTINCTROW is a synonym for DISTINCT.
7 See https://www.w3resource.com/sql/select-statement/queries-with-distinct-multiple-columns.php for examples.
8 So the epithet "optimiser" is itself sub-optimal.
9 It's very similar to "COUNT()" therefore.
10 The query cache is deprecated as of MySQL 5.7.20 and is removed in MySQL 8.0. Deprecation includes SQL_CACHE and SQL_NO_CACHE.

11 We will look at the CREATE TABLE syntax later.

12 Only available in MySQL 5.6 onwards, although PARTITION on the table is supported earlier.

13 BETWEEN is inclusive of values, not exclusive.

14 Therefore beware porting code between platforms without full testing.

15 PROCEDURE is deprecated as of MySQL 5.7.18 and is removed in MySQL 8.0. For an example see https://dev.mysql.com/doc/refman/5.7/en/procedure-analyse.html.

16 However, if you have --secure-file-priv set, then you will need to use the directory specified in the secure_file_priv variable.

17 The wildcard may vary – some engines use *.

18 Note the case insensitivity in this example – this may not always apply, depending on the SQL engine.

19 Note that not all engines support all four types: MySQL only supports three, omitting the FULL JOIN.

20 And that, in turn, is really a synonym for SHOW COLUMNS.

21 A semi-join is one where the result only contains the columns from one of the joined tables. A semi-join between two tables returns rows from the first table where there is at least one match in the second. The difference between a semi-join and a conventional join is that the rows in the first table are returned at most once. It is specified using EXISTS or IN.

22 This table is from https://dev.mysql.com/doc/refman/5.7/en/explain-output.html which also includes the JSON format.

23 NULL-safe equality, returning 1 rather than NULL if both operands are NULL and 0 rather than NULL if one operand is NULL.

24 A WINDOW clause is useful for queries in which multiple OVER clauses would otherwise define the same window. Instead, you can define the window once, give it a name and refer to the name in the OVER clauses. See https://dev.mysql.com/doc/refman/8.0/en/window-functions-named-windows.html for details.

25 See https://dev.mysql.com/doc/refman/5.7/en/create-table.html#create-table-temporary-tables, for a full description.

26 CHANGE is a MySQL extension to standard SQL. MODIFY is a MySQL extension for Oracle compatibility.

27 These examples are from https://dev.mysql.com/doc/refman/5.7/en/alter-table.html#alter-table-redefine-column.

28 Note that MySQL associates privileges with the combination of a host name and user name and not with only a user name as per standard SQL.

29 Note that Standard SQL does not have global or database-level privileges, nor does it support all the privilege types that MySQL supports.

30 See https://dev.mysql.com/doc/refman/5.7/en/load-data.html for a full description.

31 Backslash is the MySQL escape character within strings in SQL statements. The escape sequences '\t' and '\n' specify tab and newline characters respectively.

32 In .csv, lines have fields separated by commas and enclosed within double quotation marks, with an initial line of column names. The lines in such a file are terminated by carriage return/newline pairs.

33 This needs to be the full file path, so it probably resides in /var/lib/mysql-files/ and may need GRANT FILE on *.* to dbuser<n>@'localhost'.

34 These examples are from https://dev.mysql.com/doc/refman/5.7/en/create-procedure.html

35 This allows the semicolon to be used in the procedure definition without MySQL treating it as a terminator and evaluating the line.

36 An atomic transaction is an indivisible set of database operations (hence the "atom" bit) such that either all occur, or nothing occurs.

37 Watch out for date formats though!

38 Data manipulation language is a subset of SQL including such as SELECT, UPDATE, INSERT, DELETE. The other subsets are DDL (**data definition language** – CREATE, ALTER, DROP), DCL (**data control language** – GRANT, REVOKE) and TCL (**transaction control language** – COMMIT, ROLLBACK).

REFERENCES

Connolly, TM & Begg, CE, 2014, *Database Systems: A Practical Approach to Design, Implementation and Management*. New Jersey, NJ: Pearson.

Ganney P, Maw P, White M, Ganney R, ed. 2022, *Modernising Scientific Careers The ICT Competencies*, 7th edition, Tenerife: ESL.

Larson, PÅ, Blanas, S, Diaconu, C, Freedman, C, Patel, JM & Zwilling, M, 2011, High-performance concurrency control mechanisms for main-memory databases. *Proceedings of the VLDB Endowment*, 5(4), 298–309.

MySQL 5.7 Reference Manual (1) 2022 [online] Available: https://dev.mysql.com/doc/refman/5.7/en/ [Accessed 02/04/22].

MySQL 5.7 Reference Manual (2) 2022 [online]. Available: https://dev.mysql.com/doc/refman/5.7/en/explain-output.html [Accessed 02/04/22].

MySQL 5.7 Reference Manual (3) 2022 [online] Available: https://dev.mysql.com/doc/refman/5.7/en/grant.html [Accessed 02/04/22].

Stack Overflow 2012 [online]. Available: https://stackoverflow.com/questions/976185/what-are-some-of-your-most-useful-database-standards [Accessed 02/04/22].

Vallejo, E, Sanyal, S, Harris, T, Vallejo, F, Beivide, R, Unsal, O, & Valero, M, 2011, Hybrid Transactional Memory with Pessimistic Concurrency Control, *International Journal Of Parallel Programming*, 39(3), 375–396. doi:10.1007/s10766-010-0158-x.

4 Data Mining

4.1 INTRODUCTION

Data mining is *"the practice of searching through large amounts of computerized data to find useful patterns or trends"* (Merriam-Webster 2022 [online]). The overall goal of the data mining process is to extract information from a data set and transform it into an understandable structure for further use.

In this we see a similarity with the equation found in Chapter 1:

$$\text{Information} = \text{Data} + \text{Structure}$$

The term "data mining" is a misnomer, because the goal is the extraction of patterns and knowledge from large amounts of data, not the extraction (mining) of data itself. It is frequently (mistakenly) applied to any form of large-scale data or information processing as well as any application of computer decision support system, including artificial intelligence, machine learning and business intelligence.

Possibly one of the most famous pieces of data mining was the work of the Bletchley Park codebreakers, who spent their time looking for patterns in the data in order to interpret it.

As a note of caution in this field, it is worth recalling the words of the British statistician George E. P. Box: *"All models are wrong, but some are useful"* (Box and Draper 1987). There are great similarities with Statistical Analysis – the key differences are summarised in Table 4.1.

The actual data mining task is the analysis of large quantities of data to extract previously unknown, interesting patterns such as groups of data records (cluster analysis), unusual records (anomaly detection) and dependencies (association rule mining, sequential pattern mining).

> The traditional method of turning data into knowledge relies on manual analysis and interpretation. For example, in the health-care industry, it is common for specialists to periodically analyze current trends and changes in health-care data, say, on a quarterly basis. The specialists then provide a report detailing the analysis to the sponsoring health-care organization; this report becomes the basis for future decision making and planning for health-care management.
>
> (Fayyad et al. 1996)

So let us look at how we might turn data into knowledge. The **knowledge discovery in databases** (KDD)[1] process is commonly defined with the following stages:

1. Selection
2. Pre-processing
3. Transformation
4. Data mining
5. Interpretation/evaluation

DOI: 10.1201/9781003316244-4

TABLE 4.1

Key Differences between Data Analysis and Statistical Analysis

	Data Analysis	Statistical Analysis
Data Volume	Large amounts	Usually limited to a sample (i.e. a portion of the population)
Tools	Data science toolbox, e.g. programming languages like Python and R, or frameworks like Hadoop and Apache Spark	Usually mathematical-based techniques such as hypothesis testing, probability and statistical theorems
Audience	Both will frequently yield results which need to be presented in a clear way to non-data scientists	

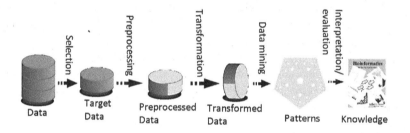

FIGURE 4.1   The steps of the KDD process.

The first four steps turn data into patterns and the final step turns patterns into knowledge and may be visualised as shown in Figure 4.1.

Possibly the most common variant of this in the data mining field is the **Cross Industry Standard Process for Data Mining** (CRISP-DM) which defines six phases:

1. Business understanding
2. Data understanding
3. Data preparation
4. Modelling
5. Evaluation
6. Deployment

The mapping between the two is shown in Table 4.2.

Of these stages/phases, three are of particular interest: Pre-processing, data mining and results validation (a sub-section of "evaluation"). We will now examine each of these in turn.

TABLE 4.2

The Mapping between the Stages/Phases in KDD and CRISP-DM

KDD	CRISP-DM
Selection	Business understanding
	Data understanding
Pre-processing	Data preparation
Transformation	Modelling
Data mining	
Interpretation/evaluation	Evaluation
	Deployment

4.2 PRE-PROCESSING

Before data mining algorithms can be used, a target data set must be assembled. As data mining can only uncover patterns actually present in the data, the target data set must be large enough to contain these patterns while remaining concise enough to be mined within an acceptable time limit. Pre-processing is the action of analysing the multivariate data sets, cleaning the target set by removing, e.g. the observations containing noise and those with missing data. It also includes correction by interpolation or cross-reference, should the data set not be large enough to survive the removal of records with missing data. If the data set is especially large, then it may be condensed by the thinning of the record set. There are many approaches to this, the simplest of which is the removal of some (but not all) similar records. It is, of course, a mistake to remove all duplicate records as the underlying pattern that is being sought will be lost due to (in an extreme case) there only being one record to represent the majority of patient journeys/results.

4.3 DATA MINING

Data mining involves six common classes of tasks. In order to examine them, let us consider a simple data set. This set consists of 23 patients who have undertaken two tests, A and B. The chart plots their scores. They are then classified as to whether they went on to develop the disease under investigation: x indicates a patient who did develop the disease, o represents a patient who did not. For the purpose of this example, the simple artificial data set in Figure 4.2 represents a historical data set that may contain useful knowledge from the point of view of early diagnosis. Note that in actual KDD applications, there are typically many more dimensions (up to several hundred) and many more data points (thousands or even millions).

Classification is the task of generalising a known structure to apply to new data. The task is to develop (or learn) a function that maps (classifies) a data item into one of several predefined classes. Examples of classification methods used as part of knowledge discovery applications include the classifying of emails as "legitimate" or "spam", identifying trends in financial markets and the automated identification of objects of interest in large image databases.

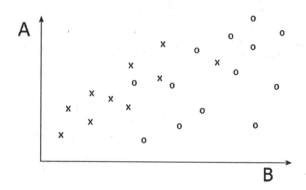

FIGURE 4.2   A simple data set for example purposes.

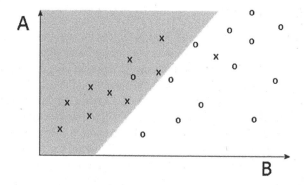

FIGURE 4.3   The example data classified by a simple partition.

Figure 4.3 shows a simple partitioning of the data into two class regions using a linear decision boundary. Note that it is not possible to separate the classes perfectly in this way.

Regression is the attempt to find a function which models the data with the least error, for estimating the relationships among data or data sets. This requires a function to be developed/learned that maps a data item to a real-valued prediction variable. This is a very common technique and examples might be estimating the probability that a patient will survive given the results of a set of diagnostic tests, predicting consumer demand for a new product as a function of advertising expenditure and predicting time series where the input variables can be time-lagged versions of the prediction variable.

Figure 4.4 shows the result of a simple linear regression where A is fitted as a linear function of B. The fit is poor because only a weak correlation exists between the two variables and therefore the classification is also weak.

Clustering is the task of discovering groups and structures in the data that are in some way or another "similar", without using known structures in the data. Such clustering is usually descriptive as the finite set of categories or clusters are used to describe the data. The clusters can be mutually exclusive and exhaustive or consist of

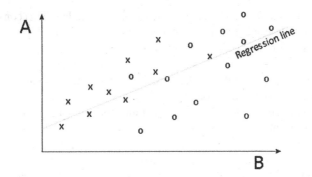

FIGURE 4.4   The example data with a regression line.

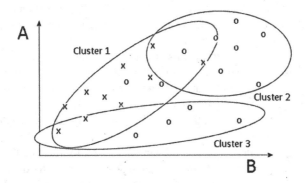

FIGURE 4.5   A possible clustering of the example data.

a richer representation, such as hierarchical or overlapping categories. Examples of clustering applications in a knowledge discovery context include discovering homogeneous subpopulations for patients in healthcare.

Figure 4.5 shows a possible clustering of the patient data set into three clusters. Note that the clusters overlap, allowing data points to belong to more than one cluster. Some techniques replace the class labels (the o and x in this example) with a third symbol in order to indicate that the class membership is no longer assumed known (i.e. the class is not considered when forming the cluster – proximity is more important).[2]

Summarisation is the provision of a more compact representation or description of the data set, including visualisation and report generation. A simple example might be tabulating the mean and standard deviations for all fields. More sophisticated methods involve the derivation of summary rules, multivariate visualisation techniques and the discovery of functional relationships between variables. Summarisation techniques are often applied to interactive exploratory data analysis and automated report generation.

Dependency modelling (also known as association rule learning) consists of finding a model that describes significant dependencies (or relationships) between variables. Dependency models exist at two levels:

1. the *structural level* of the model specifies (often in a graphical form) which variables are locally dependent on each other
2. the *quantitative level* of the model specifies the strengths of the dependencies using some numeric scale.

Probably the most common example of association rule learning is supermarkets (especially online) gathering data on customer purchasing habits to determine which products are frequently bought together and using this information for marketing purposes. In healthcare, probabilistic dependency networks use conditional independence to specify the structural aspect of the model and probabilities or correlations to specify the strengths of the dependencies. These probabilistic dependency networks are the basis of probabilistic medical expert systems that utilise databases, information retrieval and modelling of the human genome.

Change and deviation detection (also referred to as anomaly or outlier detection) is the identification of unusual data records that might be interesting or might be data errors that require further investigation. It focuses on discovering the most significant changes in the data from previously measured or normative values.

For any data-mining algorithm, there are (at a basic level) three primary components:

1. model representation – This is the language used to describe discoverable patterns. This needs to be complex enough to describe the data within an acceptable processing time. If the representation is too limited, then no amount of training time or examples can produce an accurate model for the data. It is important therefore that an algorithm designer clearly states which representational assumptions are being made by a particular algorithm.
2. model evaluation – Here, quantitative statements (or fit functions) are devised to evaluate how well a particular pattern (a model and its parameters) meets the goals of the KDD process. For example, predictive models are often judged by how well they predict outcomes on a test set. Descriptive models can be evaluated along the dimensions of predictive accuracy, novelty, utility and understandability of the fitted model. There is a close correlation between this and results validation, which we will come to shortly.
3. search –This consists of two components: parameter search and model search. Once the model representation (or family of representations) and the model-evaluation criteria are fixed, then the data-mining problem has been reduced to that of an optimisation task: find the parameters and models from the selected family that optimise the evaluation criteria. In parameter search, the algorithm searches for the parameters that optimise the evaluation criteria given some observed data and a fixed model representation. The model search is a loop over the parameter-search method. Here the model representation is changed so that a family of models is considered, parameter search being conducted on each iteration of the model.

4.3.1 Some Data-Mining Methods

There are (as you might expect) multiple data mining methods and we will consider only a few here. An important point is that each technique typically suits some problems better than others. For example, decision tree classifiers can be useful for finding structure in high-dimensional spaces and in problems with mixed continuous and categorical data (because tree methods do not require distance metrics). However, classification trees might not be suitable for problems where the true decision boundaries between classes are described, e.g. by a second-order polynomial. Therefore there is no universal data-mining method, and choosing an algorithm for an application is something of an art. In practice, a large portion of the application effort can go into properly formulating the problem (asking the right question) rather than into optimising the algorithmic details of a data-mining method.[3]

4.3.1.1 Decision Trees and Rules

Decision trees and rules that use univariate splits have a simple representational form, making the inferred model relatively easy for the user to comprehend. However, the restriction to a particular tree or rule representation can significantly restrict the functional form (and, thus, the approximation power) of the model.

Figure 4.6 shows the effect of a threshold split (a simple rule, which may come from an equally simple decision tree) applied to the B test variable for our example data set. It is clear that using such a simple threshold split (parallel to one of the feature axes) severely limits the type of classification boundaries that can be developed.

If the model space is enlarged to allow more general expressions (such as multivariate hyperplanes at arbitrary angles), then the model is more powerful for prediction but can be much more difficult to comprehend. A large number of decision trees and rule-induction algorithms exist in the fields of machine learning and applied statistics. To a large extent, they depend on likelihood-based model-evaluation methods, with varying degrees of sophistication in terms of penalising model complexity. Greedy search methods, which involve growing and pruning rules and tree structures, are typically used to explore the superexponential space of possible models. Trees

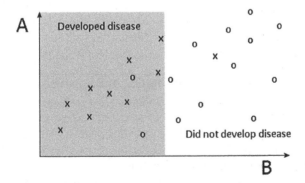

FIGURE 4.6   A threshold split applied to the example data.

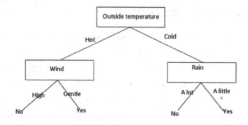

FIGURE 4.7   A simple decision tree for deciding whether or not to go running.[4]

and rules are primarily used for predictive modelling, both for classification and regression, although they can also be applied to summary descriptive modelling.

Returning to decision trees, they are generally more complex than the example in Figure 4.6 and are a type of recursive partitioning algorithm, built up of two types of nodes: decision nodes and leaves.

The decision tree starts with a node called the root[5] ("Outside temperature" in the example in Figure 4.7) and progresses through decisions until an outcome is reached (at a point called a leaf[6]). If the root is a leaf then the decision tree is trivial or degenerate and the same classification is made for all data. For decision nodes (such as "wind") a single variable is examined and the focus moves to another node based on the outcome of a comparison. The process is repeated recursively until a leaf node ("yes" or "no" in this example) is reached. At a leaf node, the majority value of training data routed to the leaf node is returned as a classification decision, or the mean-value of outcomes is returned as a regression estimate. Decision trees therefore have a great similarity with flowcharts, except that the parameters for the decision (e.g. a numeric value for blood pressure) have been built by data analysis and can be altered or even machine learned, in the same manner as a neural network (which we will look at shortly).

A single tree has the inherent flaw that it assumes that whatever comes next will follow the patterns of what has come before (which is effectively true of any trained model). An alternative method is to build an ensemble. This consists of thousands of smaller trees built from random subsections of the data. A decision is then compiled by summing the results from each of these trees. In her book "Hello World", Hannah Fry draws a comparison with the success rate of "ask the audience" versus "phone a friend" in which the former beat the latter 91% to 65%, so *"a room full of strangers will be right more often than the cleverest person you know"* (Fry 2018). It will still be unable to produce results from outside of the previous data, but it is right more frequently.[7]

4.3.1.2   Nonlinear Regression and Classification Methods

These methods consist of a family of techniques for a prediction that fits linear and nonlinear combinations of basic functions to combinations of the input variables. Basic functions used in this way include sigmoids (mathematical functions with an S-shaped curve), polynomials (mathematical functions with several variables but only using the operators addition, multiplication and integer exponentiation, e.g. x^2+xy+z^3) and splines (piecewise polynomial functions that map a curve to a set of data points by using several polynomials, each one of which fits a part of the data).

Examples of this technique include feedforward neural networks, adaptive spline methods[8] and projection pursuit regression.[9]

Nonlinear regression methods, although representationally powerful, can be difficult to interpret. For example, although the classification boundaries of Figure 4.8 might be more accurate than the simple threshold boundary of the decision tree Figure (Figure. 4.6), the threshold boundary has the advantage that the model can be expressed, to some degree of certainty, as a simple rule of the form "if B is greater than threshold, then the patient has a good prognosis."

4.3.1.3   Example-Based Methods

These methods use representative examples from the database to approximate a model. Predictions on new examples are derived from the properties of similar examples in the data whose outcome is known. Techniques include nearest-neighbour classification, regression algorithms and case-based reasoning systems.

In Figure 4.9, the class (prognosis, in this example) at any new point in the two-dimensional space is the same as the class of the closest point in the original training data set.

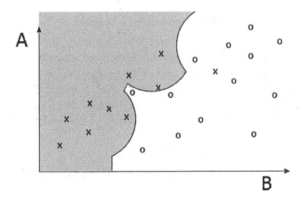

FIGURE 4.8   An example of a nonlinear decision boundary.

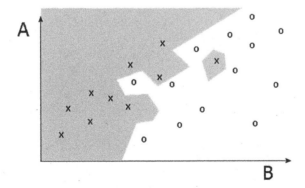

FIGURE 4.9   The use of a nearest-neighbour classifier for the patient data set.

A potential disadvantage of example-based methods (compared with tree-based or non-linear methods) is that they require a well-defined distance metric for evaluating the distance between data points. For the patient data in Figure 4.9, this is probably not a problem because tests A and B are likely to be measured in the same units. However, if variables such as the time between tests, prevalence of symptoms, gender and ethnicity were to be included, then it would be more difficult to define a metric between the variables (but not impossible – the benefit versus the predicted additional effort must then be considered).

Model evaluation is typically based on cross-validation estimates of a prediction error: parameters of the model to be estimated can include the number of neighbours to use for prediction and the distance metric itself. Like nonlinear regression methods, example-based methods are often asymptotically powerful in terms of approximation properties but, conversely, can be difficult to interpret because the model is implicit in the data and not explicitly formulated. Related techniques include kernel-density estimation and mixture modelling.

Superimposing four of the methods described above shows the importance of model evaluation. In Figure 4.10, a new data point has been added. Two of the models classify it correctly, two do not.

4.3.1.4   Probabilistic Graphic Dependency Models

Graphic models specify probabilistic dependencies using a graph structure. In its simplest form, the model specifies which variables are directly dependent on each other. Typically, these models are used with categorical or discrete-valued variables, but extensions to special cases, such as Gaussian densities, for real-valued variables are also possible. These models were initially developed within the framework of probabilistic expert systems: the structure of the model and the parameters

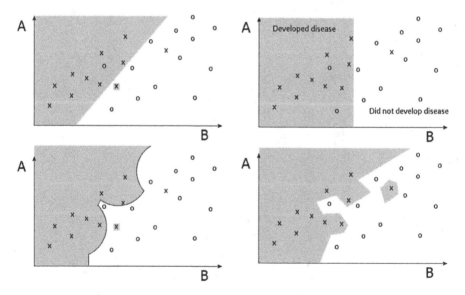

FIGURE 4.10   The addition of a new data point to four of the models.

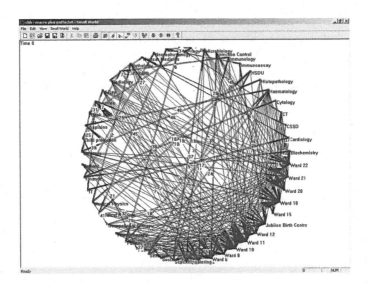

FIGURE 4.11   A graph model showing linkages between parts of a hospital and "distances" between them using external routes (Ganney 2011).

(the conditional probabilities attached to the links of the graph) were elicited from experts, such as in Figure 4.11.

More modern models learn the structure and the parameters of graphic models directly from databases. Model evaluation criteria are typically Bayesian in form,[10] and parameter estimation can be a mixture of closed-form estimates and iterative methods depending on whether a variable is directly observed or hidden. Model search can consist of greedy hill-climbing methods over various graph structures. Prior knowledge, such as a partial ordering of the variables based on causal relations, can be useful in terms of reducing the model search space to something more manageable. Graphic model induction methods are of particular interest to KDD because the graphic form of the model lends itself easily to human interpretation.

4.3.1.5   Principal Component Analysis (PCA)

PCA is a method of extracting information from data that keeps only what is most important and finds the underlying trends, using matrix algebra. The data may be high dimensional and of a random nature, which can make the patterns difficult to see – PCA attempts to identify such patterns. In an experiment, m variables are measured (where m could be a large number) and it is reasonable to expect that some of the variables may be co-dependent. With PCA, a small number (k) of new variables (often called features) are found that mostly describe the variation within the data. An important point is that k is very much smaller than m. These new variables are independent of each other, and they are created from a linear combination of the original variables. The aim is to interpret the meaning of these new variables and then to understand the original data in terms of them. With PCA, the k new variables are the best subset of new variables, in the sense that it minimises the variance of the residual noise when fitting data to a linear model. Note that PCA transforms the initial variables into new ones that are linear combinations of the original variables; it does not create new variables and attempt to fit them.

The main PCA process is:

1. The data are placed in a matrix, **D**, with the variables down the rows and the observations across the columns.
2. The means of each variable are found and subtracted from each variable, which generates the matrix **D′**.
3. The covariance matrix is constructed by $C = (1/n)\mathbf{D}'\mathbf{D}'^{\mathrm{T}}$, where n is the number of variables. This matrix has the covariance of the ith j^{th} variables in the element C_{ij}.
4. The eigenvalues and eigenvectors of the covariance matrix are found and placed in order of the size of the eigenvalues.
5. The eigenvectors that correspond to the eigenvalues whose sum is no less than 90% of the total are arbitrarily retained.[11]
6. The resulting eigenvectors are placed side by side into a matrix, **W**, which describes a new coordinate system with the axes rotated so that they align with the greatest variation of the data. The first components carry the most variation because they have larger eigenvalues.
7. The data are expressed in the new coordinate system by multiplying **D′** by the transpose of **W**, thus: $\mathbf{D}_{PCA} = \mathbf{W}^{\mathrm{T}}\mathbf{D}'$

In this way, data is moved into a new coordinate system of variables that are independent and have a lower dimension because only those variables that carry most of the variation of the data have been kept. The concept behind PCA is that the system of variables with reduced dimension carries the main trends of the data and is easier to interpret and visualise than the original data. Whilst a large number of variables may be present, through PCA, most of the features of the data can be represented in just a few variables.[12]

It is possible to perform the analysis by hand but PCA is part of software such as MATLAB and R.

The website http://setosa.io/ev/principal-component-analysis/ contains many examples, including the 17-Dimensional one in Figure 4.12.

The table in Figure 4.12 shows the average consumption of 17 types of food in grams per person per week for every country in the UK. There are some interesting variations across different food types, but the overall differences aren't particularly notable. Applying PCA gives Figure 4.13 for the first principal component and Figure 4.14 for the first two.

Both of these show Northern Ireland to be an outlier which wasn't obvious from the initial data but can now be seen using this information (re-examining the data we can see that the Northern Irish eat more grams of fresh potatoes and fewer of fresh fruits, cheese, fish and alcoholic drinks).

An important caveat to PCA is that if the original variables are highly correlated, then the solution will be very unstable. Also, as the new variables are linear combinations of the original variables, they may lack interpretation. The data does not need

	England	N Ireland	Scotland	Wales
Alcoholic drinks	375	135	458	475
Beverages	57	47	53	73
Carcase meat	245	267	242	227
Cereals	1472	1494	1462	1582
Cheese	105	66	103	103
Confectionery	54	41	62	64
Fats and oils	193	209	184	235
Fish	147	93	122	160
Fresh fruit	1102	674	957	1137
Fresh potatoes	720	1033	566	874
Fresh Veg	253	143	171	265
Other meat	685	586	750	803
Other Veg	488	355	418	570
Processed potatoes	198	187	220	203
Processed Veg	360	334	337	365
Soft drinks	1374	1506	1572	1256
Sugars	156	139	147	175

FIGURE 4.12   A sample table for PCA.

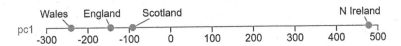

FIGURE 4.13   The first principal component from a PCA analysis of Figure 4.12.

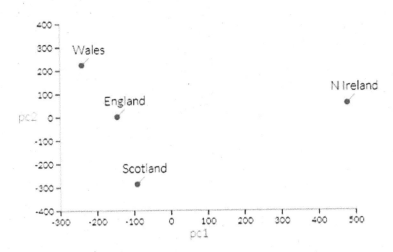

FIGURE 4.14   The first and second principal components from a PCA analysis of Figure 4.12.

to be multinormal, unless PCA is used for predictive modelling using normal models to compute confidence intervals, in which case it does.

There is a very good illustration of the use of PCA for visualisation of breast cancer microarray data on the Wikipedia page: https://en.wikipedia.org/wiki/Principal_component_analysis

4.3.1.6   Neural Networks

Neural networks are probably the major form of **Artificial Intelligence** (AI) used within healthcare. AI has its roots in the work of Simon and Newell in the 1950s and was initially used to overcome combinatorial explosion,[13] usually illustrated by an attempt to create a Turing machine to play Chess, the lower bound of the game-tree for which was estimated by Claude Shannon in 1950 to be 10^{120} (Gambiter 2008 [online]). AI in healthcare, and especially with neural networks, is generally limited to classification, especially of medical images. For example, the behold.ai system is capable of identifying "normal" chest X-Rays and achieved CE marking in 2020. Also, Siemens developed an "MRI prostate biopsy" tool, which produces a "suspicion map" of the prostate, indicating regions where abnormalities are suspected and thus reduces the need for invasive procedures. In the research arena, *"An AI system trialled at Moorfields Eye Hospital in London made the correct referral decision for over 50 eye diseases with 94% accuracy, matching the world's best eye experts."* (Moore and Nix 2021).

The fundamental building block of a neural network is a neuron, so-called as it approximates the way biological neurons process information in the brain. A computational neuron takes a set of inputs X_i and multiplies them by a set of weights W_i before adding them to a bias term b, before the application of an activation function f in a process called forward propagation. This activation function is a non-linear function that takes a numeric input and outputs a number between 0 and 1. Activation functions are used to introduce non-linearities into the network, which is a crucial aspect of neural networks. The two are compared in Figure 4.15.

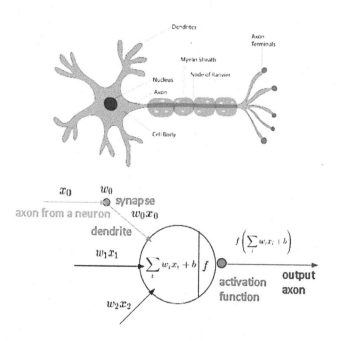

FIGURE 4.15   Drawing of a biological neuron (top) and its mathematical model (bottom).

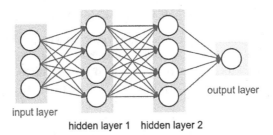

input layer

hidden layer 1   hidden layer 2

output layer

FIGURE 4.16   A 4-layer neural network with 3 inputs, 2 hidden layers of 4 neurons each and one output layer with 1 neuron.

In neural networks, neurons are organised in distinct layers and connected with each other in a graph structure whereby the outputs of neurons of one layer are the input of neurons in other layers. The most common type of layer is a **fully-connected** (FC) layer where neurons of two adjacent layers are fully pairwise connected whereas neurons of the same layer do not share any connections. The size of such a network is usually specified in terms of the total number of neurons or the number of layers. An example is shown in Figure 4.16.

To train a neural network, a training data set is required for which the intended output is known. The training procedure starts by randomly initialising the network weights W_i. A process called forward pass is then executed at each layer of the network. For FC layers, a forward pass consists of a matrix multiplication of the inputs of the layer x with the layers' weights vector W^T, followed by a bias offset and the application of an activation function, $f(W^T x + b)$. This is followed by calculation of a loss value by comparing the network's output with the intended output using a loss function J. The weights of the network W_i are then iteratively updated such that the loss function J is minimised. Neural networks can require huge data sets for training.[14]

One important consideration for neural networks is that they are very good at identifying patterns but they are not good at working out what these patterns may mean. Algorithms such as neural networks work best therefore when they explain their results so that a (human) judgement can be made. For example, it is better for an algorithm to narrow down an image to areas that may contain cancer, than to say whether or not the image does contain cancer. This helps the radiologist to focus and prevents machine errors. In facial recognition, it is better for an algorithm to report "it is 85% likely to be this person, but 80% this person or 60% this person" than to report "it is this person" and not reference the other possibilities.

In terms of model evaluation, although networks of the appropriate size can universally approximate any smooth function to any desired degree of accuracy, relatively little is known about the representation properties of fixed-size networks estimated from finite data sets. Also, the standard squared error and cross-entropy loss functions used to train neural networks can be viewed as log-likelihood functions for regression and classification, respectively. Back propagation is a parameter-search method that performs gradient descent in parameter (weight) space to find a local maximum of the likelihood function starting from random initial conditions.

4.4   DATA MINING MODELS IN HEALTHCARE

Data mining models in healthcare must always be used within the knowledge of that context. In February 2019, the Department of Health and Social Care published the guidance document "Code of conduct for data-driven health and care technology" (UK National Health Service 2021 [online]) which lists 10 principles:

- Principle 1: Understand users, their needs and the context
- Principle 2: Define the outcome and how the technology will contribute to it
- Principle 3: Use data that is in line with appropriate guidelines for the purpose for which it is being used
- Principle 4: Be fair, transparent and accountable about what data is being used
- Principle 5: Make use of open standards
- Principle 6: Be transparent about the limitations of the data used
- Principle 7: Show what type of algorithm is being developed or deployed, the ethical examination of how the data is used, how its performance will be validated and how it will be integrated into health and care provision
- Principle 8: Generate evidence of effectiveness for the intended use and value for money
- Principle 9: Make security integral to the design
- Principle 10: Define the commercial strategy

4.5   RESULTS VALIDATION

The final step of knowledge discovery from data is to verify that the patterns produced by the data mining algorithms occur in the wider data set as not all patterns found by the data mining algorithms are necessarily valid. There are, essentially, two main problems:

- spurious correlation
- a model that fits the sample but not the population

To overcome these, it is normal for the evaluation to use a test set of data, which the data mining algorithm has not been trained on.[15] The learned patterns are applied to this test set, and the resulting output is compared to the desired output.

It is common though for data mining algorithms to find patterns in the training set which are not present in the general data set. Data mining can unintentionally produce results which appear to be significant; but which do not actually predict future behaviour and cannot be reproduced on a new sample of data. This often results from investigating too many hypotheses and not performing proper statistical hypothesis testing. A simple version of this problem in machine learning is known as overfitting, but the same problem can arise at different phases of the process and thus a train/test split – when applicable at all – may not be sufficient to prevent this from happening.

If the learned patterns do not meet the desired standards, subsequently it is necessary to re-evaluate and change the pre-processing and data mining steps. If the learned patterns do meet the desired standards, then the final step is to interpret the learned patterns and turn them into knowledge. However, beware over-fitting your data: see https://www.datasciencecentral.com/profiles/blogs/how-to-lie-with-data for a discussion.

4.6 SOFTWARE

There are a lot of data mining software packages available, of which R, a programming language and software environment for statistical computing, data mining and graphics, that is part of the GNU project, has probably gained the most traction in healthcare. See https://cran.r-project.org/doc/FAQ/R-FAQ.html#What-is-R_003f for more information.

NOTES

1 "data mining" is often known as KDD.
2 Closely related to clustering is the task of probability density estimation, which consists of techniques for estimating from data the joint multivariate probability density function of all the variables or fields in the database, but that is beyond the scope of this chapter.
3 As with the choice of programming language, availability and familiarity with a toolset and technique may be the eventual deciding factor.
4 It does, of course, completely miss such cases as "a lot of hot rain" but who wants to run in that?
5 At the top, as is the case with the tree data structure we met in chapter 1.
6 At the bottom of the diagram, obviously.
7 Another example from her book involves the accuracy of pigeons to detect cancers in a medical image, with 99% accuracy when their answers were pooled.
8 Effectively a method of dividing the data into sections (the points at which the data is divided are termed "knots") and using regression to fit an equation to the data in each section.
9 A statistical model which is an extension of additive models, adapting by first projecting the data matrix of explanatory variables in the optimal direction before applying smoothing functions to these explanatory variables.
10 That is, they are probabilistic with the emphasis on expectation and are based on previous data.
11 We do this by arranging the eigenvalues in numeric order (largest first) and sum them until 90% of the total value is reached. We retain the eigenvectors that correspond to these summed eigenvalues.
12 For a fully worked example (and a problem to work through) see https://www.ncbi.nlm.nih.gov/pmc/articles/PMC3193798/.
13 The rapid growth of the complexity of a problem due to how the number of possible outcomes of the problem is affected by the input, constraints and bounds of the problem.
14 See Paul Doolan, "Radiomics and how AI can predict patients most at risk?" Scope 28:2 pp 36–37 for an investigation into reducing the size of this.
15 That is, the data is divided into two sets: test and train, as is common in neural networks.

REFERENCES

Box GEP, Draper NR. 1987, *Empirical Model-Building and Response Surfaces*, Oxford: Wiley-Blackwell. p. 424.
Fayyad U, Piatetsky-Shapiro G, and Smyth P, 1996, From Data Mining to Knowledge Discovery in Databases. AI Magazine [online] Fall 1996 pp. 37–54. Available: https://www.kdnuggets.com/gpspubs/aimag-kdd-overview-1996-Fayyad.pdf [Accessed 08/04/22].

Fry H, 2018, *Hello World: How to be Human in the Age of the Machine*, London: Doubledayp. P. 58.

Gambiter 2008, [online]. Available: http://gambiter.com/chess/computer/Shannon_number. html [Accessed 08/04/22].

Ganney, P 2011, Using Small World Models to Study Infection Communication and Control. PhD thesis. Hull University.

Merriam-Webster (2022), [online]. Available: https://www.merriam-webster.com/dictionary/ data%20mining [Accessed 08/04/22].

Moore C and Nix M, 2021, A step into the unknown: AI in medical physics and engineering, *Scope* 31:2.

Powell V and Lehe L n.d., Principal Component Analysis explained visually [online]. Available: http://setosa.io/ev/principal-component-analysis/ [Accessed 08/04/22].

UK National Health Service 2021, [online]. Available: https://www.gov.uk/government/ publications/code-of-conduct-for-data-driven-health-and-care-technology   [Accessed 08/04/22].

5 Data Analysis and Presentation

5.1 INTRODUCTION

Data analysis is *"the process of developing answers to questions through the examination and interpretation of data"* (Statistics Canada 2015 [online]). It can also enable information to be made public – some programs depend on analytical output as a major data product because, for confidentiality reasons, it is not possible to release the raw data to the public.

Prior to conducting data analysis these questions should be addressed (the answers to which are likely to form part of the eventual presentation of the analysis):

- Objectives. What are the objectives of this analysis? What issue is being addressed? What question(s) will it answer? (For research this is where the research question is formed).
- Justification. Why is this issue interesting? How will these answers contribute to existing knowledge? How is this study relevant?
- Data. What data is being used? Why is this the best source for this analysis?[1] Are there any limitations?
- Analytical methods. Which statistical techniques are appropriate? Will they satisfy the objectives?
- Audience. Who is interested in this issue/question and why?
- Is there more than one data source being used for the analysis? Are they consistent? How may they be appropriately integrated into the analysis?

5.2 APPROPRIATE METHODS AND TOOLS

An analytical approach that is appropriate for the question being investigated and the data to be analysed must be selected. When analysing data from a probability sample,[2] analytical methods that ignore the data collection design can be appropriate, provided that sufficient model conditions for analysis are met. However, methods that incorporate the sample design information will generally be effective even when some aspects of the model are incorrectly specified.

Consideration needs to be given as to how (and if) the data collection design information might be incorporated into the analysis and (if possible) how this can be done. For a design-based analysis the collection documentation may give the recommended approach for variance estimation for the data. If the data from more than one collection are included in the same analysis, then whether or not the different samples were independently selected and how this would impact the appropriate approach to variance estimation needs to be determined.

DOI: 10.1201/9781003316244-5

The data files for probability samples frequently contain more than one weight variable, particularly if the data collection is longitudinal or if it has both cross-sectional and longitudinal purposes. The collection design documentation may indicate which might be the best weight to be used in any particular design-based analysis.

When analysing data from a probability sample, there may be insufficient design information available to carry out analyses using a full design-based approach, so the alternatives require analysis. Experts on the subject matter, on the data source and on the statistical methods should be consulted if any of these is unfamiliar.

Having determined the appropriate analytical method for the data, the software choices that are available to apply to the method should be investigated. When analysing data from a probability sample by design-based methods, software specifically for survey data should be used since standard analytical software packages that can produce weighted point estimates do not correctly calculate variances for survey-weighted estimates.[3]

Commercial software is often the best choice for implementing the chosen analyses, since these software packages have usually undergone more testing than non-commercial software. It may therefore be necessary to reformat the data in order to use the selected software.

A variety of diagnostics should be included among the analytical methods if models are being fitted to the data.

Data sources vary widely with respect to missing data. The data source should document the degree and types of missing data and the processing of missing data that has been performed. At one extreme, there are data sources which seem complete – any missing units have been accounted for through a weight variable with a nonresponse component and all missing items on responding units have been filled in by imputed[4] values. At the other extreme, there are data sources where no processing has been done with respect to missing data. The work required to handle missing data can thus vary widely.

Whether imputed values should be included in the analysis requires deciding and if so, how they should be handled, as imputed values have already been fitted to a model. If imputed values are not used, consideration must be given to what other methods may be used to properly account for the effect of nonresponse in the analysis. The most extreme approach is to disregard all records which are not complete, but this may then render the sample too small to be usable.

Any caveats about how the approaches used to handle missing data could impact the results need to be included in the report.

For a discussion on handling missing data, see https://www.datasciencecentral.com/profiles/blogs/how-to-treat-missing-values-in-your-data-1.

5.3   INTERPRETATION OF RESULTS

Since many analyses in healthcare are based on observational studies rather than on the results of a controlled experiment, drawing conclusions from them concerning causality is not recommended. Alternatively, if that is the point of the analysis, the nature of the data (especially if no control group exists) should be reported.

When studying changes over time, focusing on short-term trends without inspecting them in light of medium-and long-term trends is erroneous. Frequently, short-term trends are merely minor fluctuations around a more important medium- and/or long-term trend. For example, many infections are cyclic in nature and so a short-term trend can be very misleading.

Where possible, arbitrary time reference points should be avoided. Instead, meaningful points of reference, such as the last major turning point for economic data, generation-to-generation differences for demographic statistics and legislative changes for social statistics should be used.

5.4   PRESENTATION OF RESULTS

The presentation[5] should focus on the important variables and topics. In all cases, the intended audience will influence the depth of the data to be presented as well as the progression through the information. Trying to be too comprehensive will often interfere with a strong story line for a magazine article or presentation to a board (Trust or Project), whereas a lack of depth in a peer-reviewed journal will lead to questions as to the trustworthiness of the investigation. In either case, ideas should be arranged in a logical order and in order of relevance or importance. Headings, subheadings and sidebars strengthen the organisation of an article, paper or slide.

The language should be as simple as the subject and audience permits. Depending on the target audience, a loss of precision may be an acceptable trade-off for more readable text.

Charts in addition to text and tables assist in communicating the interpretation of results. Headings that capture the meaning (e.g. "Women's earnings still trail men's") in preference to traditional chart titles (e.g. "Income by age and gender") can aid understanding and highlight the interpretation of the data that is being presented.[6] The tables and charts should not stand alone though – discussing the information in the tables and charts within the text can assist readers to better understand them. This should also avoid spurious interpretations being presented.

Rounding practices or procedures must be explained. Data should never have more significant digits than are consistent with their accuracy.

Information regarding the quality of the results should be part of the report. Standard errors, confidence intervals and/or coefficients of variation provide the reader/audience with important information about data quality.[7]

5.5   QUALITY INDICATORS

The main quality elements in data presentation are relevance, interpretability, accuracy and accessibility.

Analysis of data is relevant if there is an audience who is (or will be) interested in the results.

For high interpretability, the style of presentation must suit the intended audience. Sufficient details must be provided so that, if allowed access to the data, the results can be replicated. This is especially true for peer-reviewed articles.

Accurate data analysis requires appropriate methods and tools to be used to produce the results.

For analysis to be accessible, it must be available to people for whom the research results would be useful. The selection of conference/journal/magazine etc. is therefore important.

Much of the information in the preceding sections has been drawn from https://www.statcan.gc.ca/pub/12-539-x/2009001/analysis-analyse-eng.htm.

5.6   GRAPHICAL PRESENTATION

It is often said that "a picture is worth a thousand words"[8] and whilst the raw data can provide information, it is often better to use an illustration. For example, the connections between wards for an infection propagation model where Figure 5.1 is quicker to understand than Figure 5.2.

Likewise, Venn diagrams can describe data well, demonstrating the overlapping nature of the data, as Figure 5.3 and Figure 5.4 demonstrate as Figure 5.3 may be represented graphically as Figure. 5.4.

Geographic Information Systems (GIS) also assist in understanding the data and the inference the analyst wishes to draw from them. As public health bodies began to generate larger data sets and store them electronically, so the desire to be able to visualise this data in this way grew.

For example, consider infection propagation. GIS in its simplest form maps the infection occurrences onto a geographic map and animates using a time-based measurement, thus showing how the infection has spread and how and when it either

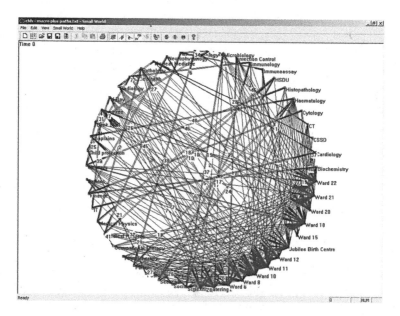

FIGURE 5.1   Graph of ward linkages.

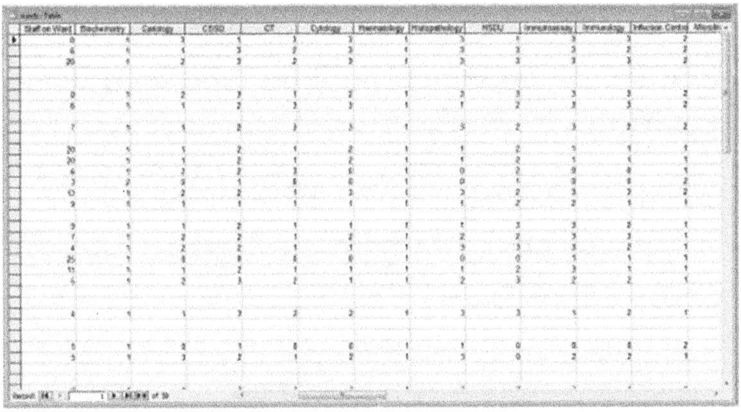

FIGURE 5.2 Database table of ward linkages.

In-house	Platform as a service	Software as a service	Infrastructure as a service	Data as a service	Backup as a service
Fabric	Fabric	Fabric	Fabric	Fabric	Fabric
Sewing machines	Sewing machines	Sewing machines	Sewing machines	Sewing machines	Sewing machines
Packaging and Shipping	Packaging and Shipping	Packaging and Shipping	Packaging and Shipping	Packaging and Shipping	Packaging and Shipping
Storage of stock	Storage of stock	Storage of stock	Storage of stock	Storage of stock	Storage of stock
Feedback & complaints	Feedback & complaints	Feedback & complaints	Feedback & complaints	Feedback & complaints	Feedback & complaints
Overflow storage	Overflow storage	Overflow storage	Overflow storage	Overflow storage	Overflow storage

Business manages

Vendor manages

FIGURE 5.3 Shorts as a service, tabulated.[9]

grew into an epidemic or died out. It is therefore predominantly a reflective tool, rather than a predictive one.

The power of a GIS comes from the ability to aggregate and visualise large data sets and thereby discover patterns in the data – in this way, GIS can be used to spot an epidemic earlier than might be achieved using conventional reporting tools. For example, a threshold might be set for a certain number of cases in a certain sized area

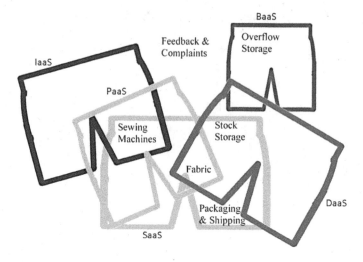

FIGURE 5.4   Shorts as a service, Venn diagram.

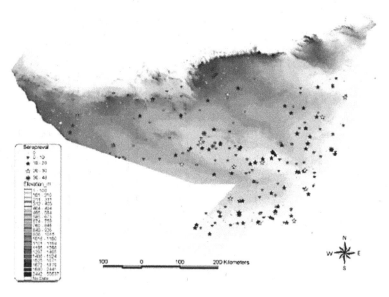

FIGURE 5.5   A GIS of the rift valley fever. (Soumare et al. 2007).

within a certain time frame. Traditionally this has been monitored by dividing a larger area into set smaller ones and counting the occurrences in each smaller area. The GIS can aggregate in many different ways, thus determining whether an area that is of the correct size but crosses one of these divisions contains sufficient cases to trigger an alert.

Figure 5.5 shows a GIS is of the Rift Valley Fever[10] spatial distribution in relation to ground elevation. This spatial distribution shows clusters of high sero-prevalence located mostly in the Nugal Valley, directing authorities to concentrate their efforts on this area. A GIS worked well due to the nomadic nature of the herds, meaning that herds could not be followed but locations could.

5.7 STANDARDS

Whilst there are no industry standards as such for data visualisation, there have been many proposed. Two such are:

- MSCUI (**Microsoft Common User Interface**), a set of data visualisation style guidelines for Office Add-ins (Microsoft 2022 [online]).
- "Data and Design – a simple introduction to preparing and visualising information" by Trina Chiasson, Dyanna Gregory et al. (Chiasson et al. 2021 [online]). It's a free, Creative Commons-licensed e-book which explains important data concepts in simple language.

The most important standard of all though is that the visualisation must assist the understanding of the data analysis.

5.8 COMMERCIAL SOFTWARE: EXCEL

Probably the most common tool for data analysis and presentation in the NHS is Excel[11] (although other spreadsheets and statistical packages do exist, this has by far the largest user base). We will now look at the main forms of analysis in this package.

5.8.1 CHARTS

Firstly, a small mathematical niggle. The figures in this section are not graphs. Figure 5.1 is a graph. The ones here are merely charts, because in mathematics a graph is a collection of vertices and the edges which connect them.[12] Graphs have been used to describe and investigate physical concepts such as road systems, electrical circuits, atomic bonds and computer networks (especially the Internet); relational concepts such as matches played between football teams; and social concepts such as "friend" and "acquaintance". A graph is therefore an abstract representation of the circuit, history or social relationship that is under study. Both graphs and charts are graphical visualisations, but only one of them is actually a graph.

So, let's now look at Excel charts. Whilst the raw tabular data can often provide sufficient information, a graphical approach is generally more appropriate. There are 6 main types of chart in Excel. We will use the table of infection rates in Table 5.1 as an example to illustrate them:

The pie chart was originally developed by Florence Nightingale (Figure 5.6 shows an example of her work).

A pie chart can display only one series of data and is a circular statistical chart, with sectors or slices representing the proportion of data. The arc length of each sector represents the quantity proportionally and so it is effective at showing proportions, but not much else, e.g. Figure 5.7.

TABLE 5.1

Infection Rates for Three Strains of Disease, Seasonally

	Winter	Spring	Summer	Autumn	Year
Strain A	17	24	20	23	84
Strain B	22	28	25	29	104
Strain C	26	34	20	32	112

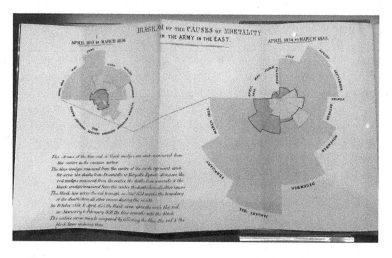

FIGURE 5.6 Diagram dated 1858 by Florence Nightingale of a coloured pie chart to illustrate causes of death in the British Army. (Originally from Notes on Matters Affecting the Health, Efficiency, and Hospital Administration of the British Army sent to Queen Victoria in 1858).

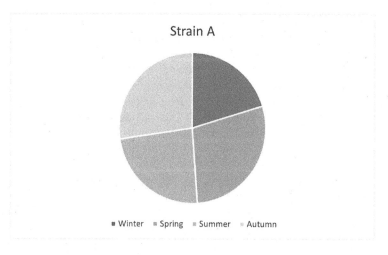

FIGURE 5.7 A pie chart.

The column chart very effectively shows the comparison of one or more series of data points. But the clustered column chart (see Figure. 5.8) is especially useful in comparing multiple data series.

The use of a different colour for each data series easily demonstrates how a single series, Strain A for example, changes over time. As the columns are clustered the three data series for each time period can be compared.

One variation of this chart type is the stacked column chart, as shown in Figure 5.9.

In a stacked column chart, the data points for each time period are stacked instead of clustered. This chart type demonstrates the proportion of the total for each data point in the series.

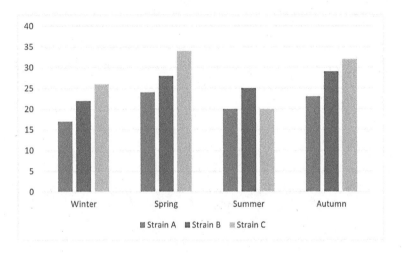

FIGURE 5.8   A clustered column chart.

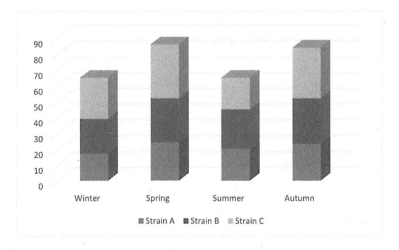

FIGURE 5.9   A stacked column chart.

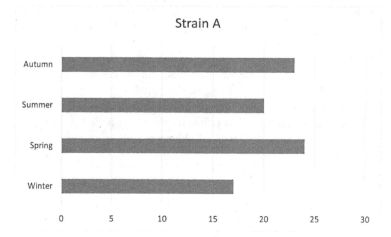

FIGURE 5.10   A bar chart.

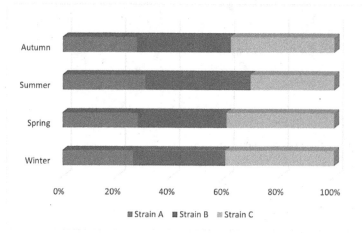

FIGURE 5.11   A 100% stacked bar chart showing the complete data but as a percentage of overall infections.

The bar chart (see Figure 5.10 and Figure 5.11) is effectively a rotated column chart. The use of a bar chart versus a column chart therefore depends on the type of data and user preference. However, bar charts do tend to display and compare a large number of series better than the other chart types.

The line chart (see Figure 5.12) is especially effective in displaying trends and is equally effective in displaying trends for multiple series (see Figure 5.13).

As this type of chart is usually used to show some changes and trends in the data it can demonstrate time-based differences well. Figure 5.14 shows how common

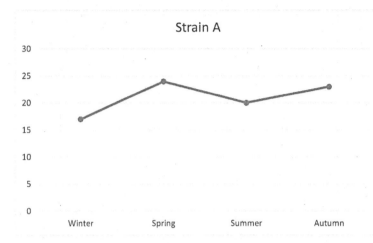

FIGURE 5.12   A line chart.

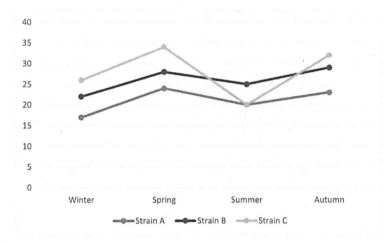

FIGURE 5.13   A line chart with multiple series.

source epidemics usually produce more new cases earlier and faster than host-to-host epidemics. Once the infected source is closed, sealed, or removed, the common source epidemic usually abates rapidly. Host-to-host epidemics are slower to grow and slower to diminish.

Area charts (see Figure 5.15 and Figure 5.16) are like line charts except that the area below the plot line is solid. Like line charts, area charts are used primarily to show trends over time or some other dimension. An area chart is used to illustrate quantitative data graphically by plotting the data points and connecting them into line segments.

A scatter chart uses cartesian coordinates to illustrate the values of two common variables for a data set. In this case, the data is represented as a collection of points.

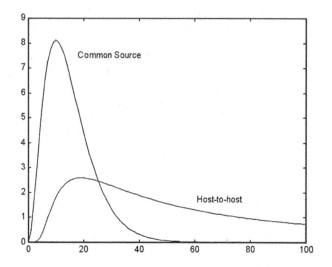

FIGURE 5.14   A line chart showing time-based differences.

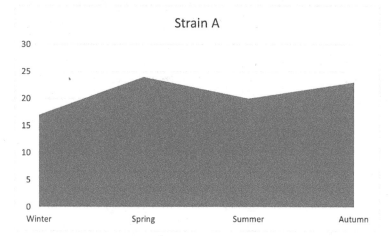

FIGURE 5.15   An area chart.

They show how the values of two or more series compare over time (or some other dimension). Scatter charts can therefore show how much one variable is affected by another.[13]

To illustrate the scatter chart, we will use the worksheet values in Table 5.2 (the charts are in Figure 5.17 and Figure 5.18).

In the simple case (a linear chart), the series pair has a positive correlation if they increase similarly and a negative correlation if one decreases as the other increases. Otherwise, they have no correlation. The chart in Figure 5.18 is more complex but shows positive correlation as both plots move in the same direction at the same time.

Scatter charts can also demonstrate cyclic behaviour as can be seen in Figure 5.19.

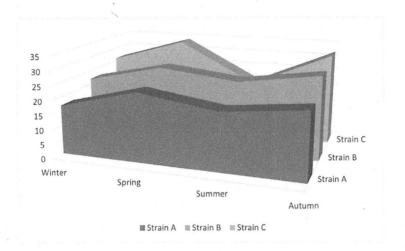

FIGURE 5.16   A 3-D area chart.

TABLE 5.2
Values for Strains A and B, by Month

	Jan	Feb	Mar	Apr	May	Jun	Jul	Aug	Sep	Oct	Nov	Dec
Strain A	10	20	30	50	60	70	55	47	70	50	35	40
Strain B	20	25	40	60	70	76	65	55	77	60	27	20

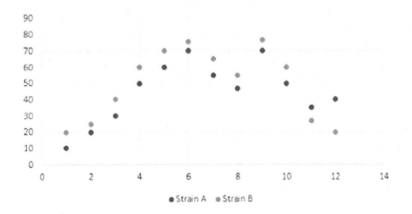

FIGURE 5.17   A scatter chart.

   This connected scatter chart (i.e. the points have been connected) shows laboratory reports of confirmed Norovirus infections in England and Wales, 1995 to 2002.
It can clearly be seen that there is seasonal variation for infection rates.

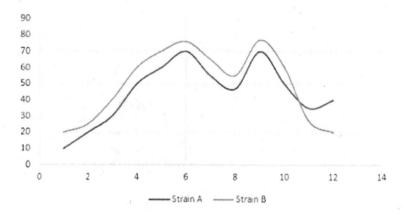

FIGURE 5.18   A scatter chart with smooth lines.

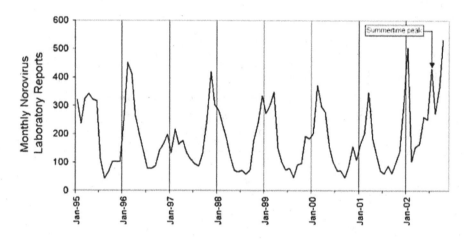

FIGURE 5.19   A connected scatter chart showing seasonal variation (Lopman et al. 2002).

Excel provides the facility to plot a trend line on a scatter chart (exponential, linear, logarithmic, polynomial, power or moving average), with options to display the equation and the R-squared value. The R-squared value is a measure of fit and the closer to 1, the better the line fits the data.

For example, a table of the sums spent on CDs etc. in various years[14] is shown in Table 5.3.

A scatter chart with a linear trend line can be seen in Figure 5.20, which is clearly a poor representation of the data, as indicated by the R^2 value of 0.0415.[15]

A polynomial of order 2, as seen in Figure 5.21 is much better.

The best fit of all is a moving average, as seen in Figure 5.22, which is as you'd expect as it doesn't try to fit one equation to the whole of the data, it just seeks the best fit locally.

TABLE 5.3
Sums of Money (in GBP) Spent on CDs etc. by the Author

Year	Total
1996	341.25
1997	473.99
1998	515.55
1999	316.98
2000	520.13
2001	571.32
2002	739.33
2003	993.32
2004	1085.62
2005	438.38
2006	104.26

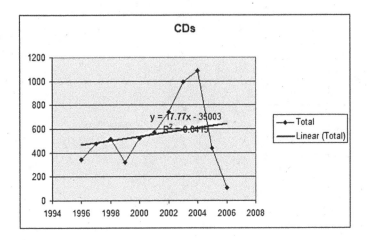

FIGURE 5.20 A scatter chart with linear trend line.

Whilst the trend line may usefully describe the data, it can also predict it. To do this in Excel, the "Year" range is extended (in this case to reach 2017). The polynomial (degree 2) in Figure 5.23 predicts a negative spend (which would be nice), whereas a degree 5 (Figure 5.24) predicts astronomic spending.

In the end, the linear model gives the only usable prediction (Figure 5.25), despite the relatively poor fit, of £821.32 and £839.09, respectively, although the real figures were £131.45 and £9.97. The moving average is unable to be extended into the future as it relies on fitting to local data – the best that can be achieved is to extend the final part of the fitted curve, thereby ignoring all preceding data.

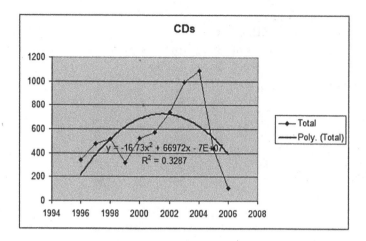

FIGURE 5.21 A polynomial order 2 trend line.

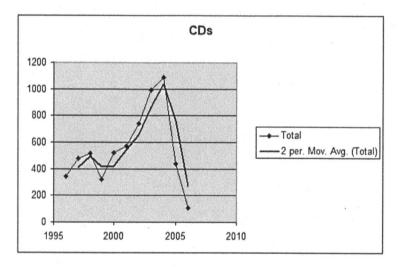

FIGURE 5.22 A moving average trend line.

Other statistical tests can be performed, such as regression, correlation and T tests, but they are more the subject of a statistics book than this one.

Whilst summary statistics (mean, standard deviation and correlation) can assist in understanding the data, they can be misleading. The famous "Anscombe's Quartet" (Figure 5.26) has four sets of data which each have the same summary statistics, but are clearly different data.

For a further set of examples, together with generation rules, see the Datasaurus Dozen (Matejka and Fitzmaurice, 2017).

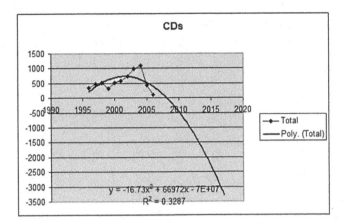

FIGURE 5.23 A polynomial trend line used for prediction.

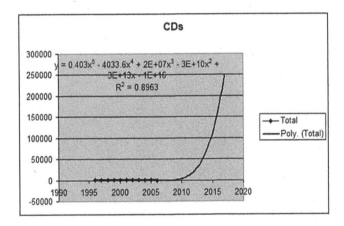

FIGURE 5.24 A polynomial trend line of degree 5 used for prediction.

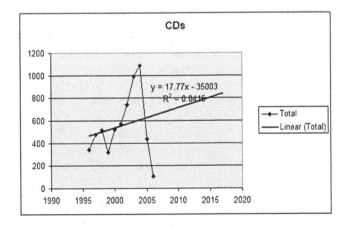

FIGURE 5.25 A linear trend line used for prediction.

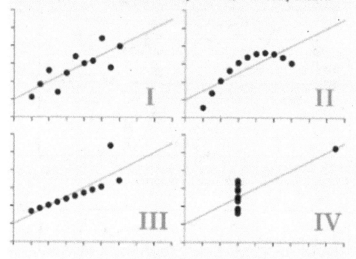

FIGURE 5.26   Anscombe's Quartet. (Anscombe 1973)

5.9   BLINDED WITH SCIENCE

Finally, it is important to convey the information truthfully. Many studies have remarked on the potential for a science-like element (such as a chart or statistic) to be given more authority than it deserves. The presentation of data analysis must therefore inform, not mislead.

Some general pointers (see "standards", above) are:

Pie charts seem friendly, but are actually hard to read. This is because humans are generally unable to make accurate estimates or comparisons of angles. Slices that are close in size are difficult to compare and the only comparison that can be made when there is a clear distinction is that one is bigger than the other.

Lengths are easier to compare than areas or volumes. Humans cannot easily compare two dimensions at once (e.g. two circles[16]), rarely making accurate estimates.

Bar and column charts are very good for comparisons, giving a simple visual representation of comparative values through the height of the bars. To judge the end points, the bars should begin at a zero baseline, as just showing the tips of the bars to exaggerate differences in the data removes the ability to make useful visual comparisons.

Line charts often show a trend. Altering the aspect ratio can distort the information the chart is conveying, as stretching the height of the chart can create fake drama, while stretching the width can underplay it.

Each part of a chart should aid understanding, not hide it. Everything presented should be there for a purpose. Excess colour (use as many as are necessary – no more), graphical clutter (such as background, borders, shading, dark grid lines and needless labels) and a plethora of special effects distract from the data.

A colour map, whilst making an image more visually appealing, may distort the interpretation of the data, especially if the colour bar itself is omitted (a more common occurrence than most would expect). A colour map such as batlow is therefore preferable to one such as rainbow or jet.

> The colour bar should be perceptually uniform to prevent data distortion and visual error. This means the perceptual colour differences between all neighbouring colours should appear the same. If two neighbouring colours have a different variation compared to other neighbouring colours (e.g., two greenish versus two yellowish-reddish colours in rainbow colour maps), the colour bar is perceptually non-uniform and not scientific.
>
> (Crameri et al. 2020)

Use the correct number of dimensions for the data. 3D is only useful when plotting the third dimension.[17] Otherwise it can skew the data and make comparisons harder.

Extra decimal places can look impressive and imply accuracy (possibly more than is actually present in the analysis). However, appropriate rounding makes numbers easier to compare and conveys information quicker. As mentioned earlier, the rounding method must be displayed.

Text assists understanding. Appropriate labels inform, so every chart and every axis need at least a title.[18]

Tables are useful for looking up individual numbers. In order to do this, they must show information in a way that makes identifying numbers and row-to-row comparisons easy. Aligning numbers to the right assists this, whereas left or centre alignment forces the eyes to jump back and forth between values, making comparisons of magnitudes difficult.[19]

Keep it simple: the intention is to inform, not to confuse. It's relatively easy to take something simple and make it complicated, but much harder to take something complex and present it in a way that is accessible. Sometimes a chart isn't the right presentation: for one or two values just showing the numbers may be best.

Charting two sets of data with one axis on the left and another on the right can be confusing and suggests a relationship that may not exist. It also takes time for the reader to understand and can lead to values being compared when it is meaningless given that the scales and units of the data are different.

For example, while Figure 5.27 makes sense, Figure 5.28 implies that all activities burn the same number of calories.

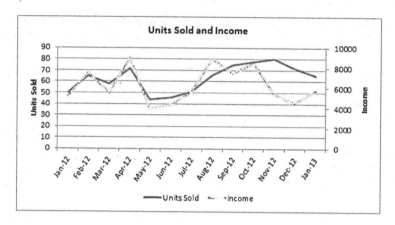

FIGURE 5.27　A double axis chart. (dedicatedexcel.com 2021 [online]).

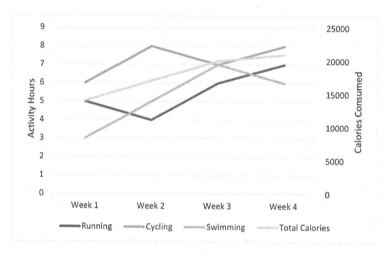

FIGURE 5.28　A less helpful double axis chart.

NOTES

1 Availability is a large factor, especially when dealing with confidential information or that covered by the GDPR.
2 A sample where each person (when dealing with people) in the population has a probability of being selected for the sample. This probability might not be the same for each person in order to balance the sample but must be known for each person selected. The sample is then used to predict the population.
3 The problem revolves around ignoring the sampling variation in w_i. See https://fmwww.bc.edu/RePEc/csug2017/Canada17_Chen.pdf for a worked example.
4 Impute: assign (a value) to something by inference from the value of the products or processes to which it contributes.
5 This may be a PowerPoint, peer-reviewed paper, discussion document, etc.
6 An exception to this is the discussion document, unless the purpose is to debate the interpretation.

7 For some amusing examples, see Dave Gorman's "Modern Life Is Goodish" where he looks at advertisements where only 7 out of 20 people agreed with the statement being presented.
8 Original attributions are hard to come by as the phrase has morphed through time. My favourite is Napoleon's "A good sketch is better than a long speech".
9 With thanks to Rachel O'Leary.
10 A disease prevalent among domesticated herd animals.
11 Especially true in Medical Physics and Clinical Engineering.
12 See also graphical databases in Section 7.3, Chapter 7.
13 However, beware variables that look correlated but aren't – see Section 4.5, Chapter 4 for details.
14 I did some DJ-ing at the time. I think you can tell where that activity peaked.
15 For a discussion of goodness of fit measures, see https://www.datasciencecentral.com/profiles/blogs/7-important-model-evaluation-error-metrics-everyone-should-know.
16 People usually underestimate the size of the bigger circles and overestimate the size of the smaller ones. Apparently this happens because we instinctively judge the lengths or widths of the shapes and not their areas.
17 In the case of the area chart above (Figure 5.16) the third dimension is the strains and using 3D makes their comparison easier.
18 The charts in this chapter may not all have a title, but do have a caption. Both would be overkill.
19 With numbers in a table, especially with a fixed number of decimal places, the magnitudes are easier to grasp with right-aligned than centre-aligned (they effectively form a type of bar chart). With a fixed number of decimal places the decimal point is in the same position for right-aligned numbers but is not so for left-aligned.

REFERENCES

Anscombe, F. J. 1973, Graphs in Statistical Analysis. *American Statistician*. 27 (1): 17–21. doi: 10.1080/00031305.1973.10478966. JSTOR 2682899.

Chiasson T, Gregory D et al., 2021, Data and Design – a simple introduction to preparing and visualising information [online]. Available: https://datadesignbook.github.io/english/index.html [Accessed 11/04/22].

Crameri, F., Shephard, G.E. and Heron, P.J., 2020, The misuse of colour in science communication. *Nat Commun*. [online] 11, 5444. Available: https://doi.org/10.1038/s41467-020-19160-7 [Accessed 11/04/22].

dedicatedexcel.com 2022, [online] Available: https://dedicatedexcel.com/how-to-create-a-chart-with-two-y-axis-in-excel-2010/ [Accessed 16/06/22].

Lopman B.A., Brown D.W. and Koopmans M, 2002, Human Caliciviruses in Europe. *Journal of Clinical Virology*. 24, 137–160.

Matejka J and Fitzmaurice M, 2017, *Same Stats, Different Graphs: Generating Datasets with Varied Appearance and Identical Statistics through Simulated Annealing*. Toronto Ontario Canada: Autodesk Research [online]. Available: https://www.autodeskresearch.com/publications/samestats [Accessed 11/04/22].

Microsoft 2022, [online]. Available: https://docs.microsoft.com/en-us/office/dev/add-ins/design/data-visualization-guidelines [Accessed 11/04/22].

Soumare B, Tempia S, Cagnolati V, Mohamoud A, Van Huylenbroeck G and Berkvens D, 2007, Screening for Rift Valley fever infection in northern Somalia: A GIS based survey method to overcome the lack of sampling frame. *Veterinary Microbiology*. [online] 121 (2007) 249–256. Available: ScienceDirect [Accessed 11/04/22].

Statistics Canada 2015, [online]. Available: https://www.statcan.gc.ca/pub/12-539-x/2009001/analysis-analyse-eng.htm [Accessed 11/04/22].

6 Boolean Algebra

6.1 INTRODUCTION

Boolean Algebra is the algebra of logic and is named after its creator, George Boole, an English mathematician born in Lincoln in 1815. There's also a crater on the Moon named after him and he wrote discourses on differential equations, but it's his "Laws of Thought" from 1854 which he is best remembered for, and it pretty much laid the foundations for the information age. In it he maintained:

> No general method for the solution of questions in the theory of probabilities can be established which does not explicitly recognise, not only the special numerical bases of the science, but also those universal laws of thought which are the basis of all reasoning, and which, whatever they may be as to their essence, are at least mathematical as to their form.
>
> (Boole 2012)

Just as the algebra we are used to will deal with numerical entities, so Boolean algebra deals with logical entities. In conventional algebra we would normally deal with x and y, where x and y can take any numerical value (e.g. 42, 4.2, 4E2, π). In Boolean algebra we deal with A and B, where A and B can take any logical value (i.e. *true* or *false*).[1]

6.2 NOTATION

There are several notations,[2] but the conventions we shall use here are as shown in Figure 6.1.

6.3 TRUTH TABLES

As logical entities can only take two values (*true* or *false*) it is possible to simply write out all possible combinations of simple expressions. This is an excellent way of validating Boolean algebra (but can be somewhat long-winded). The basic combinations are shown in Figure 6.2.

It is worth noting the order of entries in the table. By convention, false comes before true. The reason for this is that another way of representing false is with 0 and true is with 1. This would render the OR table from Figure 6.2 as shown in Figure 6.3.

Thus the A and B columns, taken as consecutive digits, count in binary from 0 to 3. In more complex truth tables this can be a useful way of checking that all the possible combinations have been written out.

DOI: 10.1201/9781003316244-6

Logical entity	A, B, C, ...
NOT A	$\overline{A}$
A AND B	A.B
A OR B	A+B
true	T
false	F

FIGURE 6.1 Boolean notation.

A	$\overline{A}$
F	T
T	F

A	B	A.B
F	F	F
F	T	F
T	F	F
T	T	T

A	B	A+B
F	F	F
F	T	T
T	F	T
T	T	T

FIGURE 6.2 NOT, AND & OR truth tables.

A	B	A+B
0	0	0
0	1	1
1	0	1
1	1	1

FIGURE 6.3 The OR table using 0 and 1 in place of F and T.

6.4 ALGEBRAIC RULES

Whilst many of the rules in Figure 6.4 are intuitive (and some seem so because of the resemblance to "normal" algebra – not particularly safe ground to be on), some results require further investigation. For example, let us consider A+(B.C)=(A+B).(A+C). The truth table for each side of this equation can be constructed and, if they are the same, then the rule is proven, as we see in Figure 6.5.

$$A + A = A$$

$$A.A = A$$

Symmetric or Commutative laws $\quad \begin{cases} A+B = B+A \\ \quad A.B = B.A \end{cases}$

Associative laws $\quad \begin{cases} A+B+C = (A+B)+C = A+(B+C) \\ \quad A.B.C = (A.B).C = A.(B.C) \end{cases}$

$$A.(B+C) = A.B + A.C$$

$$A + (B.C) = (A+B).(A+C)$$

De Morgan's laws $\quad \begin{cases} \overline{A.B} = \bar{A} + \bar{B} \\ \overline{A+B} = \bar{A}.\bar{B} \end{cases}$

$$A + true = true$$

$$A.true = A$$

$$A.false = false$$

$$A + \bar{A} = true$$

$$A.\bar{A} = false$$

$$A + A.B = A$$

$$A + \bar{A}.B = A + B$$

$$A.B + B.C + C.A = (A+B).(B+C).(C+A)$$

$$A.\bar{B} + A.C + B.C = A.\bar{B} + B.C$$

$$\bar{\bar{A}} = A$$

FIGURE 6.4 The Boolean algebraic rules.

A	B	C	$B.C$	$A+(B.C)$	$A+B$	$A+C$	$(A+B).(A+C)$
F	F	F	F	F	F	F	F
F	F	T	F	F	F	T	F
F	T	F	F	F	T	F	F
F	T	T	T	T	T	T	T
T	F	F	F	T	T	T	T
T	F	T	F	T	T	T	T
T	T	F	F	T	T	T	T
T	T	T	T	T	T	T	T

FIGURE 6.5 The truth table to investigate the rule A+(B.C)=(A+B).(A+C).

6.5 LOGICAL FUNCTIONS

When x is a real variable there are many[3] functions of x which are possible. Because Boolean variables can only take two values, it is possible to write out all functions of a number of variables.

6.5.1 FUNCTIONS OF ONE VARIABLE

For a single variable, A, there are four possible functions as shown in Figure 6.6. The order of the possible values follows the aforementioned "binary" order of functions (i.e. from 00 to 11 – FF to TT).

From Figure 6.6 we can see that f_0 and f_3 are constant functions and $f_1 = A$ and $f_2 = \bar{A}$.

6.5.2 FUNCTIONS OF TWO VARIABLES

There are 16 possible functions of two variables (2^{2^2}), as shown in Figure 6.7. Of these, f_0 and f_{15} are constant functions.

$$f_3 = A$$

$$f_5 = B$$

$$f_{12} = \bar{A}$$

$$f_{10} = \bar{B}$$

$$f_1 = A.B$$

$$f_7 = A + B$$

$$f_6 = A \, XOR \, B$$

$$f_9 = A \equiv B$$

A	f_0	f_1	f_2	f_3
F	F	F	T	T
T	F	T	F	T

FIGURE 6.6 The functions of one variable.

A	B	f_0	f_1	f_2	f_3	f_4	f_5	f_6	f_7	f_8	f_9	f_{10}	f_{11}	f_{12}	f_{13}	f_{14}	f_{15}
F	F	F	F	F	F	F	F	F	F	T	T	T	T	T	T	T	T
F	T	F	F	F	F	T	T	T	T	F	F	F	F	T	T	T	T
T	F	F	F	T	T	F	F	T	T	F	F	T	T	F	F	T	T
T	T	F	T	F	T	F	T	F	T	F	T	F	T	F	T	F	T

FIGURE 6.7   The functions of two variables.

$$f_{14} = \overline{A.B} = A\,\mathrm{NAND}\,B = A \uparrow B$$

$$f_8 = \overline{A + B} = A\,\mathrm{NOR}\,B = A \downarrow B$$

$$f_{13} = A \Rightarrow B$$

$$f_2 = A\neg \Rightarrow B$$

$$f_{11} = B \Rightarrow A$$

$$f_4 = B\neg \Rightarrow A$$

Some of these functions we have met earlier. The others are:

XOR (Exclusive OR), which is sometimes rendered EOR or EXOR and often
  uses the symbol $\oplus$. It is called "exclusive" (as opposed to "OR" which is
  inclusive) because the output of true is ambiguous when both inputs are true
  – it therefore excludes that case.
NAND is "NOT AND" and NOR is "NOT OR".
$\equiv$ (equivalent) is also known as iff and XNOR.
Implies ($\Rightarrow$) is not particularly intuitive as the meaning "if A is true then B is
  also true" is clear when A is T, but not so when A is F (which always gives
  a T output). It may also be rendered NOT A OR B.
Does not imply ($\neg \Rightarrow$) may also be rendered A AND NOT B.

6.6   SIMPLIFICATION OF LOGICAL EXPRESSIONS

Let us consider a three-input single-bit binary addition, as shown in Figure 6.8. The
sum remains in the binary position with the carry "carried" to the next position up
(as in normal addition).

A	B	C	Carry	Sum
0	0	0	0	0
0	0	1	0	1
0	1	0	0	1
0	1	1	1	0
1	0	0	0	1
1	0	1	1	0
1	1	0	1	0
1	1	1	1	1

FIGURE 6.8   A three-input single-bit binary addition.

Expressing this in logic terms (again translating 1 as T and 0 as F), we see that

$$Carry = \bar{A}.B.C + A.\bar{B}.C + A.B.\bar{C} + A.B.C \tag{6.1}$$

An expression derived directly from a truth table in this way is called the **Disjunctive Normal Form** or DNF.

This expression can be simplified algebraically:

$$\bar{A}.B.C + A.\bar{B}.C + A.B.\bar{C} + A.B.C$$

$$=\left(\bar{A}.B.C + A.\bar{B}.C + A.B.\bar{C}\right) + A.B.C$$

$$=\left(\bar{A}.B.C + A.B.C\right) + \left(A.\bar{B}.C + A.B.C\right) + \left(A.B.\bar{C} + A.B.C\right)$$

$$=\left(\bar{A} + A\right).B.C + A.\left(\bar{B} + B\right).C + A.B.\left(\bar{C} + C\right)$$

$$= B.C + A.C + A.B$$

(i.e. the expression is true if any two values are true – this also covers the special case where all three are true, as we can see from Figure 6.8).

6.7   A SLIGHT DETOUR INTO NAND AND NOR

In "functions of two variables" (above) we introduced NAND and NOR. On initial examination, NAND (a combination of NOT and AND) and NOR (a combination of

A	A	$A\uparrow A$	$\overline{A}$
F	F	T	T
T	T	F	F

FIGURE 6.9 The truth table demonstrating the equivalence of NAND and NOT.

NOT and OR) are not particularly useful as they are not intuitive logical concepts and are merely combinations of concepts we already have.

However, we can show that AND, OR and NOT can be represented as expressions using only NAND. Equally we can show that they can be represented as expressions using only NOR. An example (NOT) is given in Figure 6.9, showing that $\overline{A} = A \uparrow A$

At this point we leave the fun and games of pure mathematics to enter the word of electronics.[4]

The concept of functional completeness states that a set of operators (or, in electronics terms, a set of gates) is functionally complete if every logical expression can be realised by an expression (circuit) containing only operators (gates) from the set.

Every logical expression has a DNF and every DNF can be realised as a circuit containing only AND, OR and NOT gates. Therefore {AND, OR, NOT} is functionally complete. However, as we can represent these operators by expressions containing only NAND, {NAND} is also functionally complete. Similarly, {NOR} is functionally complete.

This makes electronics development simpler, as all logic circuits can be developed using only NAND gates, meaning you can construct circuits using **Integrated Circuits** (ICs) containing lots of NAND gates instead of a plethora of ANDs ORs and NOTs. One other benefit is that NAND gates are considerably easier to manufacture, hence making them cheaper to use.

6.8 KARNAUGH MAPS

A Karnaugh map is a very visual method of simplifying a logical expression. For an expression with i variables, there are 2^i possible combinations. A Karnaugh map is a diagram with 2^i areas, one for each combination. Figure 6.10 shows the layout for two variables.

By convention, we normally only label the variable, not the inversion. Expressions not using all the variables are represented by a group of cells. Figure 6.11 shows a Karnaugh map for 4 variables with the expression $A.B.C.$ marked with Xs.

Returning to the example of the sum and carry bits for earlier (6.1), we require to simplify

$$Carry = \overline{A}.B.C + A.\overline{B}.C + A.B.\overline{C} + A.B.C$$

Marking this up as a Karnaugh map gives the map in Figure 6.12.

By grouping adjacent cells[5] with 1,2,4,8 or 16[6] cells in each group,[7] we can simplify the expression, as shown in Figure 6.13.

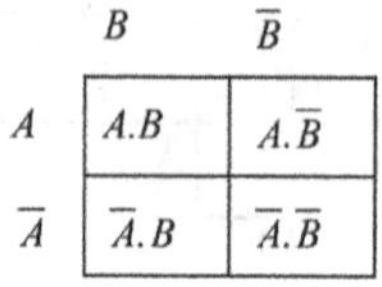

FIGURE 6.10   A Karnaugh map for two variables.

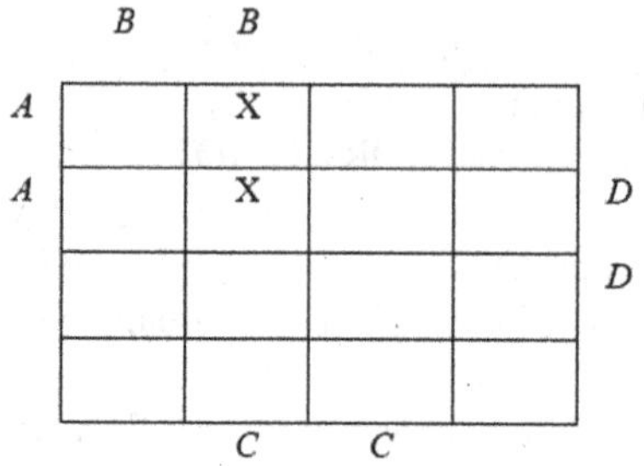

FIGURE 6.11   A Karnaugh map for 4 variables showing how a 3-variable expression is marked.

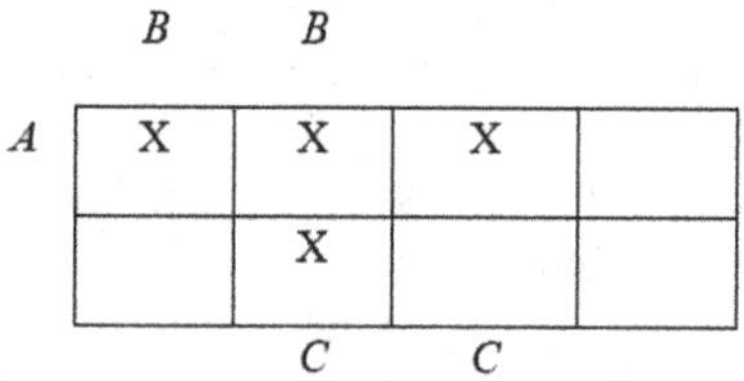

FIGURE 6.12   A Karnaugh map for the carry bit.

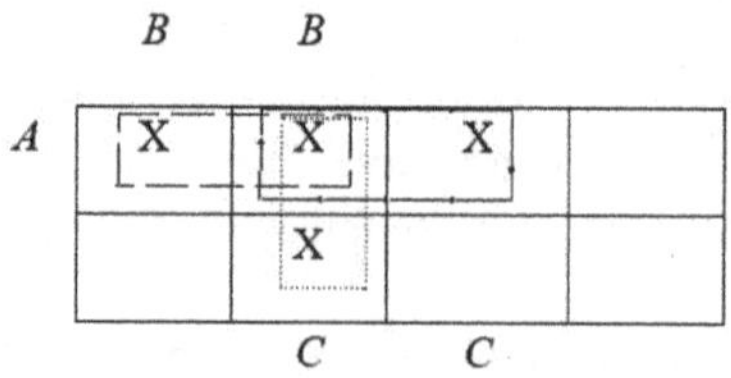

FIGURE 6.13   Groups in Figure 6.12 are identified.

Group dashed = $A.B$
Group dotted = $B.C$
Group arrows = $A.C$
This gives $Carry = A.B+B.C+A.C$ as before.

Returning to another previous example, we showed that $A+(B.C)=(A+B).(A+C)$ using a Truth Table. This can equally be done using a Karnaugh map, as

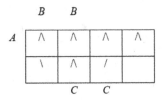

FIGURE 6.14   A Karnaugh map proof of an algebraic rule.

demonstrated in Figure 6.14, where \ indicates $(A+B)$ and / indicates $(A+C)$, we can see that the intersection of these two sets (given by $\wedge$) forms two groups: A and $B.C$. Thus $A+(B.C)=(A+B).(A+C)$ as before.

6.9   USING BOOLEAN ALGEBRA IN FORMING AND VALIDATING QUERIES

Database queries (and loop conditions in programs) are logical entities. A query such as "Find all patients aged over 50 whose spine scan is not Osteopenic" can be expressed as "SELECT patients WHERE age>50 AND (spine BMD<0.9 OR spine BMD >1.08)". In other words, we formulate an expression with parts which will equate to either *true* or *false*. Once we have done this, we can apply Boolean algebra. In particularly complex queries, it is worth expressing them as fully as possible and then simplifying using a Karnaugh map or Boolean algebra rather than trying to do it all "in your head".

A simple example of the benefit of this approach is when inverting a query. To select the set of patients not collected in the previous query, you may be tempted to write "Find all patients aged 50 or under whose spine scan is Osteopenic" or "SELECT patients WHERE age≤50 AND (spine BMD≥0.9 OR spine BMD≤1.08)" i.e. reversing the test conditions. However, we know from Boolean algebra that $\overline{A+B}=\overline{A}.\overline{B}$; therefore, the correct expression is "SELECT patients WHERE age≤50 OR (spine BMD≥0.9 AND spine BMD≤1.08)".

6.10   BINARY AND MASKING

A further use of Boolean algebra is in manipulating binary numbers. For example, if an eight-bit number is stored as a sign bit followed by a seven-digit number, the eight bits can store numbers from −127 to +127.[8]

In order to obtain the absolute value of such a number, we only need to "strip off" the topmost bit. This can be done by ANDing the number with 127, as shown in Figure 6.15 for the value −73.

Another example is the use of two bytes to store a 16-bit number. These are often stored low byte first,[9] but the important part is that we need to separate a number into two numbers. The high byte can be found by dividing by 256 (assuming we're dealing with integers so any remainder is lost). The low byte is easily found by ANDing the number with 255.

-73	11001001
127	01111111
-73 AND 127	01001001
=73	

FIGURE 6.15   Finding the absolute value of a binary number using AND.

Similar tricks can be used to extract individual bits (which may have significance in, for example, an image file header where eight Yes/No flags may be packed into one byte).

NOTES

1 There is such a thing as fuzzy logic, where other values exist, but we won't be dealing with them here as they are more probabilistic and are thus suited more to decision making, identification, pattern recognition, optimisation and control.
2 One particular variant is !A to indicate NOT A, which follows from programming syntax. Others also exist for AND and OR.
3 Possibly infinite, depending on which definition of infinity you are using.
4 Just as with complex numbers, mathematicians devised something that was fun to play with and then physicists ruined it all by finding a use for it.
5 The grouping can "wrap around".
6 i.e. powers of 2.
7 Cells can be in more than one group.
8 This is actually slightly inefficient as there are then two representations of zero, but it makes the next bit easier.
9 It's just a convention – it can be the other way round. Low byte first is known as Little Endian and the opposite is Big Endian. Both terms come from Gulliver's Travels describing debates about which end of an egg should be cracked first. In Gulliver, such debate led to war.

REFERENCES

Boole, G, 2012 [Originally published by Watts & Co., London, in 1952]. Rhees, Rush, ed. *Studies in Logic and Probability* (Reprint ed.). Mineola, NY: Dover Publications. p. 273. ISBN: 978-0-486-48826-4. Retrieved 27 October 2015, lifted from Wikipedia.

7 NoSQL

(With grateful thanks to James Moggridge at UCLH for permission to use his presentation as a basis for this chapter and especially for the examples that remain untouched).

7.1 INTRODUCTION

A NoSQL database provides a mechanism for storage and retrieval of data that is modelled in ways other than the tabular relations used in relational databases, "NoSQL" originally referring to "non SQL" or "non-relational" databases. Whilst such databases have existed since the late 1960s, the "NoSQL" term came to prominence in the early twenty-first century, triggered by the needs of Web 2.0 companies such as Facebook, Google and Amazon. NoSQL systems are also sometimes called "Not only SQL" to emphasise that they may support SQL-like query languages.

NoSQL is thus a loose umbrella term describing the means of storing data other than in a standard relational table schema. There are a variety of technologies under the NoSQL umbrella[1]:

- Document storage – e.g. Apache CouchDB, ArangoDB, BaseX, Clusterpoint, Couchbase, Cosmos DB, HyperDex, IBM Domino, MarkLogic, MongoDB, OrientDB, Qizx, RethinkDB
- Object storage – e.g. Cache (Intersystems)
- Graph – e.g. AllegroGraph, ArangoDB, InfiniteGraph, Apache Giraph, MarkLogic, Neo4J, OrientDB, Virtuoso.
- Wide Column – e.g. Bigtable (Google), Accumulo, Cassandra, Druid, HBase, Vertica.
- Key/Value – e.g. DynamoDB (Amazon), Aerospike, Apache Ignite, ArangoDB, Couchbase, FairCom c-treeACE, FoundationDB, HyperDex, InfinityDB, MemcacheDB, MUMPS, Oracle NoSQL Database, OrientDB, Redis, Riak, Berkeley DB, SDBM/Flat File dbm, Voldemort.

The reasons for adopting a NoSQL technology include:

- Unstructured data that doesn't fit a schema – forcing such data into a schema is inefficient
- Flexibility – there is no requirement for object-relational mapping and data normalisation, the models working with self-contained aggregates or **Binary Large Objects** (BLOBs).
- Scalability – an increase in data size produces slower response times. A RDBMS solves this by upgrading the hardware (more RAM, CPU, HDD

etc.) whereas a NoSQL solution distributes the database load onto additional hosts. This is horizontal rather than vertical scaling (also known as "scaling out" rather than "scaling up") and is very suited to handling big data.[2]

- Speed – better response times depending on the nature of the data. Data structures used by NoSQL databases (e.g. key-value, wide column, graph, or document) are different from those used by default in relational databases, making some operations faster in NoSQL – especially when dealing with huge volumes of data.
- Rapid development – NoSQL systems have simple APIs (**Application Programming Interface**) and tend to be text-based protocols implemented via HTTP REST with JSON (see Section 12.3, Chapter 12).
- Less dependence on Database admin skills

Many NoSQL stores compromise consistency in favour of availability, partition tolerance and speed. Barriers to the greater adoption of NoSQL stores include the use of low-level query languages,[3] lack of standardised interfaces and huge previous investments in existing relational databases. Most NoSQL stores lack true ACID (**Atomicity, Consistency, Isolation, Durability**)[4] transactions, although a few databases, such as MarkLogic, Aerospike, FairCom c-treeACE, Google Spanner,[5] Symas LMDB and OrientDB have made them central to their designs.

Instead, most NoSQL databases offer a concept of "eventual consistency" in which database changes are propagated to all nodes "eventually" (typically within milliseconds) so queries for data might not return updated data immediately or might result in reading data that is not accurate, a problem known as stale reads. Additionally, some NoSQL systems may exhibit lost writes and other forms of data loss. Fortunately, some NoSQL systems provide concepts such as write-ahead logging to avoid this.

A summary of the differences between SQL and NoSQL databases is in Table 7.1.

TABLE 7.1
A Summary of the Differences between SQL and NoSQL Databases

SQL Databases	NoSQL Databases
Relational databases.	Non-relational databases.
Only one type of SQL database is available, as the data are stored in tables.	Multiple types of NoSQL database are available, e.g. key-value stores, document databases, wide-column stores and graph databases.
The schema and the structure of the data are predefined. For the addition of new fields the schema must be altered which may cause system interruptions.	Dynamic schemas. There is no need to predefine the schema which is going to store the data. New data can be added any time, without system interruptions.
Usually, a single server hosts the database, and it is scaled vertically.	Horizontally scaling, allowing data spreading across many servers.
They use joins, which allows the design of complex queries, but can slow down response times.	Join can be only performed in collections in a complex aggregation query.

Some strengths and weaknesses of NoSQL as compared to a RDBMS are:

7.1.1 STRENGTHS

- NoSQL can be used as either a primary or an analytic data source (especially for online applications).
- It has big data capability.
- There is no single point of failure.
- NoSQL provides fast performance and horizontal scalability.
- NoSQL can handle structured, semi-structured and unstructured data equally.
- NoSQL databases don't require a dedicated high-performance server.
- It is generally simpler to implement than a RDBMS.
- NoSQL has a flexible schema design which can easily be altered without downtime or service disruption.

7.1.2 WEAKNESSES

- There are no standardisation rules.
- There are limited query capabilities (especially true for complex queries).
- NoSQL databases and tools are less mature, stable and well-understood than those related to RDBMS.
- NoSQL does not offer any traditional database capabilities, such as consistency when multiple transactions are performed simultaneously.
- As the volume of data increases it is difficult to maintain unique values as keys become difficult.
- NoSQL doesn't work as well with relational data as a RDBMS.[6]
- The learning curve is quite steep for new developers.
- NoSQL is generally open-source which enterprises often avoid due to issues with robustness, support, integration into existing infrastructure and a need to publish source code (amongst others).

We will now examine two types of NoSQL store: Document Storage and Graph.

7.2 DOCUMENT STORAGE

Document stores appear the most natural among the NoSQL database types because they're designed to store everyday documents without pre-processing and they provide complex querying and calculations on this form of data.

The central concept of a document store is the notion of a "document". While each document-oriented database implementation differs on the details of this definition, in general, they all assume that documents encapsulate and encode data (or information) in some standard formats or encodings (i.e. there may well be more than one format/encoding in use). Encodings in use include XML, YAML[7] and JSON as well as binary forms like BSON.[8] Documents are addressed in the database via a unique key that represents that document.[9] One of the other defining characteristics of a document-oriented database is that in addition to the key lookup performed by a key-value store, the database offers an API or query language that retrieves documents based on their contents.

One form of document storage is MongoDB, which is part of the MEAN[10] Stack. Documents are stored in JSON format and the system is open source, which may account for its popularity given its age relative to SQL languages.[11]

It is possible to compare MongoDB terminology with traditional RDBMS – some examples are shown in Table 7.2.

Compared to relational databases, for example, collections could be considered analogous to tables and documents analogous to records. But they are different: every record in a table has the same sequence of fields, while documents in a collection may have fields that are completely different.

The advantages and disadvantages, compared to a RDMBS are shown in Table 7.3.

As an example, consider an Ultrasound QA database in which it is required to record QA data. Fields would include:

- Site
- Date
- Person performing tests
- Systems
- Probes
- Results

TABLE 7.2

A Sample Comparison between MongoDB and a Traditional RDBMS

RDBMS	MongoDB
Table	Collection
Row	Document
Column	Field
Table Join	Embedded Documents
Primary Key	Primary Key (Default key _id provided by MongoDB itself)

TABLE 7.3

Advantages and Disadvantages of a Document Storage Database, Compared to a Relational One

Advantages	Disadvantages
Quick to set up	Relatively new
High availability	Complicated querying is harder – no joins
Scalability	Business Information and analytics tools not as mature
Flexibility	Risk of loss of consistency
Fast application development not held up by requirements to construct DB schema	Transactions not ACID compliant – no atomicity (all or nothing) beyond single document level

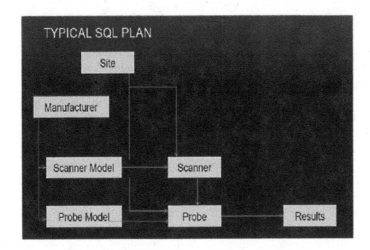

FIGURE 7.1   A typical SQL plan for an ultrasound **Quality Assurance** (QA) database.

A typical SQL plan would appear as shown in Figure 7.1 and would contain normalised data, look up tables, keys, relational tables and a results set.

The MongoDB equivalent[12] might be:

```
id: ObjectId(7df78ad8902c),
title: „UCLHQA_Report',
description: „Report on US QA for UCLH',
site: „UCLH",
author: „James „,
scanner:[ {
     sn: '2334DSFSD',
     manufacturer: 'Toshiba',
     dateTested: new Date(2011,1,20,2,15),
     performed by: 'James',
     probe[{
          sn: '2334DSFSD',
          Model: 'L38D',
          Preset: 'Abdomen',
          resolution:[{depth1-axial: '2'.2, depth1-lateral: '4.3',...}]
     }]
     probe[{
          sn: '5633FFDG',
          model: 'C94TV',
          preset: 'Gynae',
          resolution:[{depth1-axial: '2',depth1-lateral: '2',...}]],
          cystdetection:[{4at2: 'y', 2at2: 'y', 1at1: 'n'...}]
     }]
}]
scanner:[ {...}]
```

In healthcare, applications of such databases can be useful because:

- Patient reports can be relatively unstructured
- There are often requirements for new measurements
- Symptoms or patient history are often free text
- There may be a requirement to compare reports from different systems in different formats but the data itself is often comparable
- Big data can be very hard to model usefully in traditional RDBMSs.

However, the more flexible data recording is the harder it is to run accurate queries and most fields will be structured, and some will be vital/mandatory. The solution is to use a hybrid and get the best of both worlds. Postgresql, SQL Server 2016 and MySQL are all able to support NoSQL concepts.

For further thoughts on MongoDB and healthcare, see https://kousikraj. me/2012/06/05/mongo-db-and-its-application-a-case-study-of-mongodb-in-healthcare/.

7.3   GRAPHDB

A graphDB, such as Neo4J, handles relationships better than a relational DB and is built on the concept of relationship chains, such as friend of a friend of a friend and recommendations. It complies with ACID transactions yet its function is hard to replicate in traditional SQL DBs.

A graphDB stores entities as well as the relationships between those entities. The entity is stored as a node with the relationships as edges. An edge therefore gives a relationship between nodes.

This kind of database is designed for data whose relations are well represented as a graph consisting of elements interconnected with a finite number of relations between them. The type of data could be social relations, public transport links, road maps or network topologies.

GraphDBs are very good at identifying unusual patterns of behaviour – for example, credit fraud is easier to spot from a graph than from a table.

A typical SQL plan might be as shown in Figure 7.2.

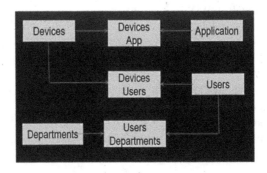

FIGURE 7.2   A typical SQL plan for a medical device database.

Giving rise to queries such as

> SELECT AppLink.LinkName, Devices.DeviceName, Users.UserName,
>   LUDept.DeptShortName
> FROM Devices INNER JOIN
> Device_AppLink ON Devices.DeviceId = Device_AppLink.DeviceId
>   INNER JOIN
> Device_User ON Devices.DeviceId = Device_User.DeviceId INNER
>   JOIN
> Users ON Device_User.UserId = Users.UserId INNER JOIN
> AppLink ON Device_AppLink.AppLinkId = AppLink.LinkId INNER
>   JOIN
> User_Dept ON Users.UserId = User_Dept.UserId INNER JOIN
> LUDept ON User_Dept.DeptId = LUDept.DeptId
> WHERE (AppLink.LinkName = 'c:\users\public\desktop\cvis tomcat.
>   lnk')

which returns

LinkName	DeviceName	UserName	DeptShortName
c:\users\public\desktop\cvis tomcat.lnk	PCCGI02128	*****	EGA
c:\users\public\desktop\cvis tomcat.lnk	PCCGI01383	*****	EGA
c:\users\public\desktop\cvis tomcat.lnk	PCCGI02741	*****	EGA
c:\users\public\desktop\cvis tomcat.lnk	PCCGI04147	*****	Service Desk
c:\users\public\desktop\cvis tomcat.lnk	PCCGI04201	*****	Service Desk
c:\users\public\desktop\cvis tomcat.lnk	PCCGI00629W7	*****	Service Desk
c:\users\public\desktop\cvis tomcat.lnk	PCCGI05270	*****	Service Desk
c:\users\public\desktop\cvis tomcat.lnk	PCCGI03876	*****	Service Desk
c:\users\public\desktop\cvis tomcat.lnk	PCCGI02111	*****	Service Desk

A GraphDB representation would be as shown in Figure 7.3, with the relationship
model as shown in Figure 7.4.

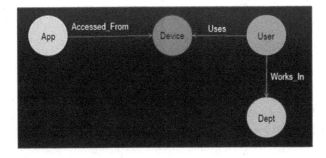

FIGURE 7.3   A GraphDB representation of a medical device database.

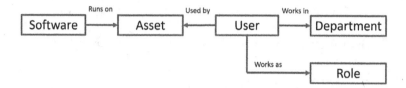

FIGURE 7.4 The relationship model for the GraphDB representation in Figure 7.3.

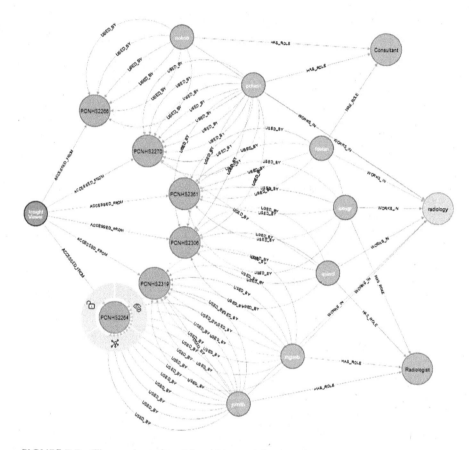

FIGURE 7.5 The graph produced from the example query.

And the same query as before would be:

MATCH (AppLink {LinkName: "c:\\users\\public\\desktop\\tomcat.
lnk"})-[:Accessed_From]->(Device)<-[:Uses]-(User)-[:Works_In]-
>(Dept) RETURN AppLink, Device,User, Dept

which produces the graph in Figure 7.5.

This is the same information as would be returned by a relational query, but displays relationships in a better form (which member of staff uses which equipment; whether they are a radiologist, etc.)

7.4   CONCLUSIONS

The "right DB" depends on the project goals. However, a relational database with a well-structured and thought out schema is always a good place to start. It is always easier to move from a structured to a less structured solution than the other way around as this problem has been solved many times.

It is worth looking at hybrid solutions and always best to choose a database technology that is well supported. Whilst NoSQL presents many possibilities, SQL is not going away as it fits too many problem spaces well.

NOTES

1 The number of vendors per technique gives a good indication as to how popular that technique is.
2 There is quite a difference between "big data" and "huge data", although you'll often hear the terms being used interchangeably. Huge data is lots of data – 6 million test results, for example. Big data is also lots of data, but the connections are vague, undefined or just very hard to conceptualise.
3 Instead of SQL, meaning for instance a lack of ability to perform ad-hoc joins across tables.
4 ACID transactions: Atomicity (all or nothing), Consistency (maintain validity), Isolation (each sequence of transaction works independently), Durability (change is persistent).
5 Though technically a NewSQL database rather than NoSQL. NewSQL databases seek to provide the scalability of NoSQL systems for online transaction processing workloads while maintaining the ACID guarantees of a traditional relational database system.
6 No surprises in that one!
7 **YAML Ain't Markup Language** – a recursive acronym (like GNU).
8 **Binary JSON**.
9 Document stores are therefore extensions of the Key/Value type of NoSQL database.
10 **Mongo, Express, Angular, Node**.
11 Whilst enterprises may not like open-source software, start-ups and small businesses/departments are especially attracted to the "free" aspect.
12 Which has more than a passing similarity to XML.

8 Network Architecture[1]

8.1 INTRODUCTION

Whilst a computer on its own is a powerful device, the possibilities and the power increase greatly when such devices are linked together to form a network. Computer networking is a major topic in computer science within healthcare and this chapter examines many of the topics within this field.

8.2 NETWORKING AND THE NETWORK ENVIRONMENT

The minimum number of devices in a network is 2.[2] The maximum number depends on the addressing method: standard IP addresses allow for 4,294,967,296[3] (which is 256[4]) but there are many ways to extend this, as we shall see.

In a hospital environment, devices are usually connected physically – i.e. with a cable. This improves reliability as well as giving larger bandwidth and higher speed. There are multiple ways of connecting devices, but the simplest is via a hub, which is essentially a connection box where all incoming signals are sent to all connected devices.

8.2.1 THE NETWORK PACKET

All networking is described in terms of packets. A network packet is a formatted unit of data carried by a packet-switched network. Computer communications links that do not support packets, such as traditional point-to-point telecommunications links, simply transmit data as a bit stream.

A packet consists of control information and user data, which is also known as the payload. Control information provides data for delivering the payload, for example: source and destination network addresses, error detection codes and sequencing information. Typically, control information is found in packet headers and trailers.

Different communication protocols use different conventions for distinguishing between the elements and for formatting the data. For example, in Point-to-Point Protocol, the packet is formatted in 8-bit bytes, and special characters are used to delimit the different elements. Other protocols like Ethernet establish the start of the header and data elements by their location relative to the start of the packet. Some protocols format the information at bit level instead of at byte level.

A good analogy is to consider a packet to be like a letter: the header is like the envelope, and the data area is whatever the person puts inside the envelope.

In the seven-layer OSI (**Open Systems Interconnection**) model of computer networking (see Section 8.8 for a full description, but for now we need only consider that the higher the layer number, the more sophisticated the model), "packet" strictly refers to a data unit at layer 3, the Network Layer. The correct term for a data unit at Layer 2, the Data Link Layer, is a frame, and at Layer 4, the Transport Layer, the correct term is a segment or datagram. For the case of TCP/IP (**Transmission Control Protocol/Internet Protocol**) communication over Ethernet, a TCP segment is carried in one or more IP packets, which are each carried in one or more Ethernet frames.

8.2.2 HARDWARE – HUB, SWITCH, ROUTER, FIREWALL

A hub (see Figure 8.1) is a simple connection box operating at the Physical Layer (layer 1) of the OSI model (see Section 8.8). Like a transport hub, it's where everything comes together. Unlike a transport hub, though, whatever comes in on one connection goes out on all other connections. It's up to the receiver to decide whether or not the message is for them. Hubs therefore work well for small networks, but get messy and slow down for larger ones. It is therefore often common to find them in small networks (e.g. at home) or in sub-networks (e.g. in an office).

A switch (see Figure 8.2) is a device that filters and forwards packets between **Local Area Network** (LAN) segments. Switches operate at the data link layer (layer 2) and sometimes the network layer (layer 3 – a higher layer and therefore more sophisticated than a hub) of the OSI Reference Model and therefore support any packet protocol. LANs that use switches to join segments are called switched LANs or, in the case of Ethernet networks, switched Ethernet LANs (as you might expect). Both hubs and switches may be referred to as "bridges".

A router (see Figure 8.3) forwards data packets along networks. A router is connected to at least two networks and is located at a gateway, the place where two or

FIGURE 8.1   A network hub. (Shutterstock).

FIGURE 8.2   A network switch. (Shutterstock).

FIGURE 8.3 A network router. (Shutterstock).

more networks connect. Routers use headers and forwarding tables to determine the best path for forwarding the packets, and they use protocols such as **Internet Control Message Protocol** (ICMP) to communicate with each other and configure the best route between any two hosts.

A large network will therefore contain all 3 of these types of devices and we will now compare them.

Routers have generally increased in sophistication, combining the features and functionality of a router and switch/hub into a single unit (and may contain a basic **Domain Name Service** (DNS)).

The functions of a router, hub and a switch are all quite different from one another, even if at times they are all integrated into a single device. We will start with the hub and the switch since these two devices have similar roles within the network.

Both the switch and the hub serve as central connections for all of the network equipment and handle a data type known as frames. Frames carry data and when one is received, it is amplified and then transmitted on to the port of the destination device. The big difference between these two devices is in the method in which frames are being delivered.

In a hub, a frame is passed along or "broadcast" to every one of its ports. It doesn't matter that the frame may be only destined for one port as the hub has no way of distinguishing which port a frame should be sent to. Passing it along to every port ensures that it will reach its intended destination. This places a lot of traffic on the network and can lead to poor network response times.

Additionally, a 10/100Mbps[4] hub must share its bandwidth across each and every one of its ports. So when only one device is broadcasting, it will have access to the maximum available bandwidth. If, however, multiple devices are broadcasting, then that bandwidth will need to be divided among all of those systems, which will degrade performance.

A switch, however, keeps a record of the **Media Access Control** (MAC) addresses of all the devices connected to it. With this information, a switch can identify which system is sitting on which port. So when a frame is received, it knows exactly which port to send it to, without significantly increasing network response times. And, unlike a hub, a 10/100Mbps switch will allocate a full 10/100Mbps to each of its ports. So regardless of the number of devices transmitting, users will always have access to the maximum amount of bandwidth. For these reasons a switch is considered to be a much better choice than a hub in many situations.

Routers are completely different devices. Where a hub or switch is concerned with transmitting frames, a router's job, as its name implies, is to route packets to other networks until that packet ultimately reaches its destination. One of the key features of a packet is that it not only contains data, but the destination address of where it's going.

A router is typically connected to at least two networks, commonly two LANs or **Wide Area Networks** (WANs) or a LAN and its **Internet Service Provider** (ISP)'s network, e.g. a home network and the Internet.

Routers might have a single WAN port and a single LAN port and are designed to connect an existing LAN hub or switch to a WAN. Ethernet switches and hubs can be connected to a router with multiple PC ports to expand a LAN. Some routers have USB ports, and more commonly, wireless access points built into them.

Besides the inherent protection features provided by the **Network Address Translation** (NAT), many routers will also have a built-in, configurable, hardware-based firewall. Firewall capabilities can range from the very basic to quite sophisticated devices. Among the capabilities found on leading routers are those that permit configuring **Transmission Control Protocol/User Datagram Protocol** (TCP/UDP) ports for games, chat services and the like, on the LAN behind the firewall. (See Section 8.6.1 for a further description of firewalls).

So, in summary, a hub glues together an Ethernet network segment, a switch can connect multiple Ethernet segments more efficiently and a router can do those functions plus route TCP/IP packets between multiple LANs and/or WANs; and much more.

It may seem that routers are therefore significantly better than bridges for connecting parts of a network together. However, bridges pass all network traffic whereas routers only handle directed traffic. Network wide broadcasts are inherently local in scope which means that they are passed along by hubs and switches, but not by routers. *"This is crucial, otherwise the global Internet would be swamped with 'broadcasts' and the world would end"* (Gibson Research Corporation 2008 [online]). The Windows "Network Neighborhood"[5] file and printer browsing depends upon network broadcasts to allow locally connected machines to find each other on the LAN – these broadcasts therefore stop at the router, so placing a shared printer on the wrong side of a router is not only poor design, it also leads to significant user frustration.

8.2.3 NETWORK TOPOLOGIES

The topology of the network can be thought of as its shape. Not its physical shape, but its logical one: much like the London Underground map shows how stations

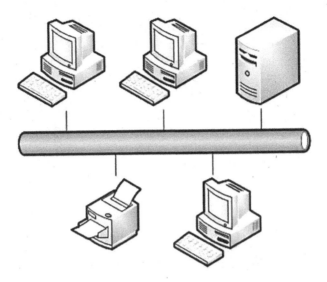

FIGURE 8.4   Bus topology.

connect, not where they are, so a network topology shows how equipment connects to each other, not where they are. The five basic topologies are bus, ring, star, tree and mesh, which we will now examine.

Bus networks (see Figure 8.4) use a common backbone to connect all devices. A single cable, the backbone, functions as a shared communication medium that devices attach or tap into with an interface connector. A device wanting to communicate with another device on the network sends a broadcast message onto the wire that all other devices see, but only the intended recipient actually accepts and processes it.

Ethernet bus topologies are relatively easy to install and don't require much cabling compared to the alternatives. However, bus networks work best with a limited number of devices. If more than a few dozen computers are added to a network bus, performance problems will be the likely result. In addition, if the backbone cable fails, the entire network effectively becomes unusable.

In a ring network, every device has exactly two neighbours for communication purposes. All messages travel through a ring in the same direction (either "clockwise" or "anticlockwise"). A failure in any cable or device breaks the loop and can take down the entire network.

To implement a ring network (see Figure 8.5), we would typically use FDDI,[6] SONET,[7] or Token Ring technology.[8]

Most small (e.g. home) networks use the star topology (see Figure 8.6). A star network features a central connection point called a "hub node" that may be a network hub, switch or (more likely) a router. Devices typically connect to the hub with **Unshielded Twisted Pair** (UTP) Ethernet.

Compared to the bus topology, a star network generally requires more cable, but a failure in any star network cable will only take down one computer's network access and not the entire LAN. (If the hub fails, however, the entire network also fails.)

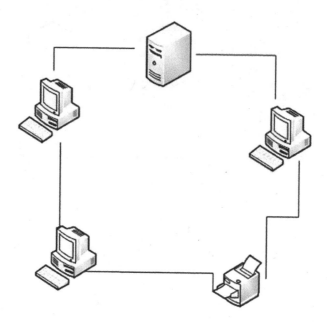

FIGURE 8.5   Ring topology.

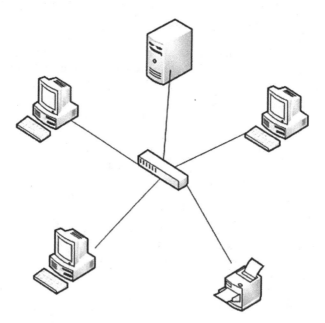

FIGURE 8.6   Star topology.

Tree topologies (see Figure 8.7) integrate multiple star topologies together onto a bus. In its simplest form, only hub devices connect directly to the tree bus, and each hub functions as the root of a tree of devices. This bus/star hybrid approach supports

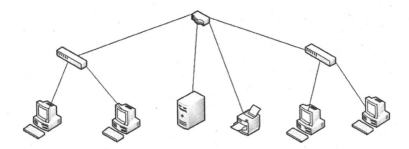

FIGURE 8.7 Tree topology.

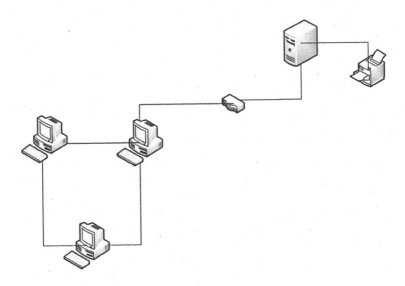

FIGURE 8.8 Mesh topology.

future expandability of the network much better than a bus (which is limited in the number of devices due to the broadcast traffic it generates) or a star (which is limited by the number of hub connection points) alone.

Mesh topologies (see Figure 8.8) involve the concept of routes. Unlike each of the previous topologies, messages sent on a mesh network can take any of several possible paths from source to destination. (Recall that even in a ring, although two cable paths exist, messages only travel in one direction.) Some WANs, most notably the Internet, employ mesh routing, specifically for the resilience that it brings.

A mesh network in which every device connects to every other is called a full mesh. As shown in Figure 8.8, partial mesh networks also exist in which some devices connect only indirectly to others.

8.3 CABLING INFRASTRUCTURE

The entire network is held together by cable, including the wireless portions as the access points need to be wired correctly. Wiring everything correctly and efficiently requires an appropriate structure/methodology.

According to the Fiber Optic Association.[9] *"Structured Cabling is the standardized architecture and components for communications cabling specified by the EIA/TIA TR42 committee and used as a voluntary standard by manufacturers to insure interoperability"* (Fiber Optic Association 2021 [online]). Examining the **Telecommunications Industry Association** (TIA) TR42 standard leads to TIA 568, in which structured cabling is even more technically defined and outlined.[10] Such technical detail is beyond the scope of this chapter (you'll be pleased to hear); however, it is worth noting that these standards define how to lay the cabling in various topologies in order to meet the needs of the network. They typically use a central patch panel from where each modular connection can be accessed as needed. Each outlet is then patched into a network switch for network use or into an IP or PBX **(Private Branch eXchange)** telephone system patch panel.

These cabling standards demand that all eight conductors in Cat5e/6/6A cable are connected, resisting the temptation to 'double-up' or use one cable for both voice and data (voice on 4 cables, data on the other four). IP phone systems, however, can run the telephone and the computer on the same wires (as they're all data at this point).

The most-quoted definition[11] is:

> Structured Cabling is defined as building or campus telecommunications cabling infrastructure that consists of a number of standardised smaller elements (structured). A properly designed and installed structured cabling system provides a cabling infrastructure that delivers predictable performance as well as has the flexibility to accommodate moves, adds and changes; maximises system availability, provides redundancy; and future proofs the usability of the cabling system.

Structured cabling is therefore an organised approach to cabling infrastructure. The opposite methodology (still in use in many data centres) is defined as "point to point" (see Figure 8.9), where patch cables (or "jumpers" – shown as dashed lines) are run directly to and from the hardware that needs to be connected.

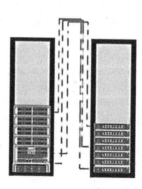

FIGURE 8.9 A point-to-point cabling methodology.

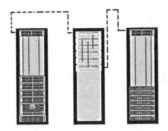

FIGURE 8.10   A structured cabling methodology.

In a structured cabling system (see Figure 8.10), a series of patch panels and trunks (multi-fibre assemblies optimised for multi-signal delivery, shown in the figure as dashed lines) are used to create a structure that allows for hardware ports to be connected to a patch panel at the top of the rack. That patch panel is then connected to another patch panel via a trunk in the **Main Distribution Area** (MDA).

The MDA (the centre rack in Figure 8.10) is the key aspect of structured cabling. This is where all moves, additions and changes can be made with short length patch cords. In this it is very like an old telephone exchange.

The main advantages of structured cabling are:

- Installation is quicker.
- Changes are much quicker as they are done in the MDA as opposed to running long patch cords from equipment racks.
- Having all ports in close proximity to one another increases patching efficiency (provided everything is correctly labelled).
- The potential for downtime is reduced as the potential for human error is drastically reduced.
- Time savings: cable and port tracing are much simpler jobs with a structured cabling system.
- Aesthetics[12]: A structured cabling system looks much cleaner than a point to point method. Since the changes are done in the MDA versus at the hardware, the hardware can be cabled up and not touched in most instances. This allows the cabling in front of the switch to remain aesthetically pleasing.

Structured cabling falls into six subsystems, as defined by ANSI/TIA-568-C.0[13] and ANSI/TIA-568-C.1[14] (note that while the terminology is very building-centric, the subsystems apply equally to a campus or multi-building site):

Entrance Facilities (EF) is the point where the external network connects with the on-premises wiring. These facilities contain the cables, network demarcation point(s), connecting hardware, protection devices and other equipment that connect to the **Access Provider** (AP) or private network cabling. Connections between the outside plant and inside building cabling are made here.

Equipment Rooms (ERs) house equipment and wiring consolidation points that serve the users inside the building or campus. These environmentally

controlled centralised spaces are usually more complex than a telecommunications room or enclosure (see below).

Backbone cabling provides interconnection between telecommunications
rooms, equipment rooms, AP spaces and entrance facilities. Backbone
cabling consists of the transmission media, main and intermediate cross-
connects and terminations at these locations. This system is mostly used in
data centres.

Telecommunications Rooms (TRs) or **Telecommunications Enclosures**
(TEs) connect between the backbone cabling and horizontal cabling including any jumpers or patch cords. A TR or TE may also provide a controlled
environment to house telecommunications equipment, connective hardware
and may serve a portion of the building/campus. The use of a TE is for a
specific implementation and not a general case. It is intended to serve a
smaller floor area than a TR (in buildings there is often a "one TR per floor"
design rule).

Horizontal cabling can be inside wiring or plenum cabling[15] and connects
telecommunications rooms to individual information outlets ("network
sockets") or work areas on the floor, usually through the wireways, conduits
or ceiling spaces of each floor. It includes horizontal cable, mechanical terminations, jumpers and patch cords located in the TR or TE and may incorporate **Multiuser Telecommunications Outlet Assemblies** (MUTOAs)
and consolidation points. A horizontal cross-connect is where the horizontal
cabling connects to a patch panel or punch up block, which is connected by
backbone cabling to the main distribution facility.

Work-area components connect end-user equipment to outlets of the horizontal
cabling system. A minimum of two telecommunications outlets (permanent links)
should be provided for each work area.[16] MUTOAs, if used, are part of the work area.

Figure 8.11 shows how these six subsystems connect together.

Regardless of copper cable type (Cat5e/Cat6/Cat6A[17]) the maximum distance is
90m for the permanent link installation and an allowance for 10m of patch cords at

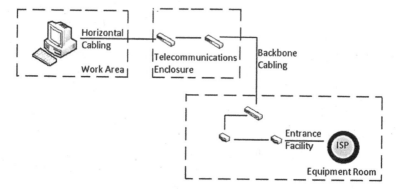

FIGURE 8.11 Structured Cabling Network.

the ends combined (usually 3m at the user end and 6m at the TR, giving a 1m buffer). Cat5e and Cat6 can both effectively run **Power Over Ethernet** (POE) applications up to 90m.[18]

It is common to colour code patch panel cables to identify the type of connection, though structured cabling standards do not require it except in the demarcation wall field. An example (Spiceworks 2013 [online]) might be:

- RED: all devices that are used for data exit points (routers, firewalls etc.)
- BLUE: between the switch port and the patch panel's back end.
- GREEN: between data servers (windows DC, exchange, Linux, DBs etc.) and the patch panel's front end.
- ORANGE: between the patch panel's front end and the backup devices (tape library etc)
- PURPLE: between the patch panel's front end and the **Voice over IP** (VoIP) **Session Initiation Protocol** (SIP) servers (Cisco CUUM, UCCX, CUPS etc.)
- BLACK: between the user device (laptop, desktop etc.) and the wall data socket.

8.4   IP ADDRESSING AND DNS

People like to communicate using names. You won't hear goalkeepers call out "1" as they rush off their line to flap hopelessly at an overhead ball, instead of "keeper". Likewise, referring to one's spouse as "number 2" isn't going to go down well. Therefore we call our devices "Linac B PC", "Endoscopy control" and so on.

Computers prefer numbers. When they're communicating, they require unique numbers. Therefore they use **Internet Protocol** (IP) addresses. An IP version 4 address is formed of 4 groups of digits, separated by dots.[19] Each group of digits can range in value from 0 to 255 – 256[20] unique numbers. The combination of these 4 groups should uniquely identify the device on the network. If it's not unique then chaos will ensue.

Of course, we still like to call our devices by names, so a network service called a DNS is usually available to translate "LinacA" into 123.45.67.89 so that the command "ping LinacA[21]" can be issued and a reply can come from 123.45.67.89 without the user having to know the IP address of LinacA.

8.4.1   IP MASK

The four-byte IP address allows us to perform grouping. A set of devices may be given addresses in the same range with only the final one or two bytes being different. Given that we have a DNS to do the translation and can therefore give our devices sensible names this may seem unnecessary, but it does allow us to segregate our network using masking.

An IP mask of 255.255.255.0 (such as in Figure 8.12) allows us to separate out the network prefix and the host number using bitwise AND.

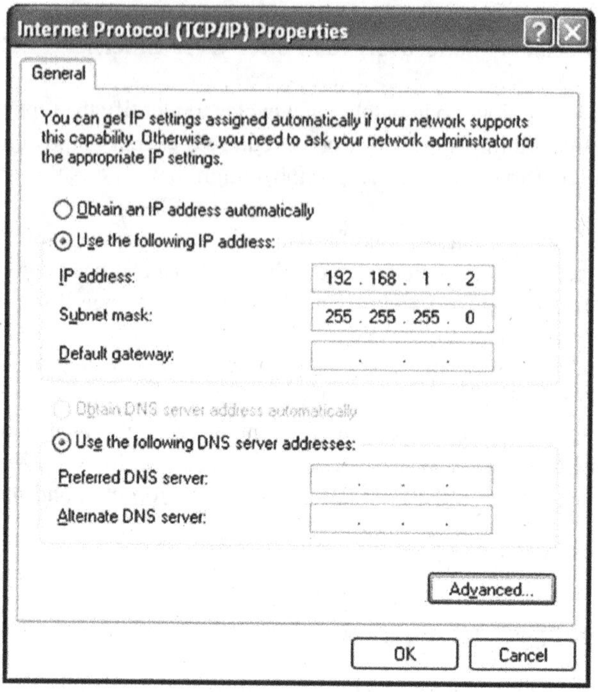

FIGURE 8.12 An IP Setup Dialog from Windows XP.

Input 1	Input 2	Output
FALSE	FALSE	FALSE
FALSE	TRUE	FALSE
TRUE	FALSE	FALSE
TRUE	TRUE	TRUE

FIGURE 8.13 Truth Table for the AND Operator.

AND does what it sounds like it should: something is true if both the first part AND the second part are true. If either part is false, the result is false. We write this in a truth table as shown in Figure 8.13.

It is common to take "1" to mean TRUE and "0" to mean FALSE, rendering our table as shown in Figure 8.14.

For example, 5 AND 3 can be written as 101 AND 011 in binary form, so using the table in Figure 8.14 to compare them bit by bit, the result is 001, or simply 1 in decimal notation.

In the example in Figure 8.12 the network address is 192.168.1 and the host number is 2.[22] Networks can be further divided to produce smaller subnets by putting

Input 1	Input 2	Output
0	0	0
0	1	0
1	0	0
1	1	1

FIGURE 8.14 Truth Table for AND using Binary Values.

more bits into the mask. For example a mask of 255.255.255.192 leaves this example with the same host number, but a host number previously of 130 would now be rendered as 2.[23]

The modern way of specifying a network mask is to specify the number of bits in it, as masks are always 1s to the left and 0s to the right and no intermingling. Thus 192.168.1.2, netmask 255.255.255.0 is written as 192.168.1.2/24 as there are 24 1s in the netmask. This is known as **Classless Inter-Domain Routing** (CIDR) notation.

It is this subnetting that allows us to write simple router rules as they can be produced for groups of IPs instead of for each one individually. By grouping similar devices together, the router rules can be such as 138.0.5.* rather than 100 separate IP addresses.

8.4.2 Ports

An IP address may contain a port number. This is a software port, not to be confused with the hardware ports we discussed earlier in the hardware section. One such example is

http://10.5.90.948:71

Here the IP address is 10.5.90.948 and the port number is 71. The port number is a 16-bit integer so can range from 0 to 65535 and it identifies a specific process to which the network message is to be forwarded when it arrives at a server. For TCP and UDP, the port number is put in the header appended to the message unit. This port number is passed logically between client and server transport layers and physically between the transport layer and the IP layer and forwarded on.

Some common port numbers[24] are (Wikipedia 2022 [online] [1]):

20: **File Transfer Protocol** (FTP) Data Transfer
21: File Transfer Protocol (FTP) Command Control
22: **Secure Shell** (SSH) Secure Login
23: Telnet remote login service, unencrypted text messages
25: **Simple Mail Transfer Protocol** (SMTP) E-mail routing
53: Domain Name System (DNS) service

80: **Hypertext Transfer Protocol** (HTTP) used in the World Wide Web
110: **Post Office Protocol** (POP3)
119: **Network News Transfer Protocol** (NNTP)
123: **Network Time Protocol** (NTP)
143: **Internet Message Access Protocol** (IMAP) Management of digital mail
161: **Simple Network Management Protocol** (SNMP)
194: **Internet Relay Chat** (IRC)
443: **HTTP Secure** (HTTPS) HTTP over TLS/SSL

8.5   IP ROUTING TABLES

A routing table is present on all IP nodes, not just on routers. The routing table stores information about IP networks and how they can be reached (either directly or indirectly). Because all IP nodes perform some form of IP routing, these tables are not exclusive to IP routers.[25] Any node that loads the TCP/IP protocol has a routing table. In this table there are a series of default entries according to the configuration of the node and additional entries can be entered either manually through TCP/IP utilities or dynamically through interaction with routers. Whilst most information will be collected dynamically, the ability to manually insert routes can aid efficiency and network resilience (so is more likely to be done on a router than a PC).

When an IP packet is to be forwarded, the routing table is used to determine:

1. The forwarding or next-hop IP address:
   For a direct delivery, the forwarding IP address is the destination IP address in the IP packet. For an indirect delivery, the forwarding IP address is the IP address of a router (the "next hop" in the route).
2. The interface to be used for the forwarding:
   The interface identifies the physical or logical interface such as a network adapter that is used to forward the packet to either its destination or the next router.

8.5.1   IP ROUTING TABLE ENTRY TYPES

An entry in the IP routing table contains the following information in this order:

Network ID: This is the network ID or destination corresponding to the route. The network ID can be a class-based, subnet or supernet network ID, or an IP address for a host route.
Network Mask: This is the mask that is used to match a destination IP address to the network ID.
Next Hop: This is the IP address of the next hop.
Interface: This contains the network interface that is used to forward the IP packet.
Metric: This is a number used to indicate the cost of the route so that the best route among possible multiple routes to the same destination can be selected. A common metric is the number of hops (routers crossed) to the

network ID. (Costs and metrics are discussed further in routing protocols, in Section 8.9.1).

Routing table entries can be used to store the following types of routes:

Directly Attached Network IDs: These are routes for network IDs that are directly attached. For directly attached networks, the Next Hop field can either be blank or contain the IP address of the interface on that network.

Remote Network IDs: These are routes for network IDs that are not directly attached but are available across other routers. For remote networks, the Next Hop field is the IP address of a router in between the forwarding node and the remote network.

Host Routes: A host route is a route to a specific IP address. Host routes allow routing to occur on a per-IP address basis. For host routes, the network ID is the IP address of the specified host and the network mask is 255.255.255.255.

Default Route: The default route is designed to be used when a more specific network ID or host route is not found. The default route network ID is 0.0.0.0 with the network mask of 0.0.0.0.

8.5.2 ROUTE DETERMINATION PROCESS

In order to determine which routing table entry is used for the forwarding decision, IP uses the following process:

1. For each entry in the routing table, a bit-wise logical AND between the destination IP address and the network mask is performed. The result is compared with the network ID of the entry and if they match then this entry is added to the list for the second stage.
2. The list of matching routes from the first stage is examined and the route that has the longest match (the route that matched the highest number of bits with the destination IP address) is chosen. The longest matching route[26] is the most specific route to the destination IP address. If multiple entries with the longest match are found (multiple routes to the same network ID, for example), the router uses the lowest metric to select the best route. If multiple entries exist that are the longest match and the lowest metric, the router is free to choose which routing table entry to use.[27]

The end result of the route determination process is thus a single route in the routing table. This chosen route provides a forwarding IP address (the next hop IP address) and an interface (the port). If the route determination process fails to find a route, then IP declares a routing error. For the sending host, an IP routing error is internally indicated to the upper layer protocol such as **Transmission Control Protocol** (TCP) or **User Datagram Protocol** (UDP). For a router, an **Internet Control Message Protocol** (ICMP) Destination Unreachable-Host Unreachable message is sent to the source host.

In order to understand this table and the route determination process, let us consider an example.

TABLE 8.1
A Windows 2000 Routing Table

Network Destination ID	Netmask	Gateway (Next Hop)	Interface	Metric	Purpose
0.0.0.0	0.0.0.0	157.55.16.1	157.55.27.90	1	*Default Route*
127.0.0.0	255.0.0.0	127.0.0.1	127.0.0.1	1	*Loopback Network*
157.55.16.0	255.255.240.0	157.55.27.90	157.55.27.90	1	*Directly Attached Network*
157.55.27.90	255.255.255.255	127.0.0.1	127.0.0.1	1	*Local Host*
157.55.255.255	255.255.255.255	157.55.27.90	157.55.27.90	1	*Network Broadcast*
224.0.0.0	224.0.0.0	157.55.27.90	157.55.27.90	1	*Multicast Address*
255.255.255.255	255.255.255.255	157.55.27.90	157.55.27.90	1	*Limited Broadcast*

8.5.3 EXAMPLE ROUTING TABLE FOR WINDOWS 2000

Table 8.1 shows the default routing table for a Windows 2000–based host (i.e. not a router). The host has a single network adapter and has the IP address 157.55.27.90, subnet mask 255.255.240.0 (/20) and default gateway of 157.55.16.1.

Firstly, let us remind ourselves of the process for route determination: bitwise AND the destination IP address and the netmask. See if it matches the Network Destination. If so, then a possible route has been found. The best option is then chosen. We will now examine each possible route in turn.

Default Route: The entry corresponding to the default gateway configuration is a network destination of 0.0.0.0 with a netmask of 0.0.0.0. Any destination IP address combined with 0.0.0.0 by a logical AND results in 0.0.0.0.[28] Therefore, for any IP address, the default route produces a match.[29] If the default route is chosen (because no better routes were found), then the IP packet is forwarded to the IP address in the Gateway column (in this case the default gateway) using the interface corresponding to the IP address in the Interface column (in this case the host's IP address).

Loopback Network[30]: The loopback network entry is designed to take any IP address of the form 127.x.y.z and forward it to the special loopback address of 127.0.0.1. This works because of the bitwise AND.

Directly Attached Network: This is the local network entry and it corresponds to the directly attached network. IP packets destined for the directly attached network are not forwarded to a router but sent directly to the destination. Note that the Gateway and Interface columns match the IP address of the node. This indicates that the packet is sent from the network adapter corresponding to the node's IP address (i.e. from this host). Note also that the netmask is the same as the host's subnet mask.

Local Host: The local host entry is a host route (network mask of 255.255.255.255) corresponding to the IP address of the host. All IP datagrams[31] to the IP address of the host are forwarded to the loopback address.

Network Broadcast: The network broadcast entry is a host route (network mask of 255.255.255.255) corresponding to the all-subnets directed broadcast address (all the subnets of the class B[32] network ID 157.55.0.0). Packets addressed to the all-subnets directed broadcast are sent from the network adapter corresponding to the node's IP address.

Multicast Address: The multicast address, with its class D network mask, is used to route any multicast IP packets from the network adapter corresponding to the node's IP address.

Limited Broadcast: The limited broadcast address is a host route (network mask of 255.255.255.255). Packets addressed to the limited broadcast are sent from the network adapter corresponding to the node's IP address.

Once all matching possible routes have been determined, the best one (i.e. the most specific – which we will look at soon) is used.

When determining the forwarding or next-hop IP address from a route in the routing table:

- If the gateway address is the same as the interface address, then the forwarding IP address is set to the destination IP address of the IP packet.
- If the gateway address is not the same as the interface address, then the forwarding IP address is set to the gateway address.

For example, (using the example routing table in Table 8.1) when traffic is sent to 157.55.16.48, the matching routes are the default and the directly attached network. The most specific route is the directly attached network as it matches 30 bits (the default matches 19). The forwarding IP address is thus set to the destination IP address (157.55.16.48) and the interface is 157.55.27.90, which is the network adapter.

When sending traffic to 157.20.0.79, only the default route matches (so it is therefore the most specific route). The forwarding IP address is set to the gateway address (157.20.16.1) and the interface is 157.55.27.90, which is the network adapter.

8.5.4   STATIC, DYNAMIC AND RESERVED IPS

There are 3 main ways of giving a device an IP address. Firstly, and most simply, is static. This is where the 4-byte number is given to that device and that device only (see Figure 8.12 where this is the case).

The second is dynamic. This is where a device requests an IP address from the DHCP[33] controller and is provided with the next spare one. This system works well with hot-spots and networks where there are more devices than IP addresses, but never has all the devices switched on at once. It's also very useful in networks where the number of devices changes frequently (e.g. a hospital network): no-one has to keep note of the addresses issued so far and no-one has to check whether a device has been decommissioned. Administratively it's the simplest method, but it too has drawbacks.

The main drawback is that some protocols have to communicate via fixed IP addresses. DICOM is one such (although later devices can often take the device name, many can't). One solution to this is to run a mixed addressing network, with some static IPs (usually in the same range so they can be easily administered) and the rest dynamic. One other solution is to use reserved IPs – in this case it is again all controlled by the DHCP controller (often part of the DNS), but when a device requests an IP it is always served the same one, which has been reserved for it. That way the network runs in dynamic mode but the needs of static addressing are met.

The problem with mixed addressing is that it requires more administration, especially to ensure that the fixed range and the static range do not overlap. On initial set-up this may be achieved by, for example, allocating addresses 0-100 to be static and addresses 101-127 to be dynamic. Problems may arise if the static range needs to be increased and the DHCP cluster (as a large organisation will have more than one DHCP server) is still dynamically issuing some IPs that ought to be static.

8.5.4.1 Two Devices with the Same IP Address

Locally this is impossible, but globally not so. If a device's IP address is only visible up to the router, then it is not visible beyond it (e.g. on the Internet) and therefore another device on the other side of the router can have exactly the same IP address and they will never conflict.

This is achieved by address translation: the router uses one set of addresses on one side (e.g. the hospital network) and another on the other side (e.g. the Internet), which it translates between dynamically. In that way, the hospital can continue to grow its IP network without worrying that it will clash with another one. The only real problems arise if more Internet addresses are required than the hospital owns, in which case it buys more (although this end is often sorted out by the ISP).

NAT is a method of remapping one IP address into another by examining the packet headers and modifying the address information within. This takes place in a router. Originally used for ease of rerouting traffic in IP networks without addressing every host, it has become a popular and essential tool in advanced NAT implementations featuring IP masquerading.

IP masquerading is a technique that hides an IP address space (e.g. your home network) behind a single (public) IP address. The outgoing address thus appears as though it has originated from the router, which must also do the reverse mapping when replies are received. The prevalence of IP masquerading is largely due to the limitations of the range of IPv4 addresses (as described at the beginning of this chapter) and the term NAT has become virtually synonymous with it.[34]

8.5.5 Where Is Your Data?

It may be on a local machine, on a local server or in "the cloud" (see Chapter 10 for a definition).

Cloud Computing is where the software and data do not reside on local servers, in the local organisation or possibly even in the same country. Whilst this frees up a lot of infrastructure and makes mobile computing more possible, it does have two main drawbacks:

The first is that a reliable network connection is essential to use the cloud.

The second is that the laws of data protection that apply to data are those that exist in the country in which the data resides: in the USA, for example, companies are allowed to sell the data they have on their network. Not to anyone, of course, but they can sell health records to insurance agencies, email addresses to marketers etc. So it is important, for health records, to know where the data is being held. In the UK there is a G-Cloud, a cloud solution hosted here for public service use, thus making it subject to the UK's data protection laws. There are also more and more assured clouds being marketed.

In early 2018 NHS Digital published guidance on data off-shoring and cloud computing for health and social care. The main points are:

- *"NHS and Social care providers may use cloud computing services for NHS data. Data must only be hosted within the UK - European Economic Area (EEA), a country deemed adequate by the European Commission, or in the US where covered by Privacy Shield.*
- *Senior Information Risk Owners (SIROs) locally should be satisfied about appropriate security arrangements (using National cyber security essentials as a guide) in conjunction with Data Protection Officers and Caldicott Guardians.*
- *Help and advice from the Information Commissioner's Office is available and regularly updated.*
- *Changes to data protection legislation, including the General Data Protection Regulation (GDPR) from 25 May 2018, puts strict restrictions on the transfer of personal data, particularly when this transfer is outside the European Union. The ICO also regularly updates its GDPR Guidance."*

(NHS Digital 2022 [online])

- NHS Digital has provided some detailed guidance documents to support health and social care organisations.

Cloud computing is becoming more trusted. In January 2020 two NHS services moved to the cloud: The NHS e-Referral Service (e-RS) and the NHS 111 Directory of Services (DoS). The quoted benefits were the saving of public money while making the services more secure and efficient (NHS Digital 2020 [online]).

A good analysis of the issues can be found at http://www.save9.com/is-it-ok-to-backup-nhs-patient-data-in-the-cloud/. It's written from a supplier perspective so has a bias towards the advantages but is essentially correct – i.e. you can use cloud services, provided you do it in the right way (compliance with IG toolkit and the DP act plus local risk assessment for example).

8.6   CONNECTING MEDICAL DEVICES TO THE HOSPITAL NETWORK

There are great advantages in connecting together ICT equipment to enable data sharing, together with the enhanced safety from a reduction in transcription errors and the increased availability and speed of access to information. However, this connectedness brings with it additional system security issues: a failure in one part may be swiftly replicated across the IT estate. There are many ways to tackle these issues and this section details some of these. It should be noted that best practice will utilise a range of security methods.

8.6.1 FIREWALLS

The first method is one of segregation, using a firewall. A firewall is, in the simplest sense, a pair of network cards (or a router) and a set of rules. A network packet arrives at one card, is tested against the rules and, if it passes, is passed to the other card for transmission. In this way, a part of a network can be protected from activity on the rest of the network by restricting the messages that can pass through it to a predefined and pre-approved set. The rules controlling this may be as simple as only allowing a predefined set of IP addresses through. Refinements include port numbers, the direction the message is travelling in, whether the incoming message is a response to an outgoing one (e.g. a web page) and specific exceptions to general rules. This is all achieved via packet filtering, where the header of the packet is examined in order to extract the information required for the rules.

The above description is of a hardware firewall. Software firewalls run on the device after the network traffic has been received. They can therefore be more sophisticated in their rules in that they can have access to additional information such as the program that made the request. Software firewalls can also include privacy controls and content filtering.[35] As the software firewall runs on the device, if the device becomes compromised then the firewall may also be compromised. The Windows 7 firewall only blocks incoming traffic, so for example it will not prevent a compromised device from sending malicious network packets. The Windows 10 firewall is similar: inbound connections to programs are blocked unless they are on the allowed list, outbound connections are allowed unless they match a rule. There are also public and private network profiles for the firewall, controlling which programs can communicate on the private network as opposed to the Internet.[36]

Firewall rules are generally set up via an interface to the firewall. For example, Figure 8.15 shows the setting of rules on a router placed between networks: 27.0.0.0/8, 192.168.0.100/24 and others (e.g. the Internet). Traffic originating from 27.0.0.0/8 destined for 192.168.0.100/24 is blocked (as it matches rule 1, which is evaluated first) whereas traffic destined for any other network (including 27.0.0.0/8) passes through, as it does not match rule 1 and rule 2 is effectively "allow anything".

#	Policy	Protocol	Destination	Port	Comment
1	Deny	Any	192.168.0.100/24	Any	Deny traffic to destination
	Allow	Any	Local LAN	Any	Wireless clients accessing LAN
	Allow	Any	Any	Any	Default rule

FIGURE 8.15   Firewall rules to prevent traffic.

#	Policy	Protocol	Source	Src port	Destination	Dst port	Comment
1	Allow	TCP	27.0.0.0/8	Any	192.168.0.200/32	80	Allow web
2	Deny	Any	Any	Any	Any	Any	Deny all
	Allow	Any	Any	Any	Any	Any	Default rule

FIGURE 8.16   Firewall rules to allow web server communication.

In Figure 8.16, the web server 192.168.0.200 needs to be accessible from 27.0.0.0/8 but neither network should be able to communicate outside. The figure shows the rules required. Any traffic from the 27.0.0.0/8 network to the web server is allowed by rule 1. Replies are also allowed as the firewall is stateful[37] and the connection has already been established by rule 1. The deny rule (rule 2) is processed second and matches all other traffic besides traffic to the web server.

8.6.2 BANDWIDTH

Bandwidth in a computer network sense is its transmission capacity, which (as it is a function of the speed of transmission) is usually expressed in bps. The most common wired bandwidths are 1 Gbps (often called "Gigabit Ethernet"), 10 Mbps (standard Ethernet) and 100 Mbps (fast Ethernet). Wireless is generally slower – 802.11g supports up to 54 Mbps, for example.[38] As bandwidth is actually the capacity, binding together several cables can increase the total bandwidth whilst not increasing the speed – although this would not normally be done in a departmental network, the point at which a hospital meets the national N3 network may be implemented this way (provided both sides of the connection can handle it – which is usually by routing pre-defined packets to specified lines, e.g. by IP address range).

It is never a good idea to reach 100% bandwidth utilisation and the average in order to avoid this may be as low as 30%, although 50% would be more common (see Figure 8.17). The amount of "spare" capacity is often termed "headroom". At 75% the throughput verses offered traffic curve starts to depart from a linear proportional increase of throughput for increase of offered traffic. At 80% the channel could be approaching overload. Much is dependent upon the traffic type - data traffic can cope with higher utilisation levels than voice as delay and jitter have more effect on the user experience for voice traffic than data traffic. Optimisation techniques such as QOS[39] can be used to prioritise voice traffic (or any other traffic that is time-critical).

The above utilisation levels are generally for non-collision-based channels. In the case of Ethernet which uses CSMA with collision detection as the access mechanism, utilisation should be much lower. An overdriven CSMA channel can result in a throughput reduction rather than an increase with increasing offered traffic. Retrys as a result of a collision lead to more retrys and more collisions and so on. Collision detection with a limitation on the number of retrys and a back off between the retrys is intended to keep the channel stable but throughput will tail off. Kleinrock (Kleinrock 1975–2011) provides good further reading.

The use of different network transport protocols can also improve throughput. For example, Skype uses TCP for video (which guarantees delivery) and UDP for audio

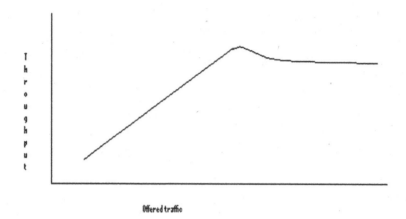

Offered traffic

FIGURE 8.17   A generalised throughput verses offered traffic curve showing a deterioration in performance when optimal levels are breached. (Taktak et al. 2019).

(which doesn't). On first inspection this appears to be the wrong way round, but TCP waits until a missed packet is delivered, whereas UDP just carries on. Thus the video freezes whereas the audio doesn't – a small glitch in the audio feed (50fps) is not going to be noticed.

Wi Fi and Bluetooth suffer from interference, and the frequency range is the same as for microwave ovens. The reason that a frequency that excites water molecules and therefore is severely attenuated by them (e.g. people) was chosen was because it was available as no-one wanted it (for precisely those reasons).

All the resilience methods outlined here require a level of redundancy: be it a copy, checksums or headroom. Thus a resilient system will always be over-engineered – in the case of bandwidth, over-engineering can remove the need for optimisation systems such as QOS, thus making the design (and therefore the support) simpler.

8.7   INFRASTRUCTURE

The core of a modern hospital's infrastructure is the supervised, air-conditioned server room, with redundant power supply provision. The users link to this facility by Ethernet and offsite electronic replication is likely. The bulk of the front line data storage will use **Storage Area Network** (SAN) or **Network Attached Storage** (NAS) and the racks of servers will often be running virtual systems. Clinical Scientist/Engineer user specialists will have remote administrator access to their systems. Critical clinical systems, such as Radiotherapy facilities will be likely to have some form of segregation, such as VLAN[40]s or firewalls. Virtual systems are particularly useful for enabling the local testing of commercial software upgrades, as it is relatively simple to destroy the server and re-create it.

8.8   THE OSI 7-LAYER MODEL

The OSI 7-layer model (see Figure 8.18) describes the transmission of messages. In sending a message, each layer (from the highest – the Application – to the lowest – the

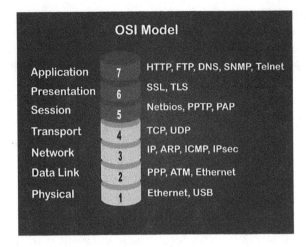

FIGURE 8.18 The seven layer OSI model. (Shutterstock).

Physical) adds a wrapper to the message, which is removed after physical transmission as the message makes its way back up the layers into the receiving application.

The **physical layer (1)**, the lowest layer of the OSI model, is concerned with the transmission and reception of the unstructured raw bit stream over a physical medium. It describes the electrical/optical, mechanical and functional interfaces to the physical medium and carries the signals for all of the higher layers. It deals with such matters as which signal state represents a binary 1, how many pins are on a connector (and what they do) and how many volts/db should be used to represent a given signal state.

The **data link layer (2)** provides error-free transfer of data frames from one node to another over the physical layer, allowing layers above it to assume virtually error-free transmission over the link. This layer is the protocol layer, providing the functional and procedural means to transfer data between network entities. The data link layer handles matters such as establishing and terminating the logical link between two nodes, transmitting/receiving frames sequentially and determining when the node "has the right" to use the physical medium (see the description of "token ring" in Figure 8.5).

The **network layer (3)** controls the operation of the subnet, deciding which physical path the data should take based on network conditions, priority of service and other factors. It is concerned with routing, subnet traffic control and logical-physical address mapping.

The **transport layer (4)** ensures that messages are delivered error-free, in sequence, and with no losses or duplications.[41] It relieves the higher layer protocols from any concern with the transfer of data between them and their peers. It handles such matters as message segmentation, message acknowledgement and session multiplexing.[42] (A session is "*a semi-permanent interactive information interchange, also known as a dialogue, a conversation or a meeting, between two or more communicating devices, or between a computer and user.*" (Wikipedia 2022 [online] [2])).

The **session layer (5)** allows session establishment between processes running on different stations. It handles session establishment, maintenance and termination, allowing two application processes on different machines to establish, use and terminate a session.[43]

The **presentation layer (6)** formats the data to be presented to the application layer. It can be viewed as the translator for the network. This layer may translate data from a format used by the application layer into a common format at the sending station, and then translate the common format to a format known to the application layer at the receiving station. It handles code conversions such as ASCII to EBCDIC,[44] bit-order and **Carriage Return/Line Feed** pairs (CR/LF). Data compression and encryption take place at this layer.

The **application layer (7)** serves as the window for users and application processes to access network services. This layer contains a variety of commonly needed functions such as resource sharing, remote printing and file access and directory services.

8.9 SCALABILITY

Networks have a habit of growing – the first network I ever put in consisted of three PCs and enabled them to share a single database so that it was visible in the two clinics and in the main office. It used a mixture of bus and ring topology in that every packet went through every PC and the bus had to be properly terminated in order to reduce errors and reflections. It took a lot of setting up and required careful planning when we added a fourth PC into it. Needless to say, that network only lasted a couple of years before it was replaced by technologies that we would recognise today – beyond "plug and play" and much more "open the box and it works".

Networks have to be able to scale – to be able to grow as the requirements for the network grow. Scaling is not mountaineering, but concerns designing a network in such a way that it can be expanded to meet future demands in a planned, graceful way. There are three main parts to this growth:

1. An increase in connected devices
2. An increase in required throughput
3. A decrease in performance (due mainly to 1 above)

It is therefore important to select a topology that will not only do the job today, but also has the opportunity to adapt to additional requirements. A simple network topology such as the tree in Figure 8.7 may be thought of as being scalable by addressing the three problems as follows:

1. Add more leaves to the tree or replace a switch/hub with one that has more ports
2. Replace hubs with switches and affected nodes with faster devices
3. Segregate the tree further, possibly adding local routing

All of these simple solutions have a limit, however – for example the number of packets flowing through the root node is only going to increase as devices increase. Scalability thus requires further thought. There are many techniques and technology for achieving this, but they all come down to planning in the end.

One such technology, **Enhanced Interior Gateway Routing Protocol** (EIGRP, which we will explore soon), is scalable in terms of hardware resources and network capacity. It achieves this by not sending out the entire routing table to their neighbours every n seconds as compared to another **Distance-Vector** (DV) protocol like **Routing Information Protocol** (RIP) which we will explore later in this chapter.

EIGRP is scalable but only really when used in conjunction with IP summarisation. Without summarisation, a link down in one part of the network has a knock-on effect on the rest of the network – assuming no feasible successor[45] is available. **Open Shortest Path First** (OSPF) (the third such protocol we will look at) on the other hand has an inbuilt hierarchy due to its area concept, but it can be quite demanding on router resources e.g. CPU/memory, so they all have their pros and cons.

IP summarisation (also called route summarisation, route aggregation or supernetting) is a technique for reducing the number of distinct IP addresses a router must maintain by grouping them together and thus representing a series of network addresses in a single summary address.[46] This saves bandwidth (only one address is passed between routers), memory (only one route is maintained) and CPU resources (packets are checked against fewer entries). Another advantage of using IP summarisation in a large complex network is that it can isolate topology changes from other routers. If a specific link is going up and down rapidly[47] a specific route may change with it, whereas a summary route does not. Therefore, routers using the summary route do not need to continually modify their tables. Of course, this summarisation is only applicable to remote routers and not to the one directly connected to the equipment.

8.9.1 RIP

RIP is one of the oldest distance-vector routing protocols which employ the hop count[48] as a routing metric. Routing loops are prevented by placing a limit on the number of hops allowed in a path from source to destination. In RIP the maximum number of hops allowed is 15, which also limits the size of networks that RIP can support. A hop count of 16 is thus considered to be an infinite distance and the route is considered unreachable.

Originally, each RIP router transmitted full updates every 30 seconds. As routing tables were small the traffic was not significant, but as networks grew in size it became evident there could be a massive traffic burst every 30 seconds, even if the routers had been initialised at random times. The theory was that, as a result of random initialisation, the routing updates would spread out in time, but this was not true in practice. Sally Floyd and Van Jacobson showed that, without slight randomisations of the update timer, the timers synchronised over time (Floyd and Jacobson 1994).

In most networking environments, RIP is not the preferred choice for routing, but it is easy to configure, because (unlike other protocols) RIP does not require any parameters.

8.9.2 OSPF

OSPF uses a **Link State Routing** (LSR) algorithm and is an **Interior Gateway Protocol** (IGP), meaning that it is a type of protocol used for exchanging routing information between gateways (routers in this case) within an autonomous system (e.g. a system of local area networks within a hospital site). OSPF is probably the most widely used IGP in large enterprise networks. **Intermediate System to**

Intermediate System (IS-IS), another link-state dynamic routing protocol, is more common in large service provider networks.

LSR protocols are one of the two main classes of routing protocols used in packet switching networks, the other being distance-vector routing protocols. The link-state protocol is performed by every node in the network that forwards packets (i.e. routers in an IP network). In LSR, every node constructs a map of the network[49] showing which nodes are connected to which other nodes. Each node then independently calculates the next best logical path from it to every possible destination in the network. This collection of best paths forms the node's routing table.[50]

OSPF therefore gathers link state information from available routers and constructs a topology map of the network. This topology is presented as a routing table to the Internet layer[51] which routes packets based solely on their destination IP address.

OSPF detects changes in the topology, such as link failures and converges on a new loop-free routing structure within seconds. It computes the shortest-path tree for each route using a method based on Dijkstra's algorithm.[52] The OSPF routing policies for constructing a route table are governed by link metrics associated with each routing interface. Cost factors may be the distance of a router (round-trip time), data throughput of a link, or link availability and reliability, expressed as simple unitless numbers.[53] This provides a dynamic process of traffic load balancing between routes of equal cost.

An OSPF network may be structured, or subdivided, into routing areas to simplify administration and optimise traffic and resource utilisation. These are denoted by 32-bit numbers similar to an IP4 address, with 0.0.0.0 being specified as the backbone through which all inter-area messages pass.

8.9.3 INTERMEDIATE SYSTEM TO INTERMEDIATE SYSTEM (IS-IS)

Like OSPF, IS-IS is an IGP, designed for use within an administrative domain or network. IS-IS is also a LSR protocol, operated by reliably flooding link state information throughout a network of routers. Each IS-IS router independently builds a database of the network's topology, using and aggregating the flooded network information. IS-IS also uses Dijkstra's algorithm for computing the best path through the network. Packets are then forwarded, based on the computed ideal path, through the network to the destination.

While OSPF was designed and built to route IP traffic and is itself a Layer 3 protocol that runs on top of IP, IS-IS is an OSI Layer 2 protocol. It therefore does not use IP to carry routing information messages. While the widespread adoption of IP may have contributed to OSPF's popularity, OSPF version 2 was designed for IPv4 and to operate with IPv6 networks, the OSPF protocol required rewriting into OSPF v3. IS-IS is neutral regarding the type of network addresses for which it can route, which allows IS-IS to easily support IPv6.

8.9.4 EIGRP

As noted earlier, EIGRP is scalable in terms of hardware resources and network capacity, which it achieves by not sending out the entire routing table every n seconds as compared to RIP. More specifically, once an initial adjacency has been formed and full routing updates exchanged, any routing change only generates a partial update. Hellos[54] (which have minimal impact on the network) are used as keepalives. EIGRP has other efficiencies such as the feasible successor concept (described in endnote [45]) and setting a maximum percentage of the bandwidth that EIGRP is allowed to use which can be useful on very slow links.

EIGRP is an advanced distance-vector routing protocol (as opposed to the LSR that OSPF and IS-IS use). It was designed by Cisco Systems as a proprietary protocol, available only on Cisco routers, but a partial functionality of EIGRP was converted to an open standard in 2013 and was published with informational status as RFC 7868 in 2016.[55]

EIGRP is a dynamic routing protocol by which routers automatically share route information and it only sends incremental updates, reducing the workload on the router and the amount of data that needs to be transmitted.

In addition to the routing table, EIGRP uses two tables to store information:

- The neighbour table, which records the IP addresses of routers with a direct physical connection to the given router.
- The topology table, which records routes that the router has learned from neighbour routing tables. Unlike a routing table, the topology table does not store all routes, but only routes that have been determined by EIGRP. The topology table also records the metrics for each of the listed EIGRP routes, the feasible successor and the successors. Routes in the topology table are marked as being either "passive" or "active". A passive route is one where EIGRP has determined the path and has finished processing. An active route is one where EIGRP is still calculating the best path. Routes in the topology table are not usable by the router until they have been inserted into the routing table. Routes in the topology table will not be inserted into the routing table if they are active, are a feasible successor, or have a higher administrative distance than an equivalent path.

EIGRP uses the **Diffusing Update Algorithm** (DUAL)[56] to improve the efficiency of the protocol and to help prevent calculation errors when attempting to determine the best path to a remote network. EIGRP determines the value of the path using five metrics: bandwidth, load, delay, reliability and **Maximum Transmission Unit** (MTU – the largest size packet or frame, specified in octets[57] that can be sent in a packet- or frame-based network such as the Internet).

8.10   WEB SERVICES: INTRODUCTION

A Web Service (sometimes written as one word, "Webservice") is platform independent software. It is designed to take 'gerbil' as input and give 'hamster' as output and do so irrespective of who or what provides the 'gerbil'. A web service written in Java, C++ etc. and hosted on a server can therefore be used by an application written in PHP/.NET/C# etc.

A good example of platform independent software is the plethora of websites that have log-in and sharing options to Facebook. It isn't practical for all those websites that integrate with Facebook to use the same technology as that of Facebook, so Facebook has created a web service and exposed it to all these websites, which then use it for authentication.

We will examine two flavours of web services: REST & SOAP. SOAP is often viewed as an "old school" technology and REST as more of a contemporary technology.[58] Web services that use the REST framework are called RESTful web services. There is no equivalent for those that use SOAP.

8.11 WEB SERVICES: REPRESENTATIONAL STATE TRANSFER (REST)

REST or RESTful web services are a way of providing interoperability between computer systems on the Internet. REST-compliant web services allow requesting systems to access and manipulate textual representations of web resources using a uniform and predefined set of stateless operations and protocol.[59]

"Web resources" were originally defined as documents or files identified by their **Uniform Resource Locator**s (URLs), but this has expanded into a more generic and abstract definition encompassing every entity that can be identified, named, addressed or handled on the Web. In a RESTful Web service, requests made to a resource's **Uniform Resource Identifier** (URI)[60] will receive a response in a defined format, such as XML, HTML or JSON. The response may confirm that some alteration has been made to the stored resource (e.g. a database), and it may provide hypertext links to other related resources or collections of resources. RESTful APIs centre around resources that are grouped into collections. For example, when browsing through directory listings and files to find image files on a website such as http://ganney.ac.uk/, a series of folders might be selected to download files from. These folders are collections of ganney.ac.uk image files. When HTTP is used (the most common case), the kind of operations available in REST include those predefined by HTTP verbs such as GET, POST, PUT and DELETE.

By using a stateless protocol and standard operations, REST systems aim for fast performance, reliability and the ability to grow, by re-using components that can be managed and updated without affecting the system as a whole, even while it is running. This latter idea does mean that documentation and version control is vitally important, especially if functionality (for example a correction factor or encryption protocol) is altered by changing a component.

The term "representational state transfer" was introduced and defined in 2000 by Roy Fielding in his doctoral dissertation and is intended to evoke an image of how such a web application behaves: it is a network of web resources (a virtual state-machine) where the user progresses through the application by selecting links (such as /ganney/paul) and operations (such as GET or DELETE (state transitions)), resulting in the next resource (representing the next state of the application) being transferred to the user for their use.

A RESTful system is subject to six constraints. By operating within these constraints, the Webservice gains desirable non-functional properties, such as performance, scalability, simplicity, modifiability, visibility, portability and reliability.

The formal REST constraints are as follows:

8.11.1   CLIENT-SERVER ARCHITECTURE

The principle behind the client-server constraint is the "separation of concerns". Separating the user interface concerns from the data storage concerns improves the portability of the user interface across multiple platforms. It also improves scalability by simplifying the server components. In addition, it allows components to evolve independently.

8.11.2   STATELESSNESS

As no client context is stored on the server between requests, each request must contain all the information necessary to service the request, and the session state is held in the client.[61] The session state can be transferred by the server (when it receives it) to another service (such as a database) to maintain a persistent state for a period and thereby allow authentication. The client then begins sending requests when it is ready to make the transition to a new state. While one or more requests are outstanding, the client is considered to be in transition.

8.11.3   CACHEABILITY

Caching is a common feature of Internet transactions as it speeds up client operations by re-using rather than re-sending data. Server responses must therefore, implicitly or explicitly, define themselves as cacheable (or not) in order to prevent clients from reusing stale or inappropriate data in response to further requests.

8.11.4   LAYERED SYSTEM

A layered system places intermediary servers into the message path.[62] Intermediary servers can improve system scalability by enabling load balancing and by providing shared caches. They can also enforce security policies, thus reducing the load on the main server.

8.11.5   CODE ON DEMAND

This constraint is the only one of the six that is optional. In order to temporarily extend or customise the functionality of a client, the server can transfer executable code – either compiled components (e.g. Java applets) or client-side scripts (e.g. JavaScript).

8.11.6   UNIFORM INTERFACE

Fundamental to the design of any REST service, the uniform interface constraint simplifies and decouples the architecture, enabling each part to evolve independently. There are four constraints for this uniform interface:

8.11.6.1   Resource Identification in Requests

Individual resources are identified in requests, so that the internal representation on the client or server need not be the one used for the transaction. For example, the server may send data from its database as HTML, XML or JSON, none of which are the server's internal representation.

8.11.6.2   Resource Manipulation Through Representations

A representation of a resource held by a client, including any metadata attached, must contain enough information to modify or delete that resource.

8.11.6.3   Self-Descriptive Messages

Each message must include enough information so that the message can be processed. For example, an Internet media type (previously known as a MIME type) may specify which parser[63] to invoke.

8.11.6.4   Hypermedia As the Engine of Application State (HATEOAS)

Once an initial URI for the REST application has been accessed (analogous to a human web user accessing the home page of a website) a REST client must be able to dynamically use server-provided links to discover all the available actions and resources that it needs. As this proceeds, the server responds with text that includes hyperlinks to other actions that are currently available. There is therefore no need for the client to be hard-coded with information regarding the structure or dynamics of the REST service as it is all found dynamically. This also aids the evolution of components principle.

These constraints of the REST architectural style affect these architectural properties:

- Component interactions can be the dominant factor in user-perceived performance and network efficiency.
- The system can scale to support large numbers of components and interactions among components. In his PhD thesis, Roy Fielding describes REST's effect on scalability as:

> REST's client–server separation of concerns simplifies component implementation, reduces the complexity of connector semantics, improves the effectiveness of performance tuning and increases the scalability of pure server components. Layered system constraints allow intermediaries— proxies, gateways and firewalls—to be introduced at various points in the communication without changing the interfaces between components, thus allowing them to assist in communication translation or improve performance via large-scale, shared caching. REST enables intermediate processing by constraining messages to be self-descriptive: interaction is stateless between requests, standard methods and media types are used to indicate semantics and exchange information, and responses explicitly indicate cacheability.

(Fielding 2000)

- Simplicity of a uniform Interface
- Modifiability of components to meet changing needs (even while the application is running)
- Visibility of communication between components by service agents
- Portability of components by moving program code with the data
- Reliability becomes the resistance to failure at the system level – what happens when components, connectors, or data fail.

HTTP-based RESTful APIs are defined with these aspects:

- There is a base URL, such as http://ganney.ac.uk/resources/
- There is an Internet media type that defines state transition data elements (e.g., Atom, microformats, application/vnd.collection+json). The current representation tells the client how to compose requests for transitions to the next available application states, which could be as simple as a URL or as complex as a Java applet.
- The standard HTTP methods are used.

8.11.7 RELATIONSHIP BETWEEN URL AND HTTP METHODS

Table 8.2 shows how HTTP methods are typically used in a RESTful API.

TABLE 8.2
HTTP Methods and REST

URL	GET	PUT	PATCH[64]	POST	DELETE
Collection, such as https://ganney.ac.uk/resources/	List the URIs of the collection's members. (Plus possibly other details)	Replace the entire collection with another collection	Not generally used	Create a new entry in the collection. The new entry's URI (assigned automatically) is usually returned by the operation	Delete the entire collection
Element,[65] such as https://ganney.ac.uk/resources/item42	Retrieve a representation of the addressed member of the collection, expressed in an appropriate Internet media type	Replace the addressed member of the collection. If it does not exist, create it	Update the addressed member of the collection	Not generally used, but does have the functionality: Treat the addressed member as a collection in its own right and create a new entry within it	Delete the addressed member of the collection

The GET method is a safe method (or nullipotent), meaning that calling it produces no side-effects: retrieving or accessing a record does not change it. The PUT and DELETE methods are idempotent, meaning that the state of the system exposed by the API is unchanged no matter how many times the same request is repeated.

Unlike SOAP-based Web services, there is no "official" standard for RESTful Web APIs. This is because REST is an architectural style, while SOAP is a protocol. REST is not a standard in itself, but RESTful implementations make use of standards, such as HTTP, URI, JSON and XML. The most common architectural constraint developers fail to meet while still describing their APIs as being RESTful[66] is the uniform interface constraint.

8.12 WEB SERVICES: SIMPLE OBJECT ACCESS PROTOCOL (SOAP)

SOAP is an XML-based messaging protocol. It defines a set of rules for structuring messages that can be used for simple one-way messaging but is particularly useful for performing **Remote Procedure Call** (RPC)-style request-response dialogues. Although HTTP is a popular transport protocol, SOAP is neither tied to it nor to any particular operating system or programming language. The only requirement for clients and servers in these dialogues is that they can formulate and understand SOAP messages. SOAP is therefore a useful building block for developing distributed applications that exploit functionality published as services over an intranet or the Internet. Matsoap is one such, based at Liverpool and links MatLab to HTTP.[67]

As an example, let us consider a simple database for patients who may need to be contacted relatively quickly. It holds a table specifying NHS number, name and telephone number. We wish to offer a service enabling other systems within the hospital to perform a lookup on this data. The service should return a name and telephone number (a two element array of strings) for a given NHS number (a 10-digit long integer[68]). A Java-style prototype for the service might be:

String[] getPatientDetails (long NHSNumber);

The SOAP approach to such a problem is to encapsulate the database request logic for the service in a method (or function) in C, VB, Java etc., then set up a process to listen for requests to the service. Such requests are in SOAP format and contain the service name and any parameters required by it. The listener process[69] decodes the incoming SOAP request and transforms it into an invocation of the method. It then takes the result of this, encodes it into a SOAP message[70] (response) and sends it back to the requester. Figure 8.19 shows the conceptual arrangement.

There are many different specific architectures possible for implementing this arrangement; one possibility is shown in Figure 8.20. Here the database system is Oracle. The service method is written in Java and connects to the database using an Oracle implementation of JDBC (see Section 16.3.3, Chapter 16). The listener process is a Java servlet running within a servlet engine such as Tomcat and is listening for those messages as an HTTP POST as the transport is HTTP over TCP/IP. The client is an excel spreadsheet using a VB Macro which utilises the Microsoft SOAP Toolkit to encode a SOAP request and decode the response received.

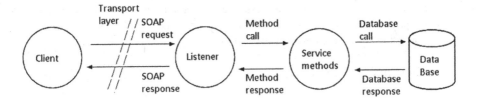

FIGURE 8.19 A SOAP message and response.

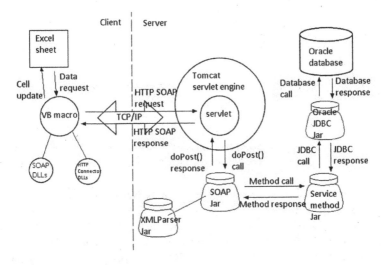

FIGURE 8.20 A sample SOAP architecture.

Note that on the client side the VB Macro relies on both the Microsoft SOAP Toolkit (the SOAP DLLs) and a HTTP Connector interface. Such HTTP Connector DLLs are typically already installed as a part of browsers such as Internet Explorer and Edge. On the server side the SOAP package relies on an XML parser to parse the SOAP messages, which in the case of Apache SOAP for Java will be Xerces.

SOAP is only one method of solving this problem. An alternative might be to allow clients direct access to a stored procedure in the database via ODBC or JDBC (see Section 16.3.3, Chapter 16). Reasons why a SOAP-based solution might be preferable include:

1. A stored procedure solution cannot send or receive rich data structures as parameters or return values. This is because parameters and return values for relational database procedure calls are limited to primitive types (integer, float, string etc.). In the example above this is not an issue as the NHS number sent is a long integer and the return values (a name and telephone number) are a pair of strings. Consider though a small extension to the

service to provide the patient's usual telephone number plus a list of other telephone numbers which are valid during certain periods. The service must now return a complex type of the form:

```
PatientContactDetail {
    String patientName;
    String phoneNumber;
    TemporaryPhoneNumber[] tempPhoneNumber;
}
```

Where the user-defined type TemporaryPhoneNumber is defined as:

```
TemporaryPhoneNumber {
    date startDate;
    date endDate;
    String phoneNumber;
}
```

Note that there can be any number (zero or more) temporary phone number records for the patient being retrieved. The prototype for the service is now:

```
PatientContactDetail getPatientDetails (long NHSNumber);
```

An ODBC or JDBC approach requires complex data structures to be flattened, but this is not possible here since there are an unknown number of TemporaryPhoneNumber records. The SOAP protocol though allows data structures of any level of complexity to be encoded.

2. With an n-tier architecture some of the business logic may be coded outside the database to which the services may require access. With a stored procedure solution, the logic could be rewritten in SQL (which may not be possible) or an openserver-style solution could be created where the calculations are passed from the stored procedure to a calculation engine which incorporates the business logic code. This may be a good solution if the business logic is not written in Java, but if it is written in java then the business logic could be included as a jar[71] between the service method jar and the JDBC jar in Figure 8.20. Whilst it isn't SOAP that provides this functionality but the servlet engine (and therefore a servlet might be written to encapsulate the business logic, database access and calculations), placing the solution in the servlet engine either requires one servlet being written per service or writing a generic servlet and creating a custom method identification and parameter encoding scheme. SOAP not only has all the method identification and parameter encoding work built-in but the protocol is a w3c standard so clients do not have to implement a custom protocol. This is particularly important when considering offering services over the Internet.

3. JDBC is only valid for Java clients. A mix of Java and non-Java clients means an inconsistent method of access to services. There may also be more

than one database (and these of different types) which a client does not need to know. A servlet-only solution (without using SOAP) is possible but for the reasons given above may not be so attractive.

In all of this, SOAP bears some similarities with the **Common Object Request Broker Architecture** (CORBA). Defined by the **Object Management Group** (OMG), CORBA is designed to facilitate the communication of systems that are deployed on diverse platforms. Like SOAP, it enables collaboration between systems on different operating systems, programming languages and computing hardware. While SOAP is XML-based, CORBA uses an object-oriented model (although the systems that use CORBA do not have to be object-oriented). CORBA uses an **Interface Definition Language** (IDL) to specify the interfaces that objects present to the outer world, and then specifies a mapping from the IDL to a specific implementation language such as C++ or Java.

CORBA thus addresses the same issues as SOAP. However, CORBA requires client stubs[72] to be compiled and distributed for each type of client in use. This is not always practical particularly when offering services to anonymous clients via the Internet. Also, CORBA's transport protocol, **Internet Inter-ORB Protocol**[73] (IIOP), is not always firewall friendly so may introduce some firewall-related obstacles.

The key distinction though is that SOAP is an XML-based protocol and consequentially particularly verbose. CORBA over IIOP has superior performance as marshalling and demarshalling in CORBA is more efficient and there is less data being transferred. The advantage that SOAP being XML-based gives is that it is human readable and writable, meaning that it is easy to read and manipulate the messages that are generated, which is extremely useful when debugging.

8.13   WEB SERVICES AND THE SERVICE WEB

A web service is a method that is callable remotely across a network. Whereas content-based services provide web pages (whether static or dynamically generated) for human consumption, web services provide data for computers. The entirety of web services available on the Internet is termed the service web.

Let us consider an example: Search engines such as Google can translate web content enabling the user to view an English language version of a web page that was written (or dynamically generated) in Spanish. The translated version is typically generated dynamically by software installed at the search engine's site. Traditionally, in order to set up a new site with similar capabilities, software to perform the translation would need to be written or procured and then plugged into the web server. Using web services it may instead be possible to offload this work to a site with a dedicated translation web service exploitable via SOAP request.[74]

The service method signature might look something like this:

```
String translate {
    String sourceLanguage;
    String targetLanguage;
    String textToTranslate;
}
```

The new search engine could then take the requested page, fragment it into markup and text, and call the web service with each fragment specifying the source and target languages. The page would then be reconstructed and passed to the browser to render.

Leaving certain details aside, such as standards for identifying languages or how the service provider might bill for usage,[75] the power of the basic idea is clear. The service web thus allows providers to concentrate on their own area of expertise and incorporate information and functionality from others' areas of expertise seamlessly.

One big challenge of this vision is indexing the service web so that services and their signatures can be easily found and exploited. The **Web Services Description Language** (WSDL) and **Universal Description, Discovery, and Integration** (UDDI) standards work together to respectively define and locate web services but the detail of these standards is beyond the scope of this chapter.[76]

8.14   SOAP MESSAGES

SOAP message syntax is described in detail at http://www.w3.org/TR/SOAP/, but we will look at a brief summary and a few examples.

A valid SOAP message is a well-formed XML document.[77] The XML prologue can be present but, if present, should contain only an XML declaration (i.e. it should not contain any **Document Type Definition** (DTD) references or XML processing instructions). It should use the SOAP envelope and SOAP encoding namespaces and have the form:

- An optional XML declaration
- A SOAP envelope (the root element) which consists of:
  - An optional SOAP header
  - A SOAP body

A SOAP-encoded RPC dialogue contains both a request message and a response message. For example, a simple service method that adds 5 to a given integer would have the method signature

> int AddFive (int number);

with the request formed as

```
<?xml version="1.0" encoding="UTF-8" standalone="no" ?>
<SOAP-ENV:Envelope
   SOAP-ENV:encodingStyle="http://ganneysoap.org/soap/encoding/"
   xmlns:SOAP-ENV="http://ganneysoap.org/soap/envelope/"
   xmlns:SOAP-ENC="http://ganneysoap.org/soap/encoding/"
   xmlns:xsi="http://www.w3.org/1999/XMLSchema-instance"
   xmlns:xsd="http://www.w3.org/1999/XMLSchema">
   <SOAP-ENV:Body>
     <ns1:AddFive xmlns:ns1="urn:GanneySoap">
     <param1 xsi:type="xsd:int">37</param1>
```

```
      </ns1:AddFive>
    </SOAP-ENV:Body>
  </SOAP-ENV:Envelope>
```

The response to this request would be formed as

```
  <?xml version="1.0" encoding="UTF-8" ?>
  <SOAP-ENV:Envelope
    xmlns:SOAP-ENV="http://ganneysoap.org/soap/envelope/"
    xmlns:xsi="http://www.w3.org/1999/XMLSchema-instance"
    xmlns:xsd="http://www.w3.org/1999/XMLSchema">
    <SOAP-ENV:Body>
      <ns1:AddFiveResponse
      xmlns:ns1="urn:GanneySoap"
      SOAP-ENV:encodingStyle="http://ganneysoap.org/soap/encoding/">
      <return xsi:type="xsd:int">42</return>
      </ns1:AddFiveResponse>
    </SOAP-ENV:Body>
  </SOAP-ENV:Envelope>
```

While most SOAP packages will take care of the SOAP message syntax details, we will now examine these messages.

The XML prologue contains only an XML declaration <?xml version="1.0" encoding="UTF-8" ?> specifying the XML version and the character encoding of the XML message.

The SOAP envelope tag <SOAP-ENV:Envelope ... > in the request message first specifies that this SOAP message's encoding style follows the schema defined at http://ganneysoap.org/soap/encoding/. Note that this is optional and will be presumed if it is not included in the response message (in this case it is). The SOAP Envelope tag also contains many namespace definitions. The namespace identifiers are standard and the SOAP specification requires that these namespaces either be defined correctly or not at all (i.e. a SOAP message with missing namespace definitions is correct and processable but one with incorrect, i.e. non-standard, definitions is incorrect and discardable). Notice that the SOAP-ENC namespace definition is missing from the response message but this does not mean the message is invalid.[78]

There is no SOAP header tag in this example. SOAP headers are optional and are typically used to transmit authentication or session management data. While authentication and session management are out of the scope of the SOAP protocol, SOAP does allow for some flexibility in SOAP messaging so that implementors can include such information.

The SOAP body tag <SOAP-ENV:Body> follows which encapsulates a single method tag porting the name of the method itself <ns1:AddFive ... > (or the same name suffixed with the word "Response" in the case of the response message). Note that the method tag is typically namespaced by the service name; in this case urn:GanneySoap to ensure uniqueness (A web service, which can contain any number of differently named methods, has a service name unique to the URL at which it is accessible).

The method tag in turn encapsulates any number of parameter tags such as the <param1 ... > tag in the request envelope. Parameter tag names can be anything at all and are typically autogenerated and have no namespace. In the response message there is only ever one parameter tag (representing the return value of the method) and it is typically named <return>.

Most of the work in a SOAP dialogue is done by the SOAP package. There are many SOAP packages available, often free of charge. Client code can thus be written in any language and on any platform for which a SOAP package is available. Amongst the numerous possibilities are:

- Visual Basic and the Microsoft SOAP Toolkit
- Java (stand-alone or Servlet/JSPs etc.) and Apache SOAP for Java
- Perl (stand-alone or CGI-Perl scripts etc.) and SOAP::Lite for Perl

It is important to remember that the choice of language, platform and SOAP package on the client-side to consume web services is entirely independent of what language, platform and SOAP package is used on the server side to provide the web services. In this way the same SOAP-based web service (deployed for example on UNIX, written in Java and exploiting Apache SOAP for Java) can be consumed by any type of client written for any platform, in any language, exploiting any SOAP package applicable to that language/platform combination. The only thing the developer needs to write is the service method logic (which may not be that simple…).

NOTES

1 Parts of this chapter (the basic concepts of a network, such as the network packet, hardware (hub, switch and router), network topologies, IP addressing (including static, dynamic and reserved) and DNS, IP masking, ports, network address translation, bandwidth and the OSI 7-layer model) are reproduced from Ganney et al. 2022.

2 Otherwise you're talking to yourself – this figure excludes virtual networking.

3 It seems like a large number, but with a world population running at 7 billion, that's only 0.5 IP addresses each.

4 bps = **bits per second**.

5 Updated to "My Network Places" from Windows 7 onwards.

6 **Fibre Distributed Data Interface** – a set of ANSI and ISO standards for data transmission on fibre optic lines in a LAN that can extend in range up to 200 km (124 miles).

7 **Synchronous Optical Network** – the American National Standards Institute standard for synchronous data transmission on optical media.

8 In a token ring, a "token" is passed around the network. The device holding the "token" is permitted to transmit – nothing else is. If a device has nothing to transmit, it passes the token on.

9 This is an American organisation (although it works internationally) hence the American spelling.

10 EIA is "**Electronic Industries Alliance**". This no longer exists and is thus often dropped from the name of the standard. EIA/TIA is often rendered ANSI/TIA to reflect its adoption.

11 This definition is cited frequently but without attribution to the original author. One such example is https://voygar.de/en/support-downloads/cabling-glossary.

12 A more aesthetically pleasing workplace is a more efficient one.

13 Generic Telecommunications Cabling for Customer Premises, which is used for generic infrastructures.

14 Commercial Building Telecommunications Cabling Standard, which is more commonly used with typical commercial building infrastructures.

15 Cable laid in the plenum space – that part of a building that can facilitate air circulation for heating and air conditioning systems, often above a false ceiling.

16 Although it can be more, for example the computer lab at UCLH specified 6 per user.

17 Cat5e supports Gigabit Ethernet (1Gbps); Cat6 supports 1Gbps at 328 feet and 10Gbps up to 33-55 metres; Cat6a supports 10Gbps at 100 metres. Cat 7 also exists but has extensive shielding making it quite stiff and requires special connectors.

18 It is worth bearing this in mind if you should decide to join two ports together at the patch panel in order to avoid running a cable across the lab: the total distance may be beyond spec.

19 IPv6 also exists, but is not widespread (yet), but does allow 2^{96} addresses.

20 Due to binary: 256 is 2^8 so each group of digits is composed of 8 bits.

21 A command that sends an "are you there" message.

22 n AND 255 = n as 255 = 11111111_2; n AND 0 = 0, so 255.255.255 uses the first three bytes to form the network address, leaving the remainder as the host number.

23 As 130 AND 192 = 2.

24 The lowest-numbered 1024 are known as the "well-known port numbers" – there is a fuller list at http://packetlife.net/media/library/23/common_ports.pdf.

25 I know I've just said this twice, but it's something people tend to forget.

26 A poor name, as it implies the route is the longest whereas it's actually the most specific and therefore probably the shortest. But that's conventions for you.

27 A router with such free will is unlikely, so it's really an algorithm that does the choosing, but the programmer was free to decide how the choice is made.

28 n AND 0 is 0 regardless of the value of n.

29 Which is why it's called the default.

30 The loopback IP address is the address used by a device to access itself and is used to test network software. This is often referred to (especially in UNIX/Linux systems) as localhost.

31 A datagram is a basic transfer unit associated with a packet-switched network.

32 A classful network divides the IP address space for IPv4 into five address classes based on the leading four address bits. Classes A, B and C provide unicast addresses for networks of three different network sizes. Class D is for multicast networking and the class E address range is reserved for future or experimental purposes.

33 Dynamic Host Configuration Protocol.

34 In order to find your "real" IP address, use the web site whatismyip.com.

35 Although there are some very famous failures in this respect, such as filtering out images containing skin tone (not helpful in medicine) and the "Scunthorpe problem" where a part of a normal word triggers a filter rule.

36 See https://www.online-tech-tips.com/windows-10/adjust-windows-10-firewall-settings/ for instructions on examining and adjusting the settings.

37 "Stateful" is the opposite of "Stateless" which we will cover in Section 8.11.2 and then in HTML. For now, "stateful" may be considered as "having memory".

38 Note that these are maximums and a wired network stands a better chance of providing the full bandwidth due to less interference. 802.11g normally only provides 20Mbps and 802.11ac, operating in the 5GHz band, has a theoretical bandwidth of 1300Mbps but only provides 200Mbps. 802.11ax (also known as Wi-Fi 6) has a maximum of 14Gbps and also provides better throughput in high-density settings, such as corporate offices, airports and dense housing situations. It requires the hardware to be correctly set up in order to achieve anything better than 800Mbps though. This, however, does position wireless better as a method of provision.

39 **Quality-of-service**, a Cisco product.

40 Virtual LAN.

41 Similar in description to the data link layer, but handling the logical data rather than physical.

42 Placing several message streams, or sessions onto one logical link and keeping track of which messages belong to which sessions.

43 It is probably simplest to envisage a session as being a connection.

44 ASCII and EBCDIC are two numerical methods of representing characters, e.g. "A" is ASCII 65 and EBCDIC 193. ASCII actually consists of four 32-character alphabets. The first (numbered zero, of course) is control characters such as horizontal tab and end of transmit block. Next is punctuation marks, then upper case letters and finally lower case letters. There are 32 due to 7-bit binary. The "spare" slots are used for additional punctuation marks with the last character (127) being delete – on paper tape that was 7 punched holes which was a deletion you can't just hit "undo" on.

45 A successor is the route to the node that has the best metric. A feasible successor is an alternative route that can be used in the event of a topology change without having to recalculate routes. The feasible successor can thus be used instantly.

46 There is a similarity with netmasking techniques therefore.

47 This activity is known as *flapping*.

48 The number of hops from source to destination.

49 This is in the form of a graph – see Section 5.8.1, Chapter 5 for a description of a mathematical graph.

50 This contrasts with distance-vector routing protocols, where each node shares its routing table with its neighbours.

51 The Internet layer of the TCP/IP model is easily compared with the network layer (layer 3) in the OSI 7-layer model. Although these two models use different classification methods, there are many similarities.

52 Dijkstra's algorithm is an algorithm for finding the shortest paths between nodes in a graph, which may represent, for example, road networks. It was conceived by computer scientist Edsger W. Dijkstra in 1956 and published three years later. It is sometimes referred to as "the LSR algorithm" although technically there are others.

53 Unless otherwise set, the link cost of a path connected to a router is determined by the bit rate (1 Gbit/s, 10 Gbit/s, etc.) of the interface.

54 A small packet, similar to ping, that is primarily used to determine the time delay in reaching a node.

55 The part that is missing is EIGRP stub, a feature that provides fast convergence in case of failure.

56 DUAL evaluates the data received from other routers in the topology table and calculates the primary (successor) and secondary (feasible successor) routes. The primary path is the path with the lowest metric. There may be multiple successors and multiple feasible successors, all of which are maintained in the topology table, but only the successors are added to the routing table and used to route packets.

57 Eight-bit bytes.

58 Although given that HTTP 1.1 was designed using REST, neither can claim to be that new.

59 A stateless protocol is a communications protocol in which no information is retained by either sender or receiver. The sender transmits a packet to the receiver and does not expect an acknowledgment of receipt. HTTP is thus stateless.

60 A string of characters used to identify a resource – http://www.ganney.ac.uk/networks is a URL, whereas http://www.ganney.ac.uk/networks/Current_Time is a URI as it identifies rather than locates the resource.

61 Session states are covered in "Databases" (see Section 3.5, Chapter 3) and "Web Programming" (see Section 11.12.1, Chapter 11).

62 A client normally can't tell whether it is directly connected to the end server or to an intermediary, so it should make no difference to the client application.

63 Parsing a sentence is separating that sentence into grammatical parts, such as subject, verb, etc. In computing a parser is a compiler or interpreter component that breaks data into smaller elements for easy translation into another, usually lower-level, language.

64 The HTTP PATCH request method applies partial modifications to a resource, as opposed to PUT, which is a complete replacement of a resource.

65 An element is part of a collection.

66 Which they can do due to the lack of an official standard to meet.

67 See https://mpcetoolsforhealth.liverpool.ac.uk/ for some examples.

68 Although an integer can normally have 10 digits, the maximum value of the first one is 2. A long integer allows 19 digits. In some implementations "int" and "long int" are the same (due to the increased number of bits in a processor), so "long long int" must be used.

69 Typically written in the same language as the service method for simplicity.

70 We will consider SOAP messages later in this chapter.

71 A jar file is a collection of Java classes (usually a library) and has the extension ".jar".

72 A client stub is a piece of code that provides entry points on the client.

73 ORB is "Object Request Broker".

74 Whilst there does not appear to be one for Google Translate, there is a SOAP interface for Bing Translate. Google can be done via third party sites such as https://www.integromat.com/en/integrations/google-translate/soap.

75 The really tricky bit.

76 See the WSDL 1.1 specification and www.uddi.org for details.

77 For more detail on XML and well-formedness visit http://www.w3.org/XML/.

78 For more detail on XML namespaces see http://www.w3.org/TR/REC-xml-names/.

REFERENCES

Fiber Optic Association 2021, [online]. Available: https://foa.org/tech/ref/premises/jargon.html [Accessed 14/04/22].

Fielding R T, 2000, Architectural Styles and the Design of Network-based Software Architectures. Doctoral dissertation, University of California, Irvine [online]. Available: https://www.ics.uci.edu/~fielding/pubs/dissertation/rest_arch_style.htm [Accessed 14/04/22].

Floyd S and Jacobson V, 1994, The Synchronization of Periodic Routing Messages, IEEE/ACM Transactions on Networking [online]. Available: https://doi.org/10.1109/90.298431 [Accessed 14/04/22].

Ganney P, Maw P, White M, Ganney R, ed. 2022, *Modernising Scientific Careers The ICT Competencies*, 7th edition, Tenerife: ESL.

Gibson Research Corporation 2008, [online]. Available: https://www.grc.com/vpn/routing.htm [Accessed 14/04/22].

Kleinrock L, 1975–2011. *Queuing Systems*. Volumes 1–3. Hoboken: John Wiley & Sons.

NHS Digital 2022, [online]. Available: https://digital.nhs.uk/data-and-information/looking-after-information/data-security-and-information-governance/nhs-and-social-care-data-off-shoring-and-the-use-of-public-cloud-services [Accessed 14/04/22].

NHS Digital 2020, [online]. Available: https://digital.nhs.uk/news-and-events/latest-news/two-national-nhs-services-move-to-the-cloud [Accessed 14/04/22].

Spiceworks 2013, [online]. Available: https://community.spiceworks.com/topic/361713-cable-color-standards [Accessed 14/04/22].

Taktak A, Ganney PS, Long D and Axell RG, 2019, *Clinical Engineering: A Handbook for Clinical and Biomedical Engineers*, 2nd edition. Oxford: Academic Press.

Wikipedia 2022a, [online] [1]. Available: https://en.wikipedia.org/wiki/Port_%28computer_networking%29 [Accessed 14/04/22].

Wikipedia 2022b, [online] [2]. Available: https://en.wikipedia.org/wiki/Session_(computer_science) [Accessed 13/08/21].

9 Storage Services[1]

9.1 INTRODUCTION

Computing has come a long way: from the Greek Antikythera analogue computer (Circa 100BCE) which was used for astronomical predictions, through the Babbage Analytical Engine (Circa 1840), Turing's Bombe (1940s), ENIAC (1945), the Apollo guidance computer (1960s) and into the PCs, tablets and embedded devices that we know today. In 1965 Gordon Moore stated Moore's law – the observation that the number of transistors in a dense integrated circuit doubles approximately every two years.[2] There is an excellent animated graph of it at https://www.reddit.com/r/dataisbeautiful/comments/cynql1/moores_law_graphed_vs_real_cpus_gpus_1965_2019_oc/.

Whilst this has increased computing power greatly, it has also increased the unwanted side-effects, most notably heat which thus requires more cooling, which involves moving parts (and in some systems water) and therefore increases reliability issues.

This chapter examines two techniques for tackling the growth of computing power and the need for storage: Virtualisation and Cloud Computing.

9.2 VIRTUAL ENVIRONMENTS

There has been a notable trend in recent years towards virtualising systems, especially for servers. In this, multiple servers are hosted on a single hardware platform, the load being distributed across all available processors.

The drivers of virtualisation are:

- CPU development has tended to lead software development
- Software applications were not using much of this new hardware power
- A single, over-resourced, physical machine can be divided into many smaller machines

This gives advantages of doing so, which include:

- Lower power consumption
- Less cooling required
- Less rack space
- Centralised administration
- Quick and efficient provisioning (just copy another **Virtual Machine** (VM))
- Easier backup and disaster recovery

There are several vendors, such as VMware (ESXi), Microsoft (Hyper-V), Citrix (XenServer), Red Hat and KVM.

DOI: 10.1201/9781003316244-9

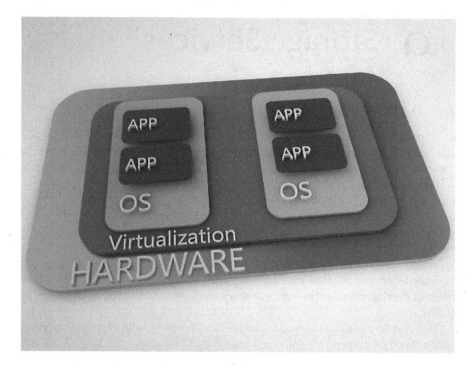

FIGURE 9.1 Schematic representation of a server hosting multiple virtual machines. (Shutterstock). Multiple apps may run under each OS.

A virtual machine, as illustrated in Figure 9.1, is not sized by providing it with more resources than it really needs, under the assumption that it'll never run out of resources and thus will work even better, as this wastes the very resources that we were trying to better utilise. Instead the "Goldilocks principle"[3] is used, for if a VM has too little resource available then it has insufficient power to get things done whereas if it has too much resource then the unused vCPU still consumes host resource due to polling and unused memory.

A VM still requires storage, but this normally comes from the organisation's SAN and thus can be added at any time (provided there's enough space on the SAN), it can be thick provisioned (like a physical server) or thin provisioned (it can use up to the amount requested).

Thick provisioning is a type of storage pre-allocation. With thick provisioning, the complete amount of virtual disk storage capacity is pre-allocated on the physical storage when the virtual disk is created. A thick-provisioned virtual disk consumes all of the space allocated to it in the datastore right from the start, so the space is unavailable for use by other virtual machines.

Thin provisioning is another type of storage pre-allocation. A thin-provisioned virtual disk consumes only the space that it needs initially and grows with time according to demand.

VMs are not a one-size-fits-all solution, though. Some applications can't be virtualised (yet) due to high IO workloads (e.g. a Trust-wide SQL Server farm containing multiple databases) and some servers may have hardware plugins (dongles etc.) that they rely on.

9.3   CLOUD COMPUTING

Cloud computing is

> a type of Internet-based computing that provides shared computer processing resources and data to computers and other devices on demand. It is a model for enabling ubiquitous, on-demand access to a shared pool of configurable computing resources (e.g., computer networks, servers, storage, applications and services), which can be rapidly provisioned and released with minimal management effort.
>
> (Wikipedia 2022 [online])

The main reasons to use the cloud are:

- System Prototyping and Evaluation
- Reduction of Costs - Saving Money
- Flexibility
- Universal access
- Up to date software
- Choice of software
- Potential to be greener and more economical

There are three types of cloud:

- **Public Cloud**: a service provider makes resources, such as applications and storage, available to the general public over the Internet. Public cloud services may be free or offered on a pay-per-usage model.
- **Private cloud**: a type of cloud computing that delivers similar advantages to public cloud, including scalability and self-service, but through a proprietary architecture. Unlike public clouds, which deliver services to multiple organisations, a private cloud is dedicated to a single organisation.
- **Hybrid cloud**: a cloud computing environment which uses a mix of on-premises, private cloud and third-party, public cloud services with orchestration between the two platforms. By allowing workloads to move between private and public clouds as computing needs and costs change, hybrid clouds give businesses greater flexibility and more data deployment options.

There are five types of cloud service:

- **Platform as a Service** (PaaS): a cloud computing model that delivers applications over the Internet. In a PaaS model, a cloud provider delivers hardware and software tools – usually those needed for application development – to its users as a service. A PaaS provider hosts the hardware and software on its own infrastructure. As a result, PaaS frees users from having to install in-house hardware and software to develop or run a new application.
- **Software as a Service** (SaaS): a third-party provider hosts software, removing the need for organisations to install and run applications on their own computers or in their own data centres. It eliminates the expense of hardware

acquisition, provisioning and maintenance, as well as software licensing, installation and support.

- **Infrastructure as a Service** (IaaS): a third-party provider hosts hardware, software, servers, storage and other infrastructure components on behalf of its users. IaaS providers also host users' applications and handle tasks including system maintenance, backup and resiliency planning.
- **Data as a Service** (DaaS): the data is hosted, offering convenient and cost-effective solutions for customer- and client-oriented enterprises.
- **Backup as a Service** (BaaS): this may be used when an organisation has outgrown its legacy storage backup and would have to go through a costly upgrade, or lacks the resources for on-premises, high-level backup. Outsourcing backup and recovery to a provider can also keep data accessible or restorable from a remote location in case of an outage or failure.

These services can be visualised as follows: Consider a business responsible for selling shorts. There are a number of ways in which such a business might be set up, each relying on outside services to a greater or lesser extent. We saw the graphical representation of this in Figure 5.3 together with another graphic showing the overlap of these services in Figure 5.4 in Chapter 5.

9.4   SECURITY AND GOVERNANCE FOR CLOUD SERVICES

Cloud services governance is a general term for applying specific policies or principles to the use of cloud computing services. The goal of cloud services governance is to secure applications and data when they are located remotely (so this can also apply to off-site hosting as well as to cloud services). Cloud services governance can be viewed as an extension of SOA (**Service Oriented Architecture**) governance, although the unique properties of a public cloud architecture – such as multi-tenancy – present slightly different concerns. Cloud services governance should complement or be integrated into existing governance processes and (as with all governance processes) is viewed as an ongoing process, not a product.

Organisations are responsible for their own information. The nature of cloud computing means that the organisation is reliant upon a third party for some element of the security of its data. The *"trust boundary"* is the point at which the responsibility passes from the owning organisation to the cloud supplier. This occurs at a different point for **Infrastructure as a Service** (IaaS), **Platform as a Service** (PaaS) and **Software as a Service** (SaaS); organisations therefore need to satisfy themselves of the security and resilience of their cloud service providers.

UK organisations that store personal data in the cloud or that use a **Cloud Service Provider** (CSP) must comply with the DPA, meaning that data processors and data controllers are accountable for the security of the personal data that they process. CSPs and organisations that use them therefore need to implement both appropriate technical and organisational measures to ensure that processing meets the GDPR's requirements and protects the rights of data subjects.

The UK government's G-Cloud framework makes it simpler[4] for the public sector to buy cloud services. Suppliers are approved by the **Crown Commercial Service**

(CCS) via the G-Cloud application process, which eliminates the need for them to go through a full tender process for each buyer.

NOTES

1   Three sections of this chapter (Introduction, Virtual Environments and Cloud Computing) are reproduced (with slight modification) from Ganney et al. 2022.
2   Originally this was every year, but he revised it to 2 years in 1975.
3   Not too much, not too little, but just right.
4   i.e. faster, cheaper and with confidence in compliance with the law.

REFERENCES

Ganney P, Maw P, White M, Ganney R, ed. 2022, *Modernising Scientific Careers The ICT Competencies*, 7th edition, Tenerife: ESL.
Wikipedia 2022, [online]. Available: https://en.wikipedia.org/wiki/Cloud_computing [Accessed 16/04/22].

10 Encryption[1]

10.1 INTRODUCTION

Assuming that all devices are secure and that the connection is also, the next issue to address is that of interception. Data may be deliberately intercepted (via packet logging) or simply mislaid. In either case, the next level of security is encryption. Successful encryption means that only the authorised receiver can read the message.

10.2 ENCRYPTION

Encryption is, of course, a very old science. It has gone from simple substitution ciphers,[2] through the complexity of the Enigma machine, to today's prime-number based techniques.

Some common terminology associated with encryption is:

- *Cryptosystem* or *cipher system* is a method of disguising messages so that only selected people can see through the disguise.
- *Cryptography* is the art of creating and using cryptosystems.
- *Cryptanalysis* is the art of breaking cryptosystems – seeing through the disguise (especially when you're not supposed to be able to do so).
- *Cryptology* is the study of both cryptography and cryptanalysis.
- *Plaintext* is the original message used as the input to the encryption system.
- *Ciphertext* is the disguised message output from the encryption system.
- The letters in the plaintext and ciphertext are often referred to as *symbols*.
- *Encryption* is any procedure to convert plaintext into ciphertext.
- *Decryption* is any procedure to convert ciphertext into plaintext.
- *Cryptology Primitives* are basic building blocks used in the encryption process. Examples of these include substitution ciphers, stream ciphers and block ciphers.

10.2.1 CIPHERS AND CRYPTOGRAPHY

A substitution cipher (such as the Caesar cipher, Atbash and ROT13) is one where each letter of the alphabet is exchanged for another. Decryption is a simple matter of reversing the substitution. For example, Atbash substitutes A for Z, B for Y etc. so decryption is actually the same as encryption, whereas Caesar displaces each letter by 3, so A becomes D, B becomes E and so on, wrapping around so Z becomes C.

A stream cipher takes a string of letters (known as a *keystream*) and combines it with the first group of letters in the message, then the next and so on.[3] For example the Vignere stream cipher substitutes each alphabetic character with a number (A=0, B=1, Z=25) and adds together the plaintext and keystream values to produce a new character (mod 26 is used, as you might expect[4]). The case of the plaintext is preserved. For example, the plaintext "Information" with keystream "structure" sees the first letter of the plaintext ("I" = 8) added to the first letter of the keystream ("s" = 18) to produce 26 = "A" (due to mod 26), being upper case as the plaintext input was. The second letter ("n" = 14) becomes "g" (13+19 = 32, Mod 26 = 6), and so on.

Block ciphers encrypt blocks of plaintext at a time using unvarying transformations on blocks of data. The Playfair cipher (one of the simplest) divides the plaintext into blocks of 2 letters and a random arrangement of alphabets in a 5x5 block is used to encrypt each of these blocks individually. Given a large enough extract of ciphertext, it might be possible to reconstruct the coding matrix, thus breaking it. Large block sizes and iterative encryption are one way of combatting this. Some of the common encryption technologies used today such as the **Advanced Encryption Standard** (AES) used worldwide, use a form of block cipher incorporating large (64 – 128-bit) blocks and pseudorandom encryption keys.

Block and stream ciphers are both examples of symmetric key cryptography, in that the same key is used for both encryption and decryption.

The alternative is asymmetric key cryptography which uses two separate keys – one of the keys encrypts the plaintext and the other decrypts it. This prevents the need for (and the associated risks with) transmitting keys either within or without the message.

A technique of sending, receiving and sending again (known as the three-pass protocol) enables secure transmission of messages without key exchange. An analogy would be the situation where a box of valuables is to be sent securely from one location to another. Sending the key to unlock the box either with it or separately would be high risk. Therefore, the sender places a padlock on the box to which only they have the key. Once sent and received (pass 1), the receiver adds their own padlock and sends it back (pass 2). The sender can now remove their padlock using their key leaving the box locked with just the receiver's lock on it. It is then sent again (pass 3) where the receiver unlocks the box to retrieve the contents. In this case, the need to send any keys is eliminated and both the encryption and decryption keys are private.

In public key cryptography, one of the keys is made public, typically the encryption key. Transmitted messages are thus encrypted using this public key of the recipient. The matching decryption key is kept private and known only to the recipient who can then use it to decrypt the ciphertext.

The primary advantage of an asymmetric cryptosystem is the increased security since private keys never need to be transmitted, unlike the symmetric key system where the key must be communicated between the sender and receiver as the same key is required for both encryption and decryption. Interception of this key can compromise the security and authenticity of future data transmissions.

10.2.2   RSA AND PGP ENCRYPTION

The most common encryption is RSA developed by Rivest, Shamir and Adelman in 1978 (Rivest et al. 1978) and relies upon the difficulty of factoring into prime numbers. It works as follows:

- Let p and q be large prime numbers and let $N = pq$. Let e be a positive integer which has no factor in common with $(p-1)(q-1)$. Let d be a positive integer such that $ed-1$ is divisible by $(p-1)(q-1)$. (10.1)
- Let $f(x) = x^e \bmod N$, where $a \bmod N$ means "divide N into a and take the remainder." (10.2)
- Let $g(x) = x^d \bmod N$. (10.3)
- Use $f(x)$ for encryption, $g(x)$ for decryption.

(Clay Mathematics Institute 2012 [online])

Therefore, in order to transmit a secure message only the numbers e and N are required. To decrypt the message, d is also required. This can be found by factoring N into p and q then solving the equation to find d. However, this factorisation would take millions of years using current knowledge and technology.[5, 6]

A simpler method is **Pretty Good Privacy** (PGP) developed by Phil Zimmerman in 1991 (Stanford University 2004 [online]) and subsequently multiply revised. In this there is a single public key (published) which is used for encryption and a single private key which is used for decryption. In PGP a random key is first generated and is encrypted using the recipient's public key. The message is then encrypted using the generated (or "session") key. Both the encrypted key and the encrypted message are sent to the recipient. The recipient then decrypts the session key using their private key, with which they decrypt the message.

So far we have only considered data transmissions. Encryption can also be used on data "at rest", i.e. on a storage device. RSA encryption is therefore usable in this context, although PGP isn't (and AES is now more common for this purpose than RSA). There are two forms of encryption: hardware and software. Both use similar algorithms but the use of hardware encryption means that the resultant storage device is portable as it requires no software to be loaded in order to be used. A device may be fully encrypted[7] or filesystem-level encrypted, which just encrypts the storage being used (often the file names are in plain text, so be careful what you call your files). Devices may use multiple keys for different partitions, thereby not being fully compromised if one key is discovered.

10.2.3 STEGANOGRAPHY, CHECKSUMS AND DIGITAL SIGNATURES

Three final concepts must be considered before we move on from encryption: steganography, checksums and digital signatures. Steganography is a process of hiding files within other files, often at bit level – image files are therefore very suitable for this.

Checksums were originally developed due to the unreliability of electronic transmission. In the simplest form, the binary bits of each part of the message (which could be as small as a byte) were summed. If the result was odd, a bit with the value 1 would be added to the end of the message. If even, then the bit would be 0. Thus, by summing the entire message's bits, the result should always be even.

As an example, consider 9. As a 7-bit number this is 0001001. It has 2 1s, so the check digit is 0, giving 00010010. It is therefore vital to know whether you are using even or odd checksums (often called parity).

Extensions to this were developed in order to detect the corruption of multiple bits and also to correct simple errors. Developed to ensure the integrity of the message due to electronic failure, these techniques can also be used to detect tampering (see Section 14.4, Chapter 14).

A full-file checksum is commonly used to ensure the reliable transmission of the file (e.g. from memory stick to PC) and is calculated using a hashing function.

> The most common checksums are MD5 and SHA-1, but both have been found to have vulnerabilities. This means that malicious tampering can lead to two different files having the same computed hash. Due to these security concerns, the newer SHA-2 is considered the best cryptographic hash function since no attack has been demonstrated on it as of yet.
>
> (Kishore 2015)

A checksum is calculated by the transmitting (or source) system and also by the receiving (or destination) system and compared. If they are the same, then the file is presumed to have been transferred without corruption.

A simple checksum example is the bank card. If you write out the 10-digit number, then replace every other number (starting with the first) with twice its value, then sum all the resultant digits,[8] you get a multiple of ten. Thus a website can quickly verify that the card number is in a valid form before contacting the card issuer for verification. The clever bit in this method is in doubling every other digit, because the most common error in typing in a string of numbers is transposition, which this catches. Just adding the digits wouldn't do this (Parker 2015).

Digital signatures verify where the information received is from. It uses a similar asymmetric cryptography technique to PGP, in that a message is signed (encrypted) using a public key and verified (decrypted) using a private key. A more complex version also uses the message, thereby demonstrating (in a similar fashion to checksums) that the message has not been altered. A valid digital signature provides three assurances: the message was created by a known sender; the sender cannot deny having sent the message and the message was not altered in transit.

Digital signatures are commonly used for software distribution, financial transactions, contract management software and in other cases where it is important to detect forgery or tampering. They are also often used to implement electronic signatures.[9] In many countries, including the United States, Algeria, Turkey, India, Brazil, Indonesia, Mexico, Saudi Arabia, Uruguay, Switzerland and in the European Union, electronic signatures have legal significance.

NOTES

1 Some sections of this chapter (RSA, PGP, steganography, checksums and digital signatures) are reproduced from Ganney et al. 2022.
2 Where each letter of the alphabet is exchanged for another. Decryption is a simple matter of reversing the substitution.
3 i.e. the keystream is shorter than the plaintext and so is used repeatedly.
4 Mod 26 divides the number by 26 and retains the remainder, so 27 Mod 26 = 1, for example.

5 It is not worth beginning to speculate on the effect of a mathematician discovering a fast factorisation method.
6 See https://sites.math.washington.edu/~morrow/336_09/papers/Yevgeny.pdf for a good description of how it all works.
7 (i.e. the entire storage, sometimes including the master boot record).
8 Remember that if you had "8" this was doubled to "16" and you now have "7" as this is "1+6".
9 A broader term that refers to any electronic data that carries the intent of a signature, but not all electronic signatures use digital signatures.

REFERENCES

Clay Mathematics Institute 2012, The RSA algorithm [online]. Available: http://www.claymath. org/posters/primes/rsa.php [Accessed 07/06/12 – link now broken, even on their website!].

Ganney P, Maw P, White M, Ganney R, ed. 2022, *Modernising Scientific Careers The ICT Competencies*, 7th edition, Tenerife: ESL.

Kishore A, 2015, *What is a Checksum and How to Calculate a Checksum* [online]. Available: https://www.online-tech-tips.com/cool-websites/what-is-checksum/ [Accessed 18/04/22].

Parker M, 2015, *Things to Make and Do in the Fourth Dimension*, London: Penguin, p. 346.

Rivest R, Shamir A, Adleman L, 1978, *A Method for Obtaining Digital Signatures and Public-Key Cryptosystems*, Communications of the ACM. 21 (2): 120–126. doi: 10.1145/359340.359342.

Stanford University 2004, [online]. Available: https://cs.stanford.edu/people/eroberts/courses/ soco/projects/2004-05/cryptography/pgp.html [Accessed 18/04/22].

11 Web Programming[1]

11.1 INTRODUCTION

Probably the most common computer interface that users experience is the web browser. This chapter looks at developing applications for this environment.

11.2 STRATEGIES FOR WEB DEVELOPMENT

Web development is not primarily an art. Neither is it purely a science. It is actually a team effort, bringing together programmers, graphic designers, content authors, interface designers, financial wizards and system testers.[2] Strategic web design therefore has a lot in common with project management, which we cover in chapter 17.

It has been said that

> Strategic design is the fusion of your organizational goals with every aspect of your design process. You aren't simply designing a user interface that looks good and is usable and accessible. You're designing an interface that will help you accomplish your organization's objectives.
>
> (Smashing Magazine [online])

This may be a little ambitious for a small departmental web site on the trust intranet, but thinking through how the site fits with the goals of the department is not a bad place to start.

There are many lists of stages in strategic design, often including items such as "analyse the competition" which are not always appropriate in the context of the NHS. Six common steps are:

1. Establish goals. What is the site trying to achieve? A website isn't predominantly a piece of art but an interface that serves a function.
2. Identify the audience. What are their demographics? A site for pre-school hearing screening will have a different look and feel to one for late onset diabetes. Factors such as age, gender, profession and technical competency influence what can (and can't) be done with the site.
3. Determine the brand image. Everything has a brand, even if nothing is being sold. Colour is vital here – is the aim to calm the audience or engage them? (NB the NHS has a set of approved colour palettes so choices may be limited – see "Design Style", next).
4. Goal-driven design direction. If the objective is to sign up volunteers for an online study, then the "sign me up" button needs to be prominent and the "about" section (describing the study) should also be highlighted. If the objective is to provide information, then usability, readability and uncluttered design are the priorities.

DOI: 10.1201/9781003316244-11

5. Measure results. Like benefits realisation in project management, this involves re-visiting the original objectives to see how well they have been met. If the objective is to provide information, then it may be worth designing in a way of capturing which pages are accessed the most.[3] A feedback form may be appropriate.

6. Kaizen. This is a Japanese philosophy which focuses on continuous improvement using small steps. A similar idea was espoused by Dave Brailsford of the British Olympic cycling team in 2012: "aggregation of incremental gain". A website is easy to tweak and therefore to incrementally improve.

11.2.1 DESIGN STYLE

There are many different approaches to website design, and the NHS has guidelines (NHS 2022 [online]) which encompass the NHS brand design (see step 3 above). Along with this comes one major factor: consistency. Microsoft built its empire on sticking to the same design rules and if a website is to be useable (and thus useful) it needs to have its own rules which it maintains. Things to consider here include:

- Colour schemes
- Positioning of controls
- Consistency of controls (e.g. if you use breadcrumb controls, make sure they don't stop half-way through the forest)
- Consistency of content (are all pages designed for reading HTML or is there a mix of pdfs/word/excel? Are all documents designed for download consistent?)

All rules can be broken but there needs to be a reason for doing so. A game website may like to have items the user has to hunt for, a bus timetable not so much. In the world of commercial websites, a frustrated customer is a lost customer: Ryanair abandoned the captcha technology long before other websites as they realised how many potential customers saw one and flicked over to another airline's site (PhocusWire [online]).

11.3   HTML

All web pages are basically formatted text, the format being HTML (**Hyper-Text Markup Language**) which essentially consists of a pair of tags[4] – a start and an end – surrounding text that is controlled by those tags. The markup tells the browser how to display the text that is so denoted. The latest standard is HTML 5 and although 5.1 and 5.2 were released, work has coalesced around the Living Standard instead (WHATWG [online]). Of course, to use the tags the user's browser has to be compliant with the standard, over which the web developer has no control.

There are some exceptions to the "pair of tags" rule in that some tags do not require an end tag but using the end tag for these does not cause a problem to the browser.

11.3.1   STATIC HTML

There are many HTML tags, some of which are listed in Table 11.1.

TABLE 11.1
Some Common HTML Tags

\<title\>	Title of page – there can only be one[5] and without it the page will not validate as HTML
\<br\>	Line break
\<p\>	Paragraph break
\<a\>	Anchor – has two main uses:
\<a href="#notes"\> \<a href="mypage2.html/"\> \<a href="mypage2.html#notes" target="_blank"\>	Jumps to specified page / position on page in specified window. The "a" is an "anchor". The first form links to a point on the same page (see next tag), the second links to a new document and the third links to a specific position in a new document opened in a new window
\<a name="notes"\>	Specifies position on page
\<b\>	Defines bold text
\<big\>	Defines big text
\<em\>	Defines emphasised text
\<i\>	Defines italic text
\<small\>	Defines small text
\<strong\>	Defines strong text
\<sub\>	Defines subscripted text
\<sup\>	Defines superscripted text
\<pre\>	Defines preformatted text (i.e. does not strip white space, tabs etc.)
\<img\>	An image file
\<frameset cols="25%,75%"\>	Divides the window into a set of frames as specified. Note that any other markup besides that describing the frames is ignored and no \<body\> tags should be used. This tag is deprecated in HTML 5 and is therefore done in CSS
\<frame src="./mypage2.html"\>	Source for one of the frames
\<table border="5"\>	Table with specified border
\<caption\>	Caption for a table
\<tr\>	Table row
\<th\>	Table heading
\<td\> \<td rowspan="2"\> \<td colspan="2"\>	Table data, which may span multiple rows or columns as specified
\<ul\>	Unordered list
\<ol\>	Ordered list
\<li\>	List item
\<dl\>	Definition list
\<dt\>	Definition term
\<dd\>	Definition
\<menu type="context" id="popup-menu"\>	Pop-up context menu
\<menuitem\>	Item for menu (new in 5.1)
\<details\>	Show or hide a block of text by clicking on an element (new in 5.1 – previously done using Javascript)
\<summary\>	Element to click on to reveal details

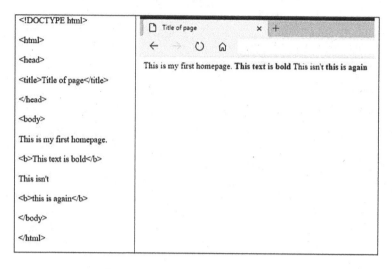

FIGURE 11.1   A basic HTML page.

HTML looks like the left-hand pane of Figure 11.1 and is rendered by the browser as the right-hand pane.

Points to note:

- All tags are in between < and > delimiters[6]
- All tags are lower case (although this is a convention and earlier versions of HTML specified upper case)
- All start tags have a corresponding end tag which is the same key word prefixed with /. Although most browsers will interpret <p> without a corresponding </p> it is contrary to the standard and the use of attributes means that the browser does not know where to stop applying them. In **Extensible HyperText Markup Language**[7] (XHTML) the <br> tag must be properly closed, like this: <br />.
- Tags are nested and do not interweave – i.e. you cannot have <a><b></a> </b>
- There are two main blocks to the page – a head and a body. The head is optional, the body is not.
- The <!DOCTYPE html> can often be omitted.
- Carriage returns in the source file are ignored, as is most white space unless specifically so marked.

HTML user agents such as Web browsers then parse this markup, turning it into a **Document Object Model** (DOM) tree, which is an in-memory representation of a document.

DOM trees contain different kinds of nodes, in particular a DocumentType node, Element nodes, Text nodes, Comment nodes and in some cases Processing Instruction nodes.

For example, this:

```
<!DOCTYPE html>
<html>
   <head>
      <title>Title of page</title>
   </head>
   <body>
      <h1>Example page</h1>
      <p>This is a <a href="demo.html">simple</a> example.</p>
      <!-- this is a comment -->
   </body>
</html>
```

would be turned into the DOM tree in Figure 11.2.

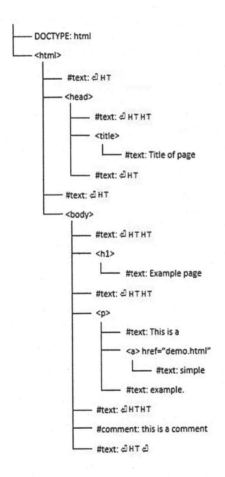

FIGURE 11.2   A DOM Tree.

The root element of this tree is the <html> element, which is always found at the root of HTML documents. It contains two elements, <head> and <body>, as well as a text node between them.

There are many more text nodes in the DOM tree than might be expected, because the source contains a number of tabs (represented here by "HT") and line breaks ("↵") that all become text nodes in the DOM. However, for historical reasons not all of the spaces and line breaks in the original markup appear in the DOM. In particular, all of the whitespace before the <head> start tag is dropped, and all of the whitespace after the <body> end tag is placed at the end of the <body>.

The <head> element contains a <title> element, which itself contains a text node with the text "Example page". Similarly, the <body> element contains an <h1> element, a <p> element and a comment.

Tags may be modified through the use of attributes (for which some are mandatory), for example <img src=image.jpg> is a mandatory attribute specifying the image to load, whereas <p align=right> is an optional attribute, producing a right-aligned paragraph. This attribute is deprecated in HTML 5,[8] for which the **Cascading Style Sheet** (CSS) syntax of <p style="text-align:right"> is now required.

<body> may also take attributes, for example <body bgcolor="#6C5BFF"> produces a page with a deep blue background and <body background="yellowpic.gif"> uses the image file as the background.

Note that due to the nesting of tags, images can be used as links, e.g.

> <a href="mypage2.html#notes" target="_blank"><img src="yellowpic.gif" alt="Alternative"></a>

Also note that lists can be nested, e.g.

> <ol>
> <li>Item 1</li>
> <li>Item 2</li>
> <ul>
> <li>Sublist</li>
> </ul>
> </ol>

As < and > are special characters in HTML (along with others) there are other codes for putting these characters onto the displayed page, as shown in Figure 11.3.

11.4   STYLE SHEETS – CSS

CSS has been a feature of HTML for many years[9] and allows the developer/designer to set an overall style for the site which is used on every page that imports the style sheet. The advantage of this is that alterations need only be made in one place to affect the entire site. Using styles clearly separates a document's presentation layout from its structure definition. The document structure is defined using HTML. The visual appearance of the page can then be changed using a style sheet.

Character	Entity Name	Entity Number
<	<	<
>	>	>
&	&	&
"	"	"
'	'	'
£	£	£

FIGURE 11.3   Codes for special characters in HTML.

CSS elements may appear in several places – they can appear in a single tag (and therefore last until the end tag), inside the <head> element of an HTML page, or in an external CSS file.

The name comes from the idea that all the styles will "cascade" into a new "virtual" style sheet by the following order of precedence (highest first):

> Inline Style (inside HTML element)
> Internal Style Sheet (inside the <head> tag)
> External Style Sheet
> Browser default

Therefore, an inline style has the highest priority, which means that it will override every style declared inside the <head> tag, in an external style sheet and in a browser default. Thus each cascades into the next to form the complete style for the page (see Table 11.2 for an example "Rendered" showing what the user will actually see).

The CSS syntax is made up of three parts: a selector, a property and a value and is written selector {property:value}. For Example, body {color:black} and p {font-family:"sans serif"}

TABLE 11.2
Style Sheet Precedence

Style Sheet	Colour	Text-Align	Font-Size
External	BlueViolet		
Internal	red	left	8pt
HTML			20pt
Rendered	red	left	20pt

The selector is normally the HTML tag to define, the property is the attribute to change and each property can take a value.

More than one property can be specified provided each property is separated with a semi-colon. As HTML disregards white space, placing each property on a new line makes them easier to read.

e.g.

h1 {text-align:center;color:Aquamarine;font-family: arial}

or

```
h2
{
text-align:left;
color:LightPink;
font-family:arial
}
```

Selectors can be grouped together with a comma. In this example all of the named header elements will be green (of which there are several shades), as noted in the comment which begins with /* and ends with */.

```
h3,h4,h5,h6
{
color: LimeGreen
}
/* turns specified headers green */
```

11.4.1 THE CLASS SELECTOR

The class selector allows different styles to be defined for the same type of HTML element. In the first example above the inline style has been used to display a particular heading but must be re-specified every time it is used. Using the class selector in the main CSS file enables reuse.

e.g. if the CSS file contains:

p.blue30marg {color:blue;margin-left: 30px}

then the html file simply needs:

<p class="blue30marg">

to result in the desired formatting.

If the tag name in the selector is omitted then the defined style may be used by all HTML elements. In this example (found in the relevant CSS file) all HTML elements with class="centre" will be centre aligned:

.centre {text-align: center}

11.4.2 APPLYING A STYLE SHEET

An internal style sheet should be used when a single document has a unique style.
Internal styles are defined in the head section by using the <style> tag, e.g.

```
<head>
<!-- The following is an internal style sheet.
It is used when a particular document has a unique style
-->
    <style type="text/css">
       hr {color: red}
       p {margin-left: 20px}
       /* This is a comment */
       body {background-image: url("./yellowpic.gif")}

       h1 {text-align:center;color:Aquamarine;font-family: arial}
       /* Placing each property on a new line makes them easier to read. */
       h2
       {
          text-align: left;
          color: LightPink;
          font-family: arial
       }
       /*Selectors can be grouped with a comma */

       h3,h4,h5,h6
       {
          color: BlueViolet
       }
    </style>
</head>
```

An inline style loses many of the advantages of style sheets by mixing content with
presentation. In order to use an inline style, a style definition can be applied to a particular tag using the style attribute. The style attribute can contain any CSS property.
This example shows how to change the colour and the left margin of a paragraph:

```
<p style="color: blue; margin-left: 30px">
This is a paragraph which has been displayed using an inline style.<br>
As you can see it is blue and has a bigger left margin...
</p>
```

An external style sheet is a text file with a .css extension and allows a style to be
applied to multiple pages. With an external style sheet, the look of an entire website
can be altered by changing this one file. Each page must link to the style sheet using
the <link> tag inside the head section, for example:

```
<head>
<!-- The following is an external style sheet -->
```

```
<link rel="style sheet" type="text/css" href="./css2.css">
<!-- The following is an internal style sheet -->
<style type="text/css">
h3 {color: red;text-align: left;font-size: 8pt}
</style>
</head>
```

11.4.3 MULTIPLE STYLE SHEETS

If some properties have been set for the same selector in different style sheets, the values will be inherited from the more specific style sheet.

For example, if the external style sheet css3.css has the property color: BlueViolet for the h3 selector and an internal style sheet has the properties h3 {color: red;text-align: left;font-size: 8pt} for the h3 selector and the html file then has the inline style "font-size: 20pt" for the h3 tag, then the final style for h3 will be color: red;text-align: left;font-size: 20pt following the order of precedence given in Section 11.4.

11.5 DYNAMIC HTML – FORMS

So far we have considered the data to be presented – we will now look at returning some. A form consists of a block of text demarked by the <form> tag. Within these tags lie the form elements, which are usually named (so we can use them), examples of which are shown in Table 11.3.

TABLE 11.3
HTML Form Elements

`<input type="text" name="firstname">`	Input of type text
`<input type="password" name="password">`	Input of type password – the input text is obscured
`<select name="Month">`	A single-element drop-down
`<option value="January">January`	A drop-down option with the value returned if selected, followed by the text to display. May is pre-selected in this example
`<option value="May" selected>May`	
`<select name="Month" size="6" multiple>`	A drop-down with six elements displayed at once and from which multiple values can be selected
`<textarea rows="10" cols="30">`	A larger text box. Text between the start and end tags is placed into the textarea initially (and this can include other HTML tags), but can be selected and changed.
`<input type="button" value="Click here!" onClick="this.form.field1. value='OK';">`	A button with an action to perform once clicked
`<fieldset>`	Groups a set of fields together, drawing a border around them
`<legend>`	Legend for a fieldset
`<input type="radio" name="response" value="male">`	A radio button, with group name and value to return if selected
`<input type="checkbox" name="bike">`	A checkbox

A form will usually have two main buttons: Submit and Reset. When the user clicks on the "Submit" button, the content of the form is sent to another file. The form's action attribute defines the name of the file to send the content to. The file defined in the action attribute usually does something with the received input.[10]

e.g.

```
<form name="input" action="http://www.w3schools.com/html/html_form_
action.asp" method="get">
<input type="reset" value="Clear form">
<input type="submit" value="Submit form">
</form>
```

Entering some details into this form and clicking the "Submit" button will send the input to a page called "http://www.w3schools.com/html/html_form_action.asp". That page will[11] display the received input.

Entering some details into the form and clicking the "reset" button will re-initialise the form and all data entered will be lost.

Information can be passed between web pages in this way. For example:

```
<form name=form1 action=page2.html>
    <input type=text name=box1>
    <input type=submit value="what does this do?">
</form>
```

If "hello" is entered into the text field and then the submit button is clicked, the page page2.html will be loaded, but the actual address used is:

```
page2.html?box1=hello
```

On page2.html, this script will display the query string:

```
<body>
    <script>
    var querystring=window.location.search.slice(1)
    document.write(querystring)
    </script>
</body>
```

11.6   DYNAMIC HTML – JAVASCRIPT

A common way of producing dynamic (i.e. non-static) effects on a web page is to use JavaScript. The button example in Table 11.3 uses JavaScript to display a message when the button is clicked but (obviously) far more complex effects are possible.

JavaScript may reside in one of three different places:

1. In the body of the web page (i.e. between the <body> tags), where the script is executed as soon as it is encountered.[12]

2. In the head of the web page (i.e. between the <head> tags), where the script is executed only when called (e.g. in response to a button press).

3. In an external .js file. This file must be referenced in the <head> portion of the page.

We will now look at some examples.

```
<html>
<body>
    <script type="text/javascript">
    <!--
    document.write("Hello World!")
    //-->
    </script>
</body>
</html>
```

This simple example writes into the web page as soon as it is encountered. The comment lines mean that browsers that are unable to interpret JavaScript will not display the code. Note that the two forward slashes at the "end of comment" line (//) is the JavaScript comment symbol. This prevents JavaScript from executing the --> tag.

```
<html>
<head>
    <script type="text/javascript">
    function sum(a,b)
    {
    A=a+b
    return A
    }
    function diff(a,b)
    {
    A=a
    A-=b
    return A
    }
    </script>
</head>
<body>
    3+2=
    <script type="text/javascript">
    document.write(sum(3,2))
    </script>
    <p>
    3-2=
    <script type="text/javascript">
    document.write(diff(3,2))
```

```
    </script>
  </body>
</html>
```

This example uses two functions pre-written into the <head> section and are called from scripts in the <body>. Note that variables are case-sensitive, so A and a are different.

JavaScript contains the usual programming constructs, such as if...else, for... next, do...while, switch() and case(). There are also a set of string handling functions such as substr() and substring() (which do different things so be sure to get the right one[13]).

Another common use of JavaScript is to refresh the screen:

```
<html>
<head>
  <script type="text/JavaScript">
  <!--
  function Refresh(timeout) {
  setTimeout("location.reload(true);",timeout);
  }
  // -->
  </script>
</head>
<body onload="JavaScript:Refresh(10000);">
  <script type="text/javascript">
    var date=new Date()
    var time=date.getSeconds()
    document.write("<b>time passing... <b>")
    document.write(time)
  </script>
</body>
</html>
```

This is, of course, a poor way to display this information as it keeps requesting data from the server (but does illustrate the use of the command). A better way would be as follows:

```
<html>
<body>
  Time passing...
  <p id='demo'></p>
  <script type="text/JavaScript">
    <!--
    // Update the timer every 10 seconds
    var x = setInterval(function() {
    // Get today's date and time
    var now = new Date();
```

```
        // Output the result in an element with id="demo"
        document.getElementById("demo").innerHTML = now.getSeconds() +
            " and counting";
        }, 10000);
        -->
    </script>
</body>
</html>
```

Note that with this method, the timer doesn't appear on screen for the first iteration (10 seconds in this case).

An external JavaScript file is referenced as follows:

```
<head>
<script src=calc.js></script>
</head>
```

Within the file are various functions, e.g.

```
function invert(){
    var f = self.document.forms[0];
    x = f.box1.value;
    if (x==0){
        alert("Please enter a number that isn't zero.");
        f.box1.focus();
        return;
    }
    else {
        f.box2.value=(1/x)
    }
    return
}
```

This function is called as

```
<input type="button" value="Invert" onClick="self.invert();">
```

Note the use of == for a test of equality.[14] There is a very useful JavaScript variable validator (to see whether your proposed variable name is valid) at https://mothereff. in/js-variables.

11.7   DYNAMIC HTML – CGI

Remember that forms have actions, e.g. <form name=form1 action=page2.html>.

Once the submit button is clicked, the form's action takes place. This will generally be to send the contents of the form to another program to process and then pass the results back, often as a new page. For this we use **Common Gateway Interface** (CGI).

Effectively, CGI programs reside in a specific directory (www/cgi-bin, for most servers, /usr/lib/cgi-bin on a Raspberry Pi) and send web page(s) to the browser when run. The CGI program takes input from the browser, reads files, writes to files (for example to record the number of "hits") and so on. As HTML is merely a text stream, the program will often use PRINT commands (such as "echo") to generate the result(s).

Some key parts are:

- The program must print Content-Type: text/html and a blank line before anything else (although that does vary from browser to browser).
- A web page that uses forms to get input codes them up before sending them.
- There are two ways of sending this coded input to the CGI program (as a shell variable or as keyboard input.)
- The CGI program must be executable (e.g. chmod +x filename)

A simple CGI example that displays the input string is as follows:
In the html:

```
<form name=form2 action=cgi-bin/formtest.cgi>
<input type=text name=box2>
<input type=submit value="does the CGI work?">
</form>
```

The formtest.cgi file (note that this must be made executable):

```
#!/bin/bash
echo -e "Content-type: text/html\n\n"
echo "<html><body>"
echo "<h1>Form test results:</h1>"
echo $QUERY_STRING
echo "</body></html>"
```

11.8   SERVER- AND CLIENT-SIDE ARCHITECTURE

With so many methods available for delivering computerised healthcare, one major question is which one to use. Here we examine one such issue: the location of the executing software.

The term "executing software" is used in order to distinguish between where the main software resides and where the data reside. Indeed, it is possible to deploy systems where the data, the software and its execution are all in separate places.

The key difference is in whether the code executes on the client device or on the server. If the former, then a loss of connection is no problem, there is a lack of contention for processing power and the code can use data from the host device without it being transmitted. Additionally, code may be re-used.

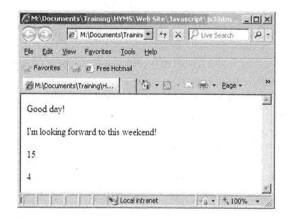

FIGURE 11.4 Web page with a time and date dependant greeting.

If the code executes on the server, then the resulting download may be smaller, confidential information residing on the server may be used and not transmitted and the server is probably a much more powerful processing device than the client. Additionally, especially for scripting, the workings of the algorithm are hidden from the end-user.

As an example, let us consider a web page with a time and date-dependant greeting, as shown in Figure 11.4.

JavaScript is client-side processing, where the web page is downloaded and the final display is generated on the client, whereas C++ is an example of a server-side technique, where the dynamic content is determined at the server and only the correct information for display is generated. The sample code for both these approaches is available at https://www.routledge.com/Introduction-to-Bioinformatics-and-Clinical-Scientific-Computing/Ganney/p/book/9781032324135.

A mixture of techniques could, of course, be deployed, so that a system may take advantage of both client- and server-side processing and minimise the disadvantages. One such might be a patient monitoring system for home use: the patient's medical history is used to generate a series of criteria, without disclosing the history. These are sent to the patient's browser and the patient enters some fresh figures. These are tested using the downloaded criteria and a result given (e.g. "normal" or "contact clinic") without transmitting the actual results.

11.9   SERVER FILES

By default, a URL that omits a page name will be served a file called index.html or index.htm, depending on the server's operating system. It is generally good practice to have one of these as the welcome page, from which all others are referenced.

Note that unless explicitly referenced, all hyperlinks will be to files in the same folder as the current page. This is especially important when attempting to reference across subfolders.

11.10   LIMITING ACCESS

On a Unix/Linux machine,[15] access can be restricted through the use of a .htaccess[16] file. This is a plain-text file with permissions of 644 or (rw-r--r--), which makes the file usable by the webserver, but prevents it from being read by a web browser. It resides in the directory for which it is controlling access.[17] The following five lines in this file will only allow access to the pages to users from the university of Clinical Engineering:

```
<Limit GET POST>
order deny,allow
deny from all
allow from .clineng.ac.uk
</Limit>
```

Any other users attempting to access these pages will receive a "403 Forbidden" error.

The .htaccess file can control the following:

- Mod_rewrite: this designates and alters how URLs and web pages on sites are displayed to the users. The following code redirects visitors to the www. version of the website and also redirects the index.html pages to the correct canonical address:

```
RewriteEngine on
RewriteCond[18] %{HTTP_HOST} ^clineng\.ac.uk$ [NC]
RewriteRule ^(.*)$ http://www.clineng.ac.uk/$1 [R=301,L]
RewriteCond %{THE_REQUEST} ^.*/index.html
RewriteRule ^(.*)index.html$ http://www.clineng.ac.uk/$1 [R=301,L]
```

- Authentication: this forces the use of a password to access parts of the site. These passwords reside in the .htpasswd file.[19] An example of such a section would be:

```
AuthUserFile /usr/local/username/safedirectory/.htpasswd
AuthGroupFile /dev/null[20]
AuthName "Please Enter Password"
AuthType Basic
Require valid-user
```

- Custom error pages: these are used in place of the defaults in the web server, in order to add a more "friendly" approach to errors. An example would be:

```
ErrorDocument 404 /friendly404.html
```

- Mime types: when the site features some application files that the server was not set up to deliver, MIME types can be added to the Apache server in the .htaccess file such as:

  AddType audio/mp4a-latm .m4a[21]

- **Server Side Includes** (SSI): One of the most common uses of SSI is to update a large number of pages with some specific data, without having to update each page individually (for example, in order to change contact information at the bottom of a page). These lines enable SSI (the first line indicates that .shtml files are valid and the second forces the server to parse them for SSI[22] commands):

  AddType text/html .shtml
  AddHandler server-parsed .shtml

Depending on the size of the web site the .htaccess file may slow down performance, but this effect is likely to be insignificant.

11.11   INTERFACING WITH A DATABASE

Databases are explored in depth in Chapter 2, so here we will only cover the connection to and a simple extraction of data from the database.

The method of connection will depend on the choice of database. For example, postgres has the command-line interface psql which can be used with the –c switch to run the command directly and then return. Hence it can easily be used in simple scripting, for example to list all patients:

  psql –c "select * from patient;"

This could be placed in a shell script called via CGI, for example.

In MySQL, the following code will select all records from the "patient" table and display them in alphabetic order of surname:

```
<?php
// Connect to database
$db = mysqli_connect('localhost','dbuser','password','appts')
or die('Error connecting to MySQL server.');
?>
<html>
<head>
</head>
<body>
<h1>PHP connect to MySQL</h1>
<h2> list of all patients</h2>
<?php
// Create query and send to database
$query = "SELECT * FROM patient ORDER BY surname";
```

```
mysqli_query($db, $query) or die('Error querying database.');
// Display results
$result = mysqli_query($db, $query);
$num = mysqli_num_rows($result);
$row = mysqli_fetch_array($result);

while ($row = mysqli_fetch_array($result)) {
echo $row['forename'] . ' ' . $row['surname'] . '<br />';
}
echo "$num listed";

// Tidy up
mysqli_close($db);
?>

</body>
</html>
```

Alternatively, the mysql command line tool could also be used (see https://dev.mysql.com/doc/refman/5.7/en/mysql.html).

The full gamut of SQL is thus available (explored further in chapter 3, "SQL").

11.12 PRIVACY AND SECURITY

In 2014, the **Open Web Application Security Project** (OWASP) undertook a survey of the top ten privacy and security risks. In 2021 these were updated to be (OWASP 2021 [online]):

1. Web application vulnerabilities
2. Operator-sided data leakage
3. Insufficient data breach response
4. Consent on everything
5. Non-transparent policies, terms and conditions
6. Insufficient deletion of user data
7. Insufficient data quality
8. Missing or insufficient session expiration
9. Inability of users to access and modify data
10. Collection of data not required for the user-consented purpose

Many of these are subject to legislation (e.g. 2, 3, 4, 6, 9 and 10 under the GDPR; 5 under Caldecott) so the penalties can be more than just embarrassing news stories. The scope for this survey was real life risks for the user (data subject) and provider (data owner) but excluded self-protection for users.[23]

Of particular relevance for the NHS, the report contained warnings such as:

- *"Internal procedures or staff are often a reason for data leakage*:
  - *Poor access management*
  - *Lack of awareness*
  - *Unnecessary copies of personal data*

- *Weak anonymization of personal data"* noting that anonoymisation can go wrong and that *"Location data, browsing behavior or device configuration can be used to identify people"*
- *"Automatic session timeout and a highly visible logout button is security state-of-the-art"* (OWASP 2015 [online]). It appears that Google, Amazon and Facebook do not implement this.

The report suggested these countermeasures:

- *"Raise Awareness among*:
  - *Product / Application Designers (business), as they decide about functionality that affects privacy*
  - *Developers / IT, as they sometimes have the choice to implement privacy friendly applications*
  - *Data Protection / Legal, as personal information is mainly processed in IT systems and IT has to be considered when implementing privacy programs*
- *Implement processes*
  - *That consider privacy in all development stages from requirements analysis to implementation (preventive)*
  - *To audit privacy measures in web applications (detective)*
- *Ask simple questions*
  - *Did you consider privacy when designing the application?*
  - *Did you address the OWASP Top 10 Privacy Risks?*
- *How are privacy incidents handled?*
  - *How is data deleted?*
  - *How do you avoid vulnerabilities in the application?*
  - *Etc.*
- *Technology examples*
  - *Avoid Data Leakage*
    - *Restrictive Access Management*
    - *Awareness campaigns*
    - *Strong anonymization techniques*
    - *Data Leakage Prevention (DLP) solutions*
  - *Improve session timeout*
    - *Configure to automatically logout after X hours / days*
    - *Obvious logout button*
    - *Educate users*
  - *Ideas for better transparency in terms & conditions*
    - *Text analyzer: readability-score.com*
    - *HTTPA: http with accountability developed by MIT*
  - *Share data with third party on click only"*

(OWASP 2015 [online])

One useful mechanism in implementing privacy and security is the web session, which we will now examine.

11.12.1 WEB SESSIONS

A web session is a data structure that an application uses to store temporary data that exists only while the user is interacting with the application. It is also specific to the user. It can therefore be used to simplify navigation around a site (remembering options selected in previous searches, for example) but also to assist in security.

A session is a key-value pair data structure, similar to a hashtable where each user has a hashkey in which their data is stored. This hashkey is the session ID.

A session data structure is shown in Figure 11.5.

Users are only able to access their own session. This may be stored on the client (by the browser, most likely in a cookie) or on the server. Even for server-managed sessions, the session ID is likely to be in a cookie, provided by the server when the web site is opened and (theoretically) destroyed when the session is over (i.e. when the browser is closed or the site left). **Java Server Pages** (JSP) will send a JSESSIONID, **Active Server Pages** (ASP) will send ASPSESSIONID and PHP will send PHPSESSID. This session ID will then be sent to the server every time a new page request is made, checked by the server to see that it is valid, and then the page request is accessed. This gets round one of the major problems with HTML: as HTML is stateless, it cannot preserve the fact that a user has already logged in and would therefore require login details on every new page.

The web page http://httpd.apache.org/docs/current/mod/mod_session.html explains how to implement sessions on an Apache server.

Session id	
'65g08e2d5'	{user_id: 42 user_name: paul }
'd58d2h46a'	{user_id: 777 user_name: arthur }
'a95ed3481'	{user_id: 999 user_name: marvin }

FIGURE 11.5   A session data structure.

A simple (and not actually very good) way of doing this is also via a CGI script, for example:

```
#!/bin/bash
echo "Content-Type: text/plain"
HTTP_SESSION="key1=foo&key2=&key3=bar"
echo
env
echo "Specifically for the session, HTTP_SESSION=:"
echo $HTTP_SESSION
```

11.12.2 Cookies

Cookies were invented by Netscape to give 'memory' to web servers and browsers. As we have noted, the HTTP protocol is stateless, meaning that once the server has sent a page to a browser requesting it, nothing is retained. Revisiting a web page is therefore treated as though this were the first visit, every time. This can lead to user frustration as information such as access to protected pages and user preferences need to be re-entered. Cookies were invented to solve this problem. There are other ways to solve it (and for protected access a server-side solution is preferable), but cookies are easy to maintain and very versatile.

A cookie is simply a small piece of text (originally a file) stored on the requesting machine. It contains:

- A name-value pair containing the data.
- An expiry date after which it is no longer valid. A cookie without an expiry date is deleted when the browser is closed. This expiry date should be in UTC (Greenwich) time.
- The domain and path of the server. An unspecified domain becomes the domain of the page that set the cookie. The purpose of the domain is to allow cookies to cross sub-domains. A cookie with domain www.bioinformatics. org will not be read by search.bioinformatics.org but setting the domain to bioinformatics.org will enable the cookie to be read by both sub-domains. A cookie cannot be set to a domain the server is not in, e.g. it is not possible to make the domain www.microsoft.com. Only bioinformatics.org is allowed, in this case. The path allows the specification of the folder where the cookie is active. Thus, a cookie that should only be sent to pages in cgi-bin has the path set to /cgi-bin. Usually the path is set to /, which means the cookie is valid throughout the entire domain.

A request for a page from a server to which a cookie should be sent causes the cookie to be added to the HTTP header. Server-side programs can then read the information and act upon it, for example by deciding whether the requestor has the right to view the page or that links should be displayed as yellow on a green background. Cookies can also be read by JavaScript, for client-side processing and mostly store user preferences in this case.

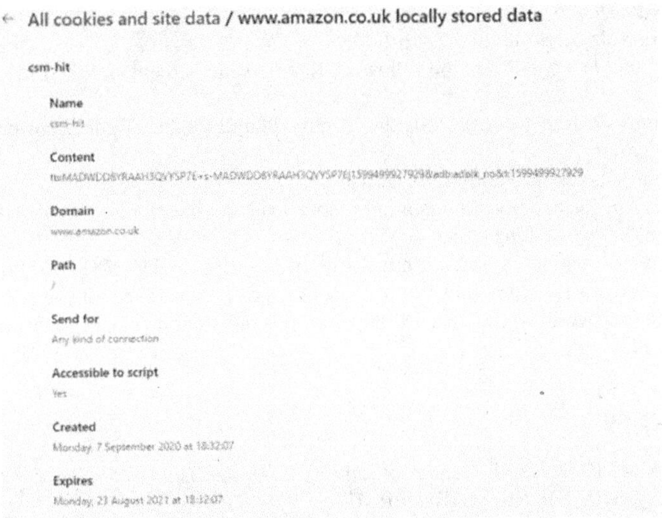

← All cookies and site data / www.amazon.co.uk locally stored data

csm-hit

Name
csm-hit

Content
fte:MADWDD8YRAAH3QVYSP7E+s-MADWDD8YRAAH3QVYSP7E|1599489927929&adbradialk_no&it:1599489927929

Domain
www.amazon.co.uk

Path
/

Send for
Any kind of connection

Accessible to script
Yes

Created
Monday, 7 September 2020 at 18:32:07

Expires
Monday, 23 August 2021 at 18:32:07

FIGURE 11.6   A cookie.

An example page is available at https://www.routledge.com/Introduction-to-Bioinformatics-and-Clinical-Scientific-Computing/Ganney/p/book/9781032324135.

Cookies are no longer stored in files but they can still be viewed via the browser. Not that they're very exciting, as Figure 11.6 demonstrates.

NOTES

1 The section "Server- and client-side architecture" is reproduced from Ganney et al. 2022.
2 In an NHS setting this is often all the same person, of course.
3 Assuming the web server won't provide that information – but access to those tools may be limited and may require an "admin" section to the site to be built.
4 In HTML and other markup languages, a tag is an element that changes the look of content or performs an action.
5 So why not call it "Highlander"?
6 A delimiter is something that marks the start and end. In this case, the character < marks the start and the character > marks the end, thereby delimiting the tag.
7 Part of the family of XML markup languages, XHTML extends HTML – the differences are discussed here: https://www.w3schools.com/Html/html_xhtml.asp.
8 For a description of deprecated tags, see https://www.w3docs.com/learn-html/deprecated -html-tags.html.
9 Since 1996 in fact.
10 If not, then the design is questionable to say the least.
11 Or did, as this particular example no longer exists.
12 There are exceptions such as the OnClick attribute for a button.
13 substr(x,y) starts at character x and extracts y characters. substring(x,y) extracts from characters x to y.
14 Other comparison operators can be found at https://www.w3schools.com/js/js_comparisons.asp
15 For Apache specifically, but also other servers implement similar technology.

16 Hypertext access.

17 It also controls access to all the sub-folders.

18 RewriteCond is a condition, the following RewriteRule is then applied if the condition is true.

19 Instructions on how to create this are at https://httpd.apache.org/docs/current/programs/htpasswd.html.

20 Not required unless group files are active.

21 This tells Apache to deliver a mp4a-latm file when it is asked for a .m4a file which it (until this point) knows nothing about.

22 For more information on SSI, see http://www.w3.org/Jigsaw/Doc/User/SSI.html.

23 They looked at risks from the point of view of the data owners' actions (e.g. a data breach revealing passwords). The risk of the user posting their personal data online wasn't considered.

REFERENCES

Ganney P, Maw P, White M, Ganney R, ed. 2022, *Modernising Scientific Careers the ICT Competencies*, 7th edition, Tenerife: ESL.

NHS 2022, [online]. Available: https://www.england.nhs.uk/nhsidentity/ [Accessed 19/04/22].

OWASP 2021, [online]. Available: https://owasp.org/www-project-top-10-privacy-risks/ [Accessed 19/04/22].

OWASP 2015, [online]. Available: https://www.owasp.org/images/c/c3/Top10PrivacyRisks_IAPP_Summit_2015.pdf [Accessed 19/04/22].

PhocusWire n.d., [online]. Available: https://www.phocuswire.com/Ryanair-abandons-Captcha-online-security-for-simpler-more-friendly-system [Accessed 19/04/22].

Smashing Magazine n.d., [online]. Available: https://www.smashingmagazine.com/2008/11/strategic-design-6-steps-for-building-successful-websites/ [Accessed 19/04/22].

WHATWG n.d., [online]. Available: https://html.spec.whatwg.org/multipage/introduction.html [Accessed 19/04/22].

12 Data Exchange[1]

12.1 INTRODUCTION

Data lies at the heart of any computerised healthcare system. We have examined several protocols and techniques for unlocking the information contained in it, but in order for that information to be successfully and efficiently utilised it must be exchanged with other systems – automatically and without re-keying. Interfaces are covered in section 13.3, Chapter 13 so in this chapter we focus on two of the most common protocols in use today: DICOM and HL7. Before we do this, though, we must first ensure that the data we exchange arrives correctly.

12.2 PARITY AND HAMMING CODES

One of the most important sets of data exchange protocols deals with the security of the data exchange: ensuring that what is received is what was transmitted.

Checksums and parity ensure that the message received is the one sent (to a level of confidence) and we will expand on the concepts here.

Simple parity checks are a single bit added to the end of a binary number, with a value such that the bits either add up to an even number (even parity) or an odd number (odd parity) depending on the protocol in use. However, this solution will not detect two errors[2] as they would cancel each other out.

To detect (and even correct) errors, we must implement Hamming codes. These were developed by Richard Hamming in 1950 (Hamming 1950). His classic [7,4] code encodes four data bits into seven bits by adding three parity bits. It can detect and correct single-bit errors. With the addition of an overall parity bit, it can also detect (but not correct) double-bit errors. There is an excellent description of how these work at https://en.wikipedia.org/wiki/Hamming_code but for this chapter a simple example will suffice.[3]

We will calculate a Hamming codeword that can correct 1-bit errors in the ASCII code for a **line feed**, (LF), 0x0A.[4]

We are going to calculate a codeword that is capable of correcting all single-bit errors in an 8-bit data element. In the codeword, there are m data bits and r redundant (check) bits, giving a total of n codeword bits.

$$n = m + r$$

The methodology of the solution is to:

1. Decide on the number of bits in the codeword
2. Determine the bit positions of the check bits
3. Determine which parity bits check which positions
4. Calculate the values of the parity bits

DOI: 10.1201/9781003316244-12

12.2.1   DECIDE ON THE NUMBER OF BITS IN THE CODEWORD

Each valid codeword of n bits contains m correct data bits. For each correct m-bit data entity, there are n bits that can be changed to give an incorrect codeword (as we are only dealing with a 1-bit error). Thus, the total number of codewords corresponding to a valid data entity is $n + 1$ (n incorrect plus 1 correct codeword). As there are 2^m valid data patterns,[5] the total number of codewords is $(n+1)2^m$. In an n-bit codeword, the possible number of patterns is 2^n, and this limits the number of correct + incorrect codes that can exist. Thus

$$\left(n+1\right)2^m \leq 2^n$$

and, since $n = m + r$,

$$\left(m+r+1\right)2^m \leq 2^{m+r}$$

so

$$\left(m+r+1\right) \leq 2^r$$

The least number of redundant bits (r) that satisfies this inequality when $m = 8$ is $r = 4$ bits. Thus, we have a 12-bit codeword. Bits 1, 2, 4 and 8 will be the check bits.[6]

12.2.2   DETERMINE THE BIT POSITIONS OF THE CHECK BITS

Using our example of 0x0A, converted to binary, the codeword will therefore be of the form cc0c 000c 1010 (where c is a check bit) and we shall use even parity. The check bit positions are the powers of 2 (positions 1,2,4,8...).

Codeword	c	c	0	c	0	0	0	c	1	0	1	0
Bit position	1	2	3	4	5	6	7	8	9	10	11	12

12.2.3   DETERMINE WHICH PARITY BITS CHECK WHICH POSITIONS

The bit positions covered by each parity bit (step 3) can be calculated by writing each bit position as a sum of the powers of 2:

1 = 1
2 = 2
3 = 1 + 2
4 = 4
5 = 1 + 4
6 = 2 + 4
7 = 1 + 2 + 4

$$8 = 8$$
$$9 = 1 + 8$$
$$10 = 2 + 8$$
$$11 = 1 + 2 + 8$$
$$12 = 4 + 8$$

12.2.4   Calculate the Values of the Parity Bits

Thus,

> Check bit 1 governs positions 1, 3, 5, 7, 9, 11[7]: Value = 0[8] (0 0 0 1 1)[9]
> Check bit 2 governs positions 2, 3, 6, 7, 10, 11: Value = 1 (0 0 0 0 1)
> Check bit 4 governs positions 4, 5, 6, 7, 12 = 0 (0 0 0 0)
> Check bit 8 governs positions 8, 9, 10, 11, 12 = 0 (1 0 1 0)

The complete codeword is therefore 0100 0000 1010.[10]

12.2.5   Using the Codeword to Correct an Error

A 1-bit error in this codeword can be corrected as follows:

First, calculate the parity bits. If all are correct, there is either no error, or errors in more than 1 bit. For a single-bit error (which is what we have calculated the code to detect), then if the parity bits show an error, add up all the erroneous parity bits, counting 1 for bit 1, 2 for bit 2, 4 for bit 4 and 8 for bit 8. The result gives the position of the erroneous bit, which can be complemented to give the correction.

For our example, let us assume the codeword is corrupted in bit 3 to give 0110 0000 1010.

Check parity (using XOR[11]):

> Bit 1[12]: B1 $\oplus$ B3 $\oplus$ B5 $\oplus$ B7 $\oplus$ B9 $\oplus$ B11 = 0 $\oplus$ 1 $\oplus$ 0 $\oplus$ 0 $\oplus$ 1 $\oplus$ 1 = 1
> Bit 2: B2 $\oplus$ B3 $\oplus$ B6 $\oplus$ B7 $\oplus$ B10 $\oplus$ B11 = 1 $\oplus$ 1 $\oplus$ 0 $\oplus$ 0 $\oplus$ 0 $\oplus$ 1 = 1
> Bit 4: B4 $\oplus$ B5 $\oplus$ B6 $\oplus$ B7 $\oplus$ B12 = 0 $\oplus$ 0 $\oplus$ 0 $\oplus$ 0 $\oplus$ 0 = 0
> Bit 8: B8 $\oplus$ B9 $\oplus$ B10 $\oplus$ B11 $\oplus$ B12 = 0 $\oplus$ 1 $\oplus$ 0 $\oplus$ 1 $\oplus$ 0 = 0

Because we are using even parity we can see that the error is in bit position 1 + 2 = 3, and bit 3 can be inverted to give the correct codeword:

$$010000001010$$

and the message has been correctly delivered.

Hamming described the problem as being one of "*packing the maximum number of points in a unit n-dimensional cube*". This may sound esoteric but it is actually (as Matt Parker points out in his book "*Things to make and do in the fourth dimension*" (Parker 2015)) the system used to solve a sudoku. A sudoku is just a grid of numbers with three mathematical overlapping patterns: one for the rows, one for the columns and one for the sections of the grid. Because of these patterns it is possible to calculate what the missing numbers are and thus correct it. This is actually very similar to the method of error correction used in mobile phone text messaging: the message is converted into a grid, some patterns are overlaid, and the message is sent, received corrupted and fixed, all without the user knowing.

It is left as an exercise to discover whether or not a corruption of a Hamming bit is also detected and repaired in this way.

12.3 JSON AND XML

Two very popular data exchange protocols are **JavaScript Object Notation** (JSON) and **Extensible Markup Language** (XML), especially for webservices. They are both human-readable and an example might be in sending credentials (username & password) to Facebook for authentication.

In XML the code is:

```
<credentials>
    <username>Ganney</username>
    <password>Rachel</password>13
</credentials>
```

And in JSON it is:

```
{"credentials":[
    {"username":"Ganney", "password":"Rachel"}
]}
```

The data that is transferred is the same in both cases – only the format is different.

XML is derived from **Standard Generalized Markup Language** (SGML). Compared to SGML, XML is quite simple. **HyperText Markup Language** (HTML) is even simpler. XML is still very verbose and is thus not well suited to many programming languages.

Comparing the two:

- Both JSON and XML are "self-describing" (i.e. human readable).
- Both JSON and XML are hierarchical (i.e. they have values within values).
- Both JSON and XML can be parsed and used by many programming languages.
- Both JSON and XML can be fetched with an XMLHttpRequest command.
- JSON is much simpler than XML, having a much smaller grammar and it maps more directly onto the data structures used in modern programming languages.
- JSON has a simpler structure leading to simpler processing.
- Although XML is extensible (and JSON isn't), JSON is not a document markup language, so there is no requirement to define new tags or attributes to represent data in it.

- JSON is easier for humans to read and write than XML. It is therefore simpler to program readers and writers for it.
- The languages JavaScript and Python have the JSON notation built into the programming language. There is a wide range of reusable software available to programmers to handle XML but this is additional software.
- XML separates the presentation of data from the structure of that data by translating the structure into a document structure, which may add complication. JSON structures are based on arrays and records which map more directly to data in programs.
- JSON is a better data exchange format. XML is a better document exchange format.
- JSON does not provide any display capabilities as it is not a document markup language, but XML can provide many views of the same data.
- XML is document-oriented whereas JSON is data-oriented, meaning that JSON can be mapped more easily to object-oriented systems.
- JSON doesn't use an end tag.
- JSON is shorter, therefore quicker to read and write.
- JSON can use arrays.

Probably the biggest difference is that XML has to be parsed with an externally linked XML parser, whereas JSON can be parsed by a standard JavaScript function. A good example of the use of JSON is in the NSHCS online curriculum (National School of Healthcare Science 2022 [online]).

The JSON syntax is a subset of the JavaScript syntax:

- Data is in name/value pairs
- Data is separated by commas
- Curly braces hold objects
- Square brackets hold arrays

The file type for JSON files is ".json" and the MIME type for JSON text is "application/json".

A common use of JSON is to exchange data to/from a web server, which is always a string. It can be parsed with JSON.parse(), converting the data into a JavaScript object. For example:

The server sends this text:

'{ "name":"Paul", "age":62, "city":"Liverpool"}'

which can be stored in a variable, say jText, and parsed using

```
var obj = JSON.parse(jText);
```

This object can then be used in the same way as any JavaScript object:

```
<p id="NameAndAge"></p>
<script>
```

```
document.getElementById("NameAndAge").innerHTML = obj.name + ","
    + obj.age;
</script>
```

Extracting the name and age as "Paul, 62" for use elsewhere in the script or web page.

12.4 DICOM

The rise of digital imaging gave rise to a host of proprietary methods of storing the data. Naturally, none of the systems could read/write to other systems. It was to overcome such problems (and more besides) that ACR and NEMA[14] proposed a standard. ACR/NEMA 300 was published in 1985 and (after several revisions and additions) was re-named DICOM (**Digital Imaging and Communications in Medicine**) in 1993. DICOM specifies not only the file format but also a networks communication protocol (based on TCP/IP) and a set of services.

DICOM, as the name implies, is used extensively in digital imaging (although it is also used for textual information exchange, such as worklists and in specialised information, such as a Radiotherapy treatment plan[15]). Standardisation also addresses one additional problem: longevity. Data format standards ensure the data can still be read in many years' time, a key consideration for medical data.

The DICOM file format is based on data sets and embeds data such as the patient identifier into this, ensuring that the image cannot be separated from the patient to whom it belongs. A DICOM data object consists of multiple data elements, each of which is tagged in order to describe the data it contains, such as Name, Date of Examination, etc.[16] We will look at a tagged file format later in this chapter.

A DICOM data object contains only one image element, although this may in turn comprise several frames. A DICOM data element consists of the tag, an optional value representation (the values for which are defined as part of the standard), the length of the data and the data itself. When a DICOM data object is exported as a DICOM file, several of the key elements are formed into the header (although they do also exist within the object – they are purely copies), along with details of the generating application. This simplifies the import of objects as the entire object need not be read prior to storage.

As mentioned above, DICOM also defines services. These are again defined in the standard, but include such as:

- Query / Retrieve
- Storage Commit
- Worklist management
- Print
- Verification

As there have been multiple versions of the DICOM standard and no application is required to implement the full standard (for example, a service may not be applicable to it) the "DICOM conformance statement" is an essential part of any system, describing the parts of DICOM that it implements (and to which version). Just as "runs on electricity" does not fully explain how to connect up a device, neither does "DICOM compliant".

DICOM is very useful for the exchange of a large quantity of data, such as a worklist or an image. It is not so useful, though, for exchanging incremental changes within a database, for which HL7 is more commonly used – we will look at that later in this chapter.

The complete DICOM standard is available online (DICOM 2022 [online]) and effectively describes two things:

1. An information model for clearly describing the real-world concepts in medical imaging.
2. A technical implementation to turn this model into bytes for storage or network exchange.

We will now consider the DICOM information model in more detail, especially parts 3 ("Information Object Definitions") and 6 ("Data Dictionary") of the standard.

12.4.1 IMAGES AS DATA

We have already seen in Figure 1.3 and Figure 1.4 how an image is formed of a set of numerical values. There we generated the image using these values, but in medical imaging the image is formed by measurements made at multiple spatial locations,[17] the image then showing the spatial relationship between these values. As with the images in those figures, those locations are often (but not always) arranged in a rectangular grid. The volume of space around each location is usually termed a pixel (2 dimensions) or voxel (3 dimensions). As with those images there may be a single (scalar) measurement at a location (e.g. a monochrome photograph or CT density) or multiple (e.g. a colour photograph or MRI **Diffusion Tensor Imaging** (DTI)[18]). The measurements may come, depending on the modality, from a single point in space (e.g. voxels in a CT image – tomographic imaging) or may be aggregated along a line (e.g. pixels in a plain film x-ray – projection imaging).

In order to clinically use the data held within an image, additional information is required. Specific to the image creation are data such as the kind of measurement contained in the image (where it fits on the various axes), the parameters used (which will depend on modality: attenuation, the echo time, the flip angle etc.), the image's orientation and spatial scaling, the body part imaged (is this the left or right leg?), and the patient's position while the image was taken.

Additional information often accompanies the image in order to assist clinical use and retrospective audit, such as where and when the image was acquired, the patient's name, identifying numbers, date of birth and diagnosis, whether the image has been archived (and if so, where and when), who requested the image be created (i.e. ordered the scan), whether a radiological report has been authorised (and what it concluded).

This additional information is called metadata – the supporting details that allow meaningful clinical use of the image. The DICOM standard specifies not only how to store the image itself (an array of pixels or voxels) but also a wide variety of these metadata.

12.4.2   Information Entities

DICOM has a strongly structured, hierarchical model for representing imaging metadata. A data dump[19] of a DICOM file header[20] may look like a series of facts relating to the image (relating to the patient, scanner, imaging configuration,[21] and hospital imaging referral) without a specific order, but this list of facts is organised into a hierarchical structure. DICOM uses **Information Entities** (IEs), each representing a distinct real-world concept and arranged into a hierarchy – six of the most important ones (which we will consider further) are shown in Figure 12.1.

In this model, the right-hand column of IEs is the hierarchy usually shown by the study browser on a scanner or PACS:

- A *Patient* represents a person being imaged
- A *Study* represents a set of related images and data, often associated with a single scanning visit
- A *Series* represents a stack of images acquired together, with many properties in common, such as the slices of a 3D CT volume or a single T1-weighted multislice MRI acquisition
- An *Image* contains an array of pixels – more generally this is called an Instance, as it may be a non-image object such as a structured report or annotation

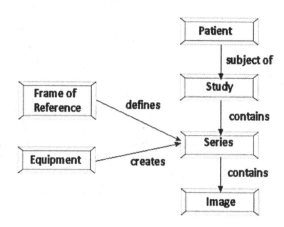

FIGURE 12.1   Information entities. Each arrow means "one or more".

Each series then has two other associated IEs, shown on the left-hand side of Figure 12.1:

- An item of *Equipment* is the system (often a scanner) creating the series
- A *Frame of Reference* represents a patient positioned in a single co-ordinate system: when several series share a frame of reference, pixels with matching co-ordinates will be at same spatial position within the patient.[22]

12.4.3 INFORMATION OBJECT DEFINITIONS

As IEs are abstract concepts, DICOM has an **Information Object Definition** (IOD) for each specific use case, such as MRI or CT images. An IOD is an object-oriented abstract data model used to specify information about Real-World Objects, providing a common view of the information to be exchanged. DICOM information objects are definitions of the information to be exchanged – effectively they are templates into which a new image is placed. Each image type, and therefore information object, has specific characteristics – an MRI image requires different descriptors to a CT image, for example.[23] These information objects are identified by unique identifiers, which are registered by NEMA.

An IOD does not represent a specific instance of a Real-World Object, but rather a class of Real-World Objects which share the same properties. An IOD used to generally represent a single class of Real-World Objects is called a Normalised Information Object (containing the attributes of certain individual IEs without context). An IOD which includes information about related Real-World Objects (including content from both the specific IE of interest (the Image) and the other IEs which provide context) is called a Composite Information Object, defining a collection of attributes across the whole hierarchy of IEs. For example, devices handling an MR Image will normally also need details of the patient, study, equipment etc. This is the type of IOD most often encountered in imaging data and is illustrated by the MR Imaging IOD in Figure 12.2. Note that Attributes are grouped into modules and macros. The composite IOD contains the image IE plus the five other IEs which give it context. Each of these is further broken down into a list of attributes: some generic (such as slice thickness in the image IE, or patient name in the patient IE) and others specific to the MR use case (such as flip angle).

It is relatively obvious that this would lead to a lot of repetition, which the DICOM standard avoids by collecting related attributes typically used together into *modules*. Both IODs and modules are defined in Part 3 of the standard[24]: Part 3 Annex A defines composite IODs, showing which modules should be included in each IE and Part 3 Annex C defines modules, showing which attributes they contain, and explaining their use and meaning.

Let us now consider an example. The MR Image IOD is defined in Part 3, section A.4 with the modules that comprise it defined in A.4.3 as shown in Table 12.1.

The "Usage" column specifies whether the module is Mandatory (M), User-defined (i.e. optional) (U) or Conditional (C) for which the conditions are specified. The "Reference" column cross-references to other definitions which provide the list

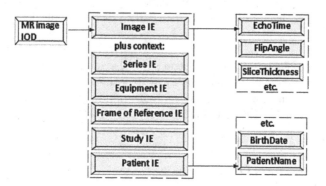

FIGURE 12.2 MR Image IOD.

TABLE 12.1

MR Image IOD Modules from https://dicom.nema.org/medical/dicom/current/output/html/part03.html#sect_A.4.3

IE	Module	Reference	Usage
Patient	Patient	C.7.1.1	M
	Clinical Trial Subject	C.7.1.3	U
Study	General Study	C.7.2.1	M
	Patient Study	C.7.2.2	U
	Clinical Trial Study	C.7.2.3	U
Series	General Series	C.7.3.1	M
	Clinical Trial Series	C.7.3.2	U
Frame of Reference	Frame of Reference	C.7.4.1	M
Equipment	General Equipment	C.7.5.1	M
Acquisition	General Acquisition	C.7.10.1	M
Image	General Image	C.7.6.1	M
	General Reference	C.12.4	U
	Image Plane	C.7.6.2	M
	Image Pixel	C.7.6.3	M
	Contrast/Bolus	C.7.6.4	C - Required if contrast media was used in this image
	Device	C.7.6.12	U
	Specimen	C.7.6.22	U
	MR Image	C.8.3.1	M
	Overlay Plane	C.9.2	U
	VOI LUT	C.11.2	U
	SOP Common	C.12.1	M
	Common Instance Reference	C.12.2	U

of attributes contained in the module. The Patient module (defined in C.7.1.1) has 43 attributes together with 12 additional tables, containing information from "Patient's Name" to "De-identification Method Code Sequence" and a full set for veterinary use. Instead we will follow the "MR Image" module (C.8.3.1), which contains 51 attributes and 2 included tables. A few of these are shown in Table 12.2.

TABLE 12.2

Extract from MR Image Module Attributes from https://dicom.nema.org/ medical/dicom/current/output/html/part03.html#sect_C.8.3.1

Attribute Name	Tag	Type	Attribute Description
Image Type	(0008,0008)	1	Image identification characteristics. See Section C.8.3.1.1.1 for specialization.
Scanning Sequence	(0018,0020)	1	Description of the type of data taken. Enumerated Values: SE Spin Echo IR Inversion Recovery GR Gradient Recalled EP Echo Planar RM Research Mode **Note** *Multi-valued, but not all combinations are valid (e.g., SE/GR, etc.).*
MR Acquisition Type	(0018,0023)	2	Identification of data encoding scheme. Enumerated Values: 2D frequency x phase 3D frequency x phase x phase

12.4.4 ATTRIBUTES

An attribute sits in the DICOM hierarchy below IE and IOD (IE->IOD->Attribute) and consists of several properties:

- A **Name**, as shown in Table 12.2. DICOM implementations should use the standardised name wherever possible for consistency.
- A **Keyword**, which is a version of the name simplified to have no spaces, punctuation, or confusing possessives (for example the attribute called *Patient's Name* has keyword form *PatientName*). This is intended for programmatic implementations, such as DICOM-handling libraries.[25]
- A **Tag**, which is a numeric identifier for the attributes: it consists of two four-digit hexadecimal[26] numbers: the first is the group and the second the element.

- A **Type**, indicating whether the value is mandatory or optional:
  - Type 1 attributes must be present, with a valid value
  - Type 2 attributes must be present, but if no meaningful value exists they may be zero-length for "no value"
  - Type 3 attributes are optional
  - Type 1C or 2C attributes are conditional, with the conditions specified in the standard. (They are not optional: if the conditions are met, they are treated as type 1 or 2 attributes and must be present; if the conditions are not met, they must not be present.)
- A **Value Representation** (VR), indicating the type of data held in the attribute; for example, a string, integer or date. (This is analogous to *data type* in programming and we met in Chapter 2.)
- A **Value Multiplicity** (VM), indicating how many values this attribute contains: values can be scalar or vector. The VM may be a single number, a range, or an unbounded range like *1-n*. (This is analogous to *array length* which we met in Chapter 1.)

All of these properties (aside from Type), for each standard attribute, are listed in the *DICOM Data Dictionary* in Part 6 of the standard. For example, the "Patient's Name" attribute has the Keyword PatientName, Tag of (0010,0010), VR of "Person Name (PN)" and VM of 1. Type is not listed here as it can vary between IODs – it may be mandatory for a CT image but optional for another modality. Type is therefore contained in the module definition, as shown in Table 12.2.

12.4.4.1　Value Representations
The full list of DICOM VRs is defined in part 5, section 6.2. Each consists of two uppercase characters. A few common ones are shown in Table 12.3.

12.4.4.2　Sequence Attributes
The VR of "SQ" (Sequence of Items) allows for the nesting of attributes. As each attribute is uniquely defined by its tag, it can only appear once in a DICOM dataset. Sequence attributes allow attributes to appear multiple times, as each SQ attribute contains multiple child datasets and each child dataset contains multiple attributes. Of course, each attribute can only appear in once in each child dataset as it must be locally unique. For example, the attribute "Other Patient IDs Sequence" (0010,1002) may contain other identifiers for the patient, for example issued by other institutions or by departmental systems and is illustrated in Figure 12.3. Each additional ID has several attributes, including the ID and its issuer. This data structure allows each additional ID to be stored, without violating the uniqueness requirement.

The patient/study/series has to be unique (i.e. it forms the primary key – the importance of which is described in Chapter 2). This should be globally unique in case the image is transferred to another institution, but in practice this isn't done[27] so the probability of data collision is high.

TABLE 12.3

Extract from DICOM Value Representations from https://dicom.nema.org/medical/dicom/current/output/html/part05.html#sect_6.2

VR Name	Definition	Character Repertoire	Length of Value
AS Age String	A string of characters with one of the following formats -- nnnD, nnnW, nnnM, nnnY; where nnn shall contain the number of days for D, weeks for W, months for M, or years for Y. Example: "018M" would represent an age of 18 months.	"0"-"9", "D", "W", "M", "Y" of Default Character Repertoire	4 bytes fixed
DS Decimal String	A string of characters representing either a fixed point number or a floating point number. A fixed point number shall contain only the characters 0-9 with an optional leading "+" or "-" and an optional "." to mark the decimal point. A floating point number shall be conveyed as defined in ANSI X3.9, with an "E" or "e" to indicate the start of the exponent. Decimal Strings may be padded with leading or trailing spaces. Embedded spaces are not allowed. **Note** *Data Elements with multiple values using this VR may not be properly encoded if Explicit-VR Transfer Syntax is used and the VL of this attribute exceeds 65534 bytes.*	"0"-"9", "+", "-", "E", "e", "." and the SPACE character of Default Character Repertoire	16 bytes maximum
FL Floating Point Single	Single precision binary floating point number represented in IEEE 754:1985 32-bit Floating Point Number Format.	not applicable	4 bytes fixed
SQ Sequence of Items	Value is a Sequence of zero or more Items, as defined in Section 7.5.	not applicable (see Section 7.5)	not applicable (see Section 7.5)

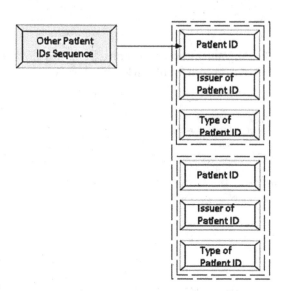

FIGURE 12.3   Sequence attributes for Other Patient IDs.

12.4.4.3   Private Attributes

In an attribute's tag, the *group* number (the first part of the tag from Table 12.2) indicates where responsibility lies for defining the attribute's use:

- If the group is *even*, the attribute must be listed in the DICOM standard
- If the group is *odd*,[28] the attribute is *private*, declared and used by a particular vendor[29] This technical encoding of attributes allows DICOM readers to ignore private attributes whose meaning is unknown to them, while still reading all standard attributes in a dataset. There is no requirement for vendors to document the purpose of their private attributes.[30]

Because there is no central registry of private tags,[31] it would be possible for two vendors to use the same tag for different purposes. In order to avoid these conflicts, the DICOM standard has a mechanism for reserving blocks for private attributes using a *Private Creator Tag*.

Private creator tags are put in the bottom section of the group: for group *spqr*[32] they must be between (*spqr*,0010) and (*spqr*,00FF). The last two digits of this element define the other elements reserved: for example, if the private creator is written to (*spqr*,0012), then all tags (*spqr*,1200) to (*spqr*,12FF) in this particular DICOM dataset are reserved for this vendor, as illustrated in Table 12.4 where 12 is this reserved value.

Despite the private tags being unpublished, simple experimentation shows that TA is the acquisition time and TP the table position. However, table position isn't always at tag (0051,1212): rather it's always at (0051,xx12) with xx set by the last two digits of the private creator tag containing the string "SIEMENS MR HEADER". This "xx"

TABLE 12.4
A Real Private Block

(0051,0012)	"SIEMENS MR HEADER"
(0051,120A)	"TA 06:05"
(0051,1212)	"TP 0"

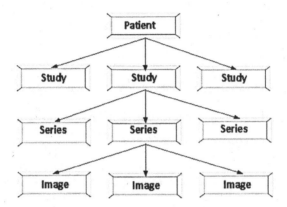

FIGURE 12.4 Object hierarchy – there will be multiple IEs at each level. The ones of specific interest are in the centre of the figure.

notation in descriptions of element numbers is often used to show that a private attribute conforms with this use.

Older versions of the standard lacked the private creator reservation mechanism, so older images (or ones taken on older equipment) may still contain odd-numbered groups without identified creators.

12.4.4.4 Unique Identifiers

A composite IOD includes both the specific IE of interest (which has been the image in our examples) together with all the IEs that give it context, illustrated in Figure 12.4.

The same context IEs will therefore be sent numerous times, once for each image. It is essential that these copies are consistent (i.e. contain the same values for each data transfer) and uniquely identified (i.e. distinguishable from the other objects at the same level of hierarchy).

Consistency is enforced by ensuring that all attributes of a context IE are the same across all child objects. For example, two consecutive slices from a series must have the same patient name attributes.

Uniqueness is enforced by defining a single primary key at each level of hierarchy. For a patient this is the patient ID, which is normally the hospital number or NHS number. A patient may appear multiple times in a database, but this image identifier must be unique so for all other levels, a DICOM **Unique Identifier**[33] (UID) is assigned, with a form like this:

    1.2.826.0.1.1844281.0.52945.31191.20140

The rules for forming a UID are:

- They can be up to 64 characters long
- They consist only of numeric digits "0-9" and "."
- Each group of digits is separated by "." and must be an integer with no leading zeroes (unless it is the number 0)
- Every UID must be globally unique[34]

DICOM uses UIDs in two ways:

- **Instance UIDs** identify a specific object: for example, a particular study (Dr Ganney's X-Ray on 3rd May 2022), series, or image. New, globally-unique values are constantly being generated within the hospital.
- **Class UIDs** identify the type of an object; they are a concise electronic way of saying "this object conforms to the CT Image IOD defined in the standard". Class UIDs are all defined in the standard or (for private classes) declared in a particular vendor's DICOM documentation.

These terms have the same meaning in DICOM as object-oriented programming: a *class* is a kind or category of object, and an *instance* is a particular individual object containing actual data.[35]

All the standard DICOM Class UIDs begin with the prefix 1.2.840.10008, and they are all registered in part 6 annex A of the standard. DICOM uses class-style UIDs to represent a number of coded values.

12.4.4.5 Attribute Example: Orientation

We will now examine a specific example of a DICOM attribute. Tomographic image IODs typically include the Image Plane Module from part 3 section C.7.6.2, which defines the basic geometric properties of a single slice, as shown in Table 12.5.

The first three attributes are mandatory (type 1) and provide enough information to locate every pixel in the 3D co-ordinate system of the patient (which we will come to shortly). The other attributes add optional additional information.

The DICOM co-ordinate system is a **Left-Posterior-Superior** (LPS) system in which x increases to the patient's left side, y increases towards the patient's back (posterior), and z increases towards the patient's head (superior), as illustrated in Figure 12.5.

The Image Orientation attribute in Table 12.5 is defined in this co-ordinate system and is illustrated in Figure 12.5. It has a VR of DS (Decimal String), and VM of 6, which is treated as 2 vectors of 3 elements: unit vectors along the row and column directions of the grid of voxels, pointing away from the corner of the first voxel in the image. This 'corner voxel' must always appear in the upper-left corner of the screen when the image is displayed.

TABLE 12.5

Image Plane Module Attributes from https://dicom.nema.org/medical/dicom/current/output/html/part03.html#sect_C.7.6.2

Attribute Name	Tag	Type	Attribute Description
Pixel Spacing	(0028,0030)	1	Physical distance in the patient between the center of each pixel, specified by a numeric pair – adjacent row spacing (delimiter) adjacent column spacing in mm. See Section 10.7.1.3 for further explanation.
Image Orientation (Patient)	(0020,0037)	1	The direction cosines of the first row and the first column with respect to the patient. See Section C.7.6.2.1.1 for further explanation.
Image Position (Patient)	(0020,0032)	1	The x, y, and z coordinates of the upper left hand corner (center of the first voxel transmitted) of the image, in mm. See Section C.7.6.2.1.1 for further explanation.
Slice Thickness	(0018,0050)	2	Nominal slice thickness, in mm.
Spacing Between Slices	(0018,0088)	3	Spacing between adjacent slices, in mm. The spacing is measured from the center-to-center of each slice. If present, shall not be negative, unless specialized to define the meaning of the sign in a specialized IOD, e.g., as in the Section C.8.4.15.
Slice Location	(0020,1041)	3	Relative position of the image plane expressed in mm. See Section C.7.6.2.1.2 for further explanation.

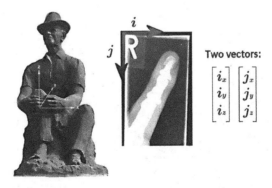

FIGURE 12.5 The DICOM co-ordinate system.

One useful feature of this co-ordinate system is in calculating a voxel's position in the patient coordinate system, as shown in Equation 12.1.

$$\begin{bmatrix} x \\ y \\ z \\ 1 \end{bmatrix} = \begin{bmatrix} i_x \cdot \Delta_c & j_x \cdot \Delta_r & 0 & p_x \\ i_y \cdot \Delta_c & j_y \cdot \Delta_r & 0 & p_y \\ i_z \cdot \Delta_c & j_z \cdot \Delta_r & 0 & p_z \\ 0 & 0 & 0 & 1 \end{bmatrix} \begin{bmatrix} m \\ n \\ 0 \\ 1 \end{bmatrix} \tag{12.1}$$

Where m and n are the indices of the voxel in the image array; Δ_r is the pixel spacing between rows and Δ_c is the pixel spacing between columns (from Pixel Spacing in Table 12.5), p is the 3-component vector (from Image Position in Table 12.5), and i and j are the two 3-component unit vectors (from Image Orientation in Table 12.5 and illustrated in Figure 12.5).

12.4.5 STANDARD ORIENTATIONS

Radiology commonly uses three anatomical planes called sagittal, axial, and coronal (see Figure 12.6). In order for representation on screen to be unambiguous, the correct labelling of axes and the standardisation of the displayed orientation is required. Sagittal slices are viewed from the patient's left, in a standing position; coronal in the same position, from the patient's front; and axial slices from the feet, with the patient on their back. Therefore if the correct choice of corner voxel (always displayed top-left of the screen) is made and the Image Orientation vectors (from Table 12.5) are set, on-screen display and labelling of the standard planes will always be correct.

The numbers in Figure 12.6 describe the three planes in the three LPS directions for i, then for j. So axial is increasing x for i and increasing y for j, whereas coronal is increasing x for i and decreasing z for j.

Earlier we noted that a key goal of DICOM is to allow safe operation across big, multi-vendor networks of imaging devices. The orientation attribute is a good example of how DICOM achieves this.[36] Because of symmetries in the human body,

first image voxel
(upper-left on display)

sagittal	axial	coronal
[0 1 0 0 0 -1]	[1 0 0 0 1 0]	[1 0 0 0 0 -1]

FIGURE 12.6 The three anatomical planes.

left-right errors are easy to make in medical imaging: in the worst case, these can lead to treatment on the wrong side of the patient's body. A "never event" such as "wrong site surgery" (which includes foot instead of hand and wrong patient) happens about 100 times per year, about one third of which are on the wrong side. By defining a standard, ambiguity-free representation for orientation, DICOM greatly reduces this risk.[37]

12.4.6 DICOM Associations

As mentioned earlier, DICOM specifies services as well as data storage. DICOM associations only use five services:

C-ECHO (equivalent to "ping")
C-STORE – send an object
C-FIND – query what the store has
C-GET – retrieve an object
C-MOVE – move and store an object

A conversation (between a workstation and the archive) looks like this (with -> indicating the direction of the message):

WS->INDEX: C-FIND PatientName="Ganney"
INDEX->WS: list of matches
WS->INDEX: C-MOVE UID=1.2.4.5 TO WS
INDEX: knows what objects are stored on CACHE1
INDEX->CACHE1: C-MOVE UID=1.2.4.5 TO WS
CACHE1->WS: C-STORE

It can be seen from this that C-MOVE is more common than C-GET.

There are, of course, other data formats – usually for specialised purposes, such as NIfTI-1 for MRI – but DICOM is the most generalised and therefore the most common.

12.4.7 DICOM-RT

An important variant use of DICOM is DICOM-RT, for radiotherapy. All the image types that might go into a radiotherapy planning system (such as CT, PET and MR) are in DICOM format. The metadata for each image type will be slightly different, but that doesn't matter – DICOM means that the planning system can understand the content of the headers and present images to the user in the correct format.

There are 5 main sub-formats of DICOM for radiotherapy: RT structure, RT-Plan, RT-Image, RT-Dose and RT-Treatment record, as summarised in Table 12.6.

They link together as shown in Figure 12.7. This figure shows the DICOM-RT objects as an extension of the DICOM standard. Note that RT Plan, which contains all RT-related information, is an important object that is needed from the start of

TABLE 12.6
Sub-formats of DICOM-RT (Sibtain et al. 2012)

DICOM-RT Object	Main Property	Example Contents
RT structure	Patient anatomical information	PTV, OAR, other contours
RT-Plan	Instructions to the linac for patient treatment	Treatment beam details e.g. gantry, collimator and couch angles; jaw and MLC positions.
RT-Image	Radiotherapy image storage/ transfer	Simulator, portal image
RT-Dose	Dose distribution data	Patient dose distributions (in 3D), Dose volume histograms
RT-Treatment	Details of treatment delivered to patient	Date and time of treatment, MU delivered, actual linac settings.

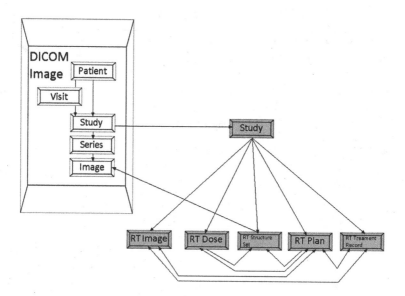

FIGURE 12.7 DICOM-RT objects.

treatment planning to the completion of treatment and is thus related to all other objects.

They are used in this way: assume we have a set of CT images on a planning system. Once outlines are drawn, usually the target volumes and critical organs, we have a set of contours on which a plan can be designed. This set of contours is stored in the DICOM-RT-Structure set. The patient record now contains a set of DICOM images and a DICOM-RT-Structure set, which contains details on the number of contours drawn, number of points in each contour and their names. This set defines a set of areas of significance, such as body contours, tumour volumes (e.g. gross target volume, clinical target volume, planning target volume), OARs, and other regions of interest.

Next some beams are added to the plan, which generates the RT-Plan, which contains details of each treatment beam, such as its name, jaw settings, energy, monitor units etc. In the DICOM-RT standard, information about the structures of interest is contained in DICOM-RT-Structure set and dose distribution in the DICOM-RT RT-Dose file, which requires the coordinates for placing their positions in relation to each other. Thus, the RT-Plan object refers only to the textual information in treatment plans, whether generated manually or by a treatment planning system.

A dose calculation is now done and this generates the DICOM-RT RT-Dose file, which contains details of the dose calculation matrix geometry, dose volume histogram etc.

Digitally reconstructed radiographs may be produced during the planning and placed in the DICOM-RT RT-Image file. Verification images taken using an electronic portal imaging device also generate images in RT-Image format. Note that CT images generated with CT simulators are considered to be ordinary CT scans and do not use the RT- extensions. In contrast to a DICOM image object, RT Image includes not only image information, but also the presentation of the image (i.e. the position, plane and orientation of the image and the distance from the machine source to the imaging plane). The RT-Image file may (if necessary) include the table position, isocentre position and patient position, together with the type of device used to limit the radiotherapy beam (e.g. a multi-leaf collimator).

Once the plan is ready for treatment, some or all of these files may be sent to the linac control system. The RT-Plan is essential in this respect but some of the others are optional and the functionality available may be vendor specific. Every time the patient has treatment, the treatment parameters used are stored by the record and verify system and at the end of treatment the full treatment record is stored in the DICOM-RT RT-Treatment file (Figure 12.8).

One often overlooked benefit of such standardisation is in upgrading and replacing systems: a new system is far more likely to be able to read old data, hence data retention can be achieved without having to keep a sample of the old system just in case the old data needs to be accessed.

12.5   HL7 (HEALTH LEVEL SEVEN)

HL7 is a data exchange format. In order for computer systems to be able to process data produced by other systems, a data standard must be agreed. Even in proprietary systems from the same manufacturer the format of the data has to be consistent so that different modules can access that data and understand its meaning.

The simplest form of this is positional meaning. In this standard, the data will always consist of the same items in the same order (and each item may be of a fixed length). For example, the data below may be converted for transmission to an external system via the process shown in Figure 12.9 to Figure 12.11.

Whilst the advantages of this system for data sharing are obvious,[38] so are the limitations.[39]

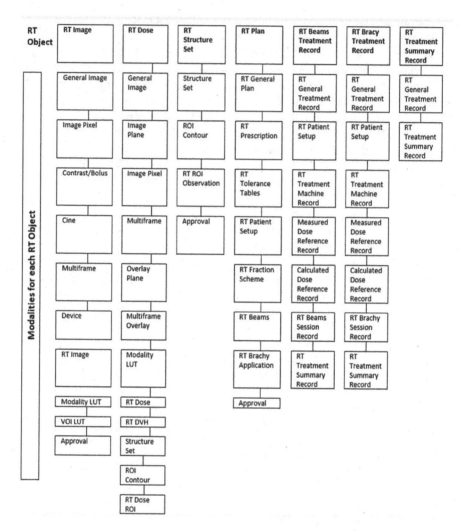

FIGURE 12.8   The seven DICOM-RT objects and their associated modules.

In order to overcome these, we might introduce a header to the data stream that describes the data that is to follow. Such a header might, for this example, be:

4^Surname^9^Forename^7^Address^22^e-mail^22^

This header first states the number of items per record (and thus also the number of items in the header), then the name and length of each field in turn, all separated by a special character. Extensions to this to describe the type of data (numeric, currency, textual, boolean etc.) are also possible.[40] It is important to know the data type as 3/4 may be a fraction, a date, a time signature or just text. Phone numbers require the

Surname	Forename	Address	e-mail
Snail	Brian	Magic Roundabout	brian@roundabout.com
Flowerpot	Bill	Garden	bill@weed.co.uk
Cat	Bagpuss	Shop Window	bagpuss@catworld.net
Miller	Windy	The Windmill, Trumpton	wmiller@chigley.ac.uk

FIGURE 12.9   A set of data, arranged in a table. The items in the first row are field names (but may also be thought of as column headings).

Surname	9
Forename	7
Address	22
Email	22

FIGURE 12.10   The maximum length of the data in each field.

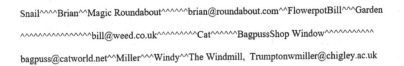

Snail^^^^Brian^^Magic Roundabout^^^^^^brian@roundabout.com^^FlowerpotBill^^^Garden

^^^^^^^^^^^^^^^^^bill@weed.co.uk^^^^^^^^^Cat^^^^^^BagpussShop Window^^^^^^^^^^^

bagpuss@catworld.net^^Miller^^^Windy^^The Windmill, Trumptonwmiller@chigley.ac.uk

FIGURE 12.11   The data converted into one long data stream. Note that unused characters are rendered with ^ and not as spaces, as these appear in the data itself.

leading zero as they are not really numbers but text (to answer that question, Matt Parker suggests (Parker 2019) asking what half of the value means – if it's meaningless, it's probably not a number) and MARCH5 may be a date or a gene. Simply importing this data into Excel (which physicists would never do…) strips the meaning and replaces them with erroneous data. Data from Figure 12.11, complete with header, is shown in Figure 12.12.

DICOM is a tagged file format. A tagged version of the example in Figure 12.11 is shown in Figure 12.13. It can be seen from this example that a tagged format is not particularly suitable for data sets which are comprised of multiple rows, as the tags appear in each row and therefore are repeated throughout the data stream. However, it is very flexible and efficient for single data item files (e.g. an image or a music file) as only the tags that have values are required to be present and additional tags can easily be incorporated.

4^Surname^9^Forename^7^Address^22^e-mail^22^Snail^^^^Brian^^Magic Roundabout^^^^^

brian@roundabout.com^^FlowerpotBill^^^Garden^^^^^^^^^^^^^^^^bill@weed.co.uk^^^^^^^^

FIGURE 12.12   The first two records of the data stream from Figure 12.11, complete with header.

Surname^9^Snail^^^^Forename^7^Brian^^Address^22^Magic Roundabout^^^^^^e-mail^22^

brian@roundabout.com^^ Surname^9^Flowerpot Forename^7^Bill^^^ Address^22^Garden^^^

^^^^^^^^^^^^ e-mail^22^ bill@weed.co.uk^^^^^^^^^EOF

FIGURE 12.13   The first two records of the data stream from Figure 12.11 in a tagged format.[41]

We now need to consider the exchange of incremental changes within a database. This case is very common in healthcare where many clinical systems take a "feed" from the PAS and return results to it. Here the PAS exchanges small changes such as date of appointment, time of arrival at reception etc., as well as large ones such as a new patient being registered (although, in a database of over a million patients this may also be considered "small").

Such incremental changes are achieved through a messaging interface and the most common standard adopted for these is HL7. Note that this is not "Health Language Seven" as some translate it – the "Level 7" refers to the position it occupies in the OSI 7-layer model – the application layer (see section 8.8, Chapter 8).

HL7 is administered by Health Level Seven International (HL7 International 2022 [online]), a not-for-profit **American National Standards Institute** (ANSI)-accredited organisation. Although HL7 develops conceptual standards, document standards and application standards, it is only the messaging standard that we will consider here.

Version 2 of HL7 was established in 1987 and went through various revisions (up to 2.7). Version 2 is backwards-compatible, in that a message that adheres to 2.3 is readable in a 2.6-compliant system.

Version 3 (the latest version) appeared in 2005 and, unlike v2, is based on a formal methodology (the **HL7 Development Framework**, or HDF) and object-oriented principles. The HDF *"documents the processes, tools, actors, rules, and artifacts relevant to development of all HL7 standard specifications, not just messaging."* (HL7 Development Framework Project 2002 [online]) As such it is largely **Universal Modelling Language** (UML) compliant, although there are currently exceptions.

The cornerstone of the HL7 v3 development process is the **Reference Information Model** (RIM). It is an object model created as part of the Version 3 methodology, consisting of a large pictorial representation of the clinical data. It explicitly represents the connections that exist between the information carried in the fields of HL7 messages. An HL7 v3 message is based on an XML encoding syntax. As such it is far more versatile than v2, but with an attendant overhead. Due to the installed userbase of v2 and the difference in message structure, v3 is not yet in widespread use. A very good comparison of the two formats can be found at http://www.ringholm.de/docs/04300_en.htm, of which a small part is reproduced in Figure 12.14 and Figure 12.15.

A hospital imaging network will therefore connect together in a way similar to that in Figure 12.16.

PID|||555-44-4444||EVERYWOMAN^EVE^E^^^^L|JONES|19620320|F|||153 FERNWOOD

DR.^^STATESVILLE^OH^35292||(206)3345232|(206)752-121||||AC555444444||67-A4335

^OH^20030520<cr>

FIGURE 12.14 A section of a HL7 v2.4 message, detailing the patient a test is for. (Ringholm 2007 [online])

```
<recordTarget>
  <patientClinical>
  <id root="2.16.840.1.113883.19.1122.5" extension="444-22-2222"
    assigningAuthorityName="GHH Lab Patient IDs"/>
  <statusCode code="active"/>
    <patientPerson>
      <name use="L">
        <given>Eve</given>
        <given>E</given>
        <family>Everywoman</family>
      </name>
      <asOtherIDs>
        <id extension="AC555444444" assigningAuthorityName="SSN"
          root="2.16.840.1.113883.4.1"/>
      </asOtherIDs>
    </patientPerson>
  </patientClinical>
</recordTarget>
```

FIGURE 12.15 A similar section to the message in Figure 12.14, using HL7 v3. (Ringholm 2007 [online])

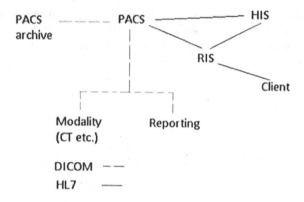

FIGURE 12.16   A generalised imaging network.

A further layer of data exchange is **Cross-enterprise Document Sharing** (XDS) which allows structured data documents of any type to be shared across platforms. Key elements in this are the document source (e.g. an EPR), the document repository (a shared store) and the document registry (essentially an index on the documents). Because the data is structured, the registry is able to index not only the title or metadata from within a header, but the data contained within the document itself. This makes searching the registry more powerful. XDS is of particular interest in a healthcare setting where the source material may be produced from a large range of systems and devices (e.g. for a PACS or an EPR).

12.6   FAST HEALTHCARE INTEROPERABILITY RESOURCES (FHIR)

FHIR is the global industry standard for passing healthcare data between systems. It is free, open, and designed to be quick to learn and implement. FHIR are part of an international family of standards developed by HL7, combining the best features of HL7's v2, v3 and CDA[42] products[43] together with the latest web standards (XML, JSON, HTTP, OAuth, etc.). There is a strong focus on implementability. The information models and APIs developed using this standard provide a means of sharing health and care information between providers and their systems regardless of the setting that care is delivered in.

NHS Digital is making extensive use of the HL7 FHIR standard, doing so across the different areas that FHIR supports, which include:

- FHIR ReSTful APIs
- FHIR Documents
- FHIR Messages

Products developed by NHS Digital make use of a consistent set of FHIR profiles and specifications, a number of which are already published. Current specifications include Transfer of Care and CareConnect.

```
<Patient xmlns="http://hl7.org/fhir">
  <id value="glossy"/>
  <meta>
    <lastUpdated value="2014-11-13T11:41:00+11:00"/>
  </meta>
  <text>
    <status value="generated"/>
    <div xmlns="http://www.w3.org/1999/xhtml">
      <p>Henry Levin the 7th</p>
      <p>MRN: 123456. Male, 24-Sept 1932</p>
    </div>
  </text>
  <extension url="http://example.org/StructureDefinition/trials">
    <valueCode value="renal"/>
  </extension>
  <identifier>
    <use value="usual"/>
    <type>
      <coding>
        <system value="http://hl7.org/fhir/v2/0203"/>
        <code value="MR"/>
      </coding>
    </type>
    <system value="http://www.goodhealth.org/identifiers/mrn"/>
    <value value="123456"/>
  </identifier>
  <active value="true"/>
  <name>
    <family value="Levin"/>
    <given value="Henry"/>
    <suffix value="The 7th"/>
  </name>
  <gender value="male"/>
  <birthDate value="1932-09-24"/>
  <careProvider>
    <reference value="Organization/2"/>
    <display value="Good Health Clinic"/>
  </careProvider>
</Patient>
```

Resource
Identity &
Metadata

Human
Readable
Summary

Extension
with URL to
definition

Standard
Data:
• MRN
• Name
• Gender
• Birth Date
• Provider

FIGURE 12.17 The FHIR structure. (FHIR 2022 [online])

The HL7 UK FHIR Reference Server is available at https://fhir.hl7.org.uk/ and includes FHIR profiles designed to be used across the country.

Profiles used within NHS digital created solutions are published on the Reference Server at https://fhir.nhs.uk/.

The simple example in Figure 12.17 shows the important parts of a resource: a local extension (third section), the human readable HTML presentation (second section), and the standard defined data content (bottom section).

Further information on the FHIR is at http://www.hl7.org/fhir/index.html

NOTES

1 The sections "DICOM-RT", "HL7" and "FHIR" are reproduced from Ganney et al. 2022.
2 But would detect 3, 5 or 7 errors.
3 This example is based on the one at http://www.cs.ucc.ie/~jvaughan/archres/notes/hamming-example.pdf, with the errors in that article corrected and annotations added.
4 This hexadecimal value is 10 in decimal, 1010 in binary.
5 Permutations of m binary values.

6 As these are the powers of 2 – the use of this will become apparent as the example progresses.

7 The bit position which includes 1 in its power of 2 sum. This follows for the other 3 check bits.

8 The parity bit (NB we are using even parity), calculated using the bit values in brackets.

9 The bit values from the governed positions, ignoring those used as governing positions.

10 The calculated parity values have now been dropped into their appropriate check digit positions.

11 XOR is "Exclusive OR" and is rendered here by the symbol $\oplus$ for brevity. It has the function:

0 XOR $0 = 0$.
0 XOR $1 = 1$.
1 XOR $0 = 1$.
1 XOR $1 = 0$.

12 The result is found by working across the line, each result feeding into the next.

13 The reader is left to ponder the wisdom of using one's spouse's name as a password.

14 American College of Radiologists and National Electrical Manufacturers Association.

15 Not, as you may suppose, an image, but a series of instructions to a linac for patient treatment.

16 Two very common tagged file formats in current use which may assist in the understanding of DICOM are TIFF and MP3.

17 This measurement will depend on the imaging modality. It may be, for example, the attenuation of an X-ray, the reflection of different wavelengths of light (a standard photograph) or proton density (an MRI image).

18 DTI shows the direction of diffusion: in the brain this will reveal the pathways.

19 By loading the file into a text editor, for example.

20 Possible because it's tagged text.

21 There is one named "BIPED" that records the number of legs of the subject – in case you're scanning in a veterinary hospital.

22 A wrist scan has the hand over the head, a neck scan doesn't, so the frame of reference is important.

23 For an MR image, the Image IE contains some MR-specific parameters such as echo time and flip angle. These are not relevant for a CT image, where Peak kiloVoltage (KVP), revolution time, etc. would be present instead.

24 The entire DICOM standard is known as NEMA standard PS3, so part 3 is sometimes referenced as PS3.3, part 4 as PS3.4 etc.

25 It is without punctuation so it can be standardised – it is a similar principle to database field naming.

26 Hexadecimal is number base 16 and uses the digits 0-9 and A-F. DICOM element numbers can thus be rendered like 0010 and 07FE. In reality, the numbers are two 32-bit binary numbers rendered in hexadecimal for ease of use. Each hexadecimal digit therefore consists of 4 bits and it is far simpler to use "D" than "1101".

27 The administrative overhead of compiling and updating such a registry is somewhat prohibitive.

28 A hexadecimal number is odd if it ends with an odd-numbered numeric or letter digit: 1, 3, 5, 7, 9, B, D, F.

29 With the exception that the lowest-numbered odd groups 0001, 0003, 0005 and 0007 may not be used for private tags.

30 Some do and it's helpful when this happens.

31 Another prohibitively difficult registry.

32 Where, s, p, q and r represent hexadecimal digits.

33 UIDs have VR UI.
34 Theoretically, global uniqueness is accomplished by registering prefixes to institutions, then appending a further unique suffix for each new object based on some combination of the equipment identifiers, time, and other random or incrementing groups. In practice, many implementations generate UIDs entirely from random or time-based input and rely on the vanishingly small likelihood of collision (and the even smaller likelihood that a colliding UID would ever be shared into the same system).
35 A class UID is "what kind of thing" whereas an instance is "what specific thing".
36 Recall that Image Orientation (Patient) has a VR of "DS" (decimal string) and VM of 6 (a unit vector along the image row and column).
37 Wrong-site surgery and wrong-person surgery are effectively the same event and there are about 50-100 per year across the NHS.
38 Simplicity is the main one – the simpler the system the more likely it is to be implemented.
39 For example, consider adding just one field to the table, or increasing the length of a field.
40 The dBase 3 file format is a good place to start.
41 Note the "EOF" indicating the end of the stream. Also that the order of the fields is now unimportant as they are prefixed by their tag in all cases.
42 Clinical Document Architecture.
43 Thus providing an evolutionary development path from HL7 Version 2 and CDA: these standards can co-exist and use each other.

REFERENCES

DICOM 2022, [online]. Available: http://dicom.nema.org/standard.html [Accessed 23/04/22].
FHIR 2022, [online]. Available: http://www.hl7.org/fhir/summary.html [Accessed 23/04/22].
Ganney P, Maw P, White M, ed. Ganney R, 2022, *Modernising Scientific Careers The ICT Competencies*, 7th edition, Tenerife: ESL.
Hamming, R W, 1950, Error detecting and error correcting codes. *Bell System Technical Journal* 29 (2): 147–160.
HL7 Development Framework Project, Project Charter, HL7 2002, [online]. Available: https://www.hl7.org/documentcenter/public/wg/mnm/docs/HDF%20Project%20Charter.doc [Accessed 23/04/22].
HL7 International 2022, [online]. Available: www.hl7.org, or www.hl7.org.uk for the UK version [Accessed 23/04/22].
National School of Healthcare Science 2022, [online]. Available: https://curriculumlibrary.nshcs.org.uk/ [Accessed 23/04/22].
Parker M, 2019, *Humble Pi: A Comedy of Maths Errors*, London: Allen Lane.
Parker, M, 2015, *Things to Make and Do in the Fourth Dimension*, London: Penguin.
Ringholm. 2007, [online]. Available: http://www.ringholm.de/docs/04300_en.htm [Accessed 09/05/22].
Sibtain A, Morgan A & MacDougall N, 2012, *Physics for Clinical Oncology (Radiotherapy in Practice)*, Oxford: OUP.

13 Hospital Information Systems and Interfaces[1]

13.1 INTRODUCTION

Hospital information systems are in use in almost every aspect of healthcare – indeed, it is almost impossible to think of an area that does not have at least a small system in use. The ease of use and especially the speed of retrieval of data have brought many benefits to patient care. This chapter examines some of the issues around these systems, beginning with one that was very common in paper-based healthcare record keeping: how long to keep records for.

13.2 DATA RETENTION

Data retention times are one of the most confusing aspects of NHS data. There are multiple pages of guidance (110 pages in the guidance that was current until 29/7/16, this went to an Excel spreadsheet with 118 entries and is currently a pdf with 38 pages of guidance, plus many others of supporting information (NHSX 2021 [online] (1))) and it can take several attempts to find the entry that best describes the data being retained. As records become more electronic, there is a temptation to retain data indefinitely but this code gives instructions on disposal.

One important point of data retention is that when records identified for disposal are destroyed, a register of these records needs to be kept. (See the Records Management Code of Practice for Health and Social Care 2016 for further information (NHSX 2021 [online] (2))).

An interesting question is what happens should data be kept beyond the recommended retention period. If it contains personal data then the GDPR's principle e (storage limitation) applies and the data must either be deleted or anonymised, otherwise there is the possibility of a fine. The exception to this is when it is being kept for public interest archiving, scientific or historical research, or statistical purposes.

If the data does not contain personal data then there appear to be no ramifications to keeping data beyond the retention limits.

13.3 HOSPITAL INFORMATION SYSTEMS AND INTERFACES

The major hospital information system in use is the **Patient Administration System** (PAS), sometimes called an **Electronic Patient Record** (EPR).[2] This system will handle all patient registrations, demographics, clinic appointments and admissions (planned and emergency). It will normally be the "gold standard" for data, in that it holds the "most correct" version, especially the demographics and GP contact information.

An EPR will have a heavy processing load. For this reason, it is common practice to have a second data repository which is a copy of the data, created via an overnight scheduled task. Reports may thereby be run against this data without incurring a performance hit on the main system. The data is at worst 24 hours out of date, but for most administrative reporting (as opposed to clinical reporting) this is acceptable. A notable exception is bed state reporting, which is therefore normally a feature of the main system.

The functionality of the EPR is greatly enhanced via interfaces to clinical systems. This will usually be via HL7 messages (see Section 12.5, Chapter 12) and may be outbound (where the data flow is from the EPR to the clinical system), inbound (the other direction) or two way. Demographics are usually outbound; test results are usually inbound and bed state information may be two way.

As the EPR will interface to many other systems, the use of an interface engine is usual. This may be thought of as a sophisticated router where outbound messages from all connected systems will be received and then passed to all relevant[3] downstream systems for processing, usually for keeping databases synchronised. It can also process these messages so that codes used by the EPR may be converted into ones used by the downstream system, for example.

An interface engine will normally include a large cache so that downstream systems that go offline (e.g. for upgrade work) are able to collect the relevant messages when they come back on line. Therefore, acknowledgement messages form an important part of the interface engine protocol.

The most recent move in the UK has been towards an **Electronic Health Record System** (EHRS) which contains all the information in an EPR but with additional data such as that from medical devices and specialised systems: it is a longitudinal record bringing together patient data such as demographics and medical history, treatment information, diagnostic information, operational information[4] and clinical care information. In doing so, it provides a fuller picture of the patient pathway and provides opportunity for data mining such as operational and clinical analytics.

13.4   EQUIPMENT MANAGEMENT DATABASE SYSTEMS

An equipment management database system is, as the name implies, a database of equipment around which a system has been constructed to manage that equipment. As such, it shares a lot of features with asset management database systems but contains additional features to aid in the management of equipment, which for the focus of this book means medical equipment.

An asset management database will contain information such as the equipment name, manufacturer, purchase cost and date, replacement cost (and planned replacement date if known), location and a service history (if appropriate). Reports may then be run against this data to produce information on capital expenditure and assistance in forecasting future expenditure. It may also be used to highlight unreliable equipment, commonalities of purchasing (in order to assist in contract negotiations and bulk buying) and equipment loss due to theft or vandalism.

An equipment management database system will include all of these features but will also include information so that the equipment may be managed appropriately.

Probably the most common example of this additional functionality is the **Planned Preventative Maintenance** (PPM).[5] In order to implement PPMs, all equipment managed must be assigned a service plan. In its simplest form, this is a set time frame: for example, a set of scales that requires calibration once a year. A more complex plan might include different tests or calibrations to be performed at different times. This type of service plan will generally still include a fixed time interval but will also include a service rotation. On the first rotation, it might be visually inspected to ensure that all seals are still in place. On the second rotation a calibration might be added and on the third a **Portable Appliance Test** (PAT) renewal might be undertaken in addition to inspection and calibration. The system will keep track of all work that is due (producing worklists for a particular time period), the service rotation that is due, parts and labour used in the maintenance and so on.

Inspecting, calibrating and testing all of a hospital's medical equipment is not a minor task: most such services spread the load across the entire year, which the management system is also able to assist in planning by reporting on peaks and troughs in the anticipated workload, adjusting for planned staff absences and potential peaks and troughs in emergency repairs (from historical data).

The system may also assist with functions such as the management of contracts (where equipment is maintained by a third party), the recording of repairs undertaken (thus determining when equipment has reached the end of its economic life) and the recording of medical device training.

In this latter example, all staff of the hospital are recorded along with the equipment they have been trained to use. Thus, training can be kept current and a mechanism for preventing the unsafe use of equipment implemented.

In order to achieve all this functionality, an equipment management database system will need to be interfaced to other systems, such as a capital asset management system, a human resources system and a contracts management system.

13.5   DEVICE TRACKING SYSTEMS

Electronic tracking devices are a growing element of hospital practice and their role is considered here in terms of potential, rather than system design. The benefits of tracking devices are numerous, from the reduction in time spent looking for equipment (estimated as being 2.5% of nurse time by UCLH in 2011[6]), to theft prevention. Other potential benefits are improved servicing due to having a reliable equipment asset register and the ability to run a more efficient equipment library.

Broadly, there are four types of technologies to consider, as described in Table 13.1. Each requires the asset to have a tag attached, which communicates with the central system. The tags may vary in size, depending on the amount of data required and the distance from the transponder that it is required to communicate with. This means that such a system may be deployed for multiple purposes. For example, it may describe bed occupancy, the stocking of a pharmacy cabinet or track the whereabouts of babies and/or vulnerable patients, as well as staff in lone working situations.

The required accuracy may also determine the technology: it is possible to "geofence" the organisation so that tracked equipment is known to be within a "fenced area" but where it is within that area is unknown (Table 13.1).

TABLE 13.1

The Four Main Technologies for Asset Tracking

Technology	Description	Pros and Cons
Wi-Fi (Wireless Network)	This system uses an organisation's Wi-Fi network (the same network that supports wireless computing devices). Tags transmit a Wi-Fi signal, which is picked up by Wi-Fi access points – if the signal is picked up by 4 or more access points, the location of the device can be triangulated in three dimensions. This may be supplemented by additional ultrasound "beacons" to more accurately locate devices within a particular room.	There is a claimed accuracy of location to 2 m (although there exists some scepticism about this, and it is probably only possible with the addition of ultrasound beacons). Makes use of an existing Wi-Fi so no additional network is required. Fully scalable for patients, assets and staff. It requires many more Wi-Fi access points than is needed just for wireless computing. Produces an additional load on the network. Tags are expensive (circa £65 each).
Radio Frequency Identification (RFID)	A network of RFID receivers is positioned across the organisation, connected to the LAN. Tags (either active or passive) communicate with these receivers, which can track their location. Passive tags respond to a signal from gateways (from where they also draw their power to do so), whereas active ones will transmit data to a nearby receiver and have on-board power. This is a proven technology, being used in many industries, including retail, where cheap passive tags are attached to stock, and their location/ presence is tracked	Tags are very cheap (passive tags are pennies each). The granularity of the mapping is not very precise. An additional network (of RFID transponders) is required.
RFID with **Infra-Red (IR)**	This uses RFID technology, supplemented by infra-red transmitters/detectors. IR can be used to track devices to a very granular level (bed space, or even a drawer), and information is then communicated centrally via an RFID network.	Accurate location of items at a very granular level. Fully scalable for patients, assets and staff. Additional network needs to be installed of RFID transponders. The solution is quite complex in that there are several technologies involved. The risk of failure is therefore higher.
Bluetooth	Bluetooth is a very short range communication protocol and so relies upon the tag being in the proximity of another device so enabled. It is very good for alerts (e.g. the tag has been separated from another device).	Whilst this technology has been around for a long time, it is the advent of the Apple AirTag, which links into Apple's Find My network that has made this viable in a hospital context. As it does not rely on fixed hardware such as RFID transponders, the granularity is dependent on the number of devices also on the Find My network. AirTags are small, contain their own power supply (a CR2032 battery) and cost about £100 for a pack of four – plus the cost of something to mount it in.

Location tracking may, of course, be implemented as a side-effect of meeting another need: for example, connecting devices to a Wi-Fi network in order to remotely distribute drug libraries to infusion pumps. This requires triangulation to be added, but the infrastructure is already in place.

13.6 INTERFACES

An interface is a point where two systems, subjects, organisations, etc. meet and interact. In computing, an interface is a shared boundary across which two or more separate components of a computer system exchange information. The exchange can be between software, computer hardware, peripheral devices, humans or combinations of these.

We can therefore see that there are several types of interface which we might be interested in:

- User interface, where the user interacts with the software.
- Hardware interface, where electronic components interact.
- Software interface, where software components[7] interact.
- Systems interface, where software systems interact in order to pass information (synchronisation and interrogation).

NOTES

1 This chapter is reproduced from Ganney et al. 2022.
2 Technically there is a difference: A PAS consists of demographics and appointments, whereas an EPR contains the full medical record including correspondence. As PAS functionality increases, the distinction becomes blurred.
3 The relevance is normally determined via a set of rules.
4 Appointments as opposed to surgery.
5 Sometimes simply referred to as "servicing".
6 Internal document, not externally published.
7 Bear in mind that an operating system is software.

REFERENCES

Ganney P, Maw P, White M, ed. Ganney R, 2022, *Modernising Scientific Careers The ICT Competencies*, 7th edition, Tenerife: ESL.
NHSX 2021a, [online] (1). Available: https://www.nhsx.nhs.uk/media/documents/NHSX_Records_Management_CoP_V7.pdf [Accessed 10/05/22].
NHSX 2021b, [online] (2). Available: https://www.nhsx.nhs.uk/information-governance/guidance/records-management-code/ [Accessed 10/05/22].

14 Backup[1]

14.1 INTRODUCTION

A reliable data centre, be it for a web presence, a clinical database or a document repository requires resilience to be built into the design. As with security, there are many ways of achieving this and the best designs will incorporate a mixture of these. In this chapter, we consider four such techniques.

14.2 REPLICATION

Replication, either of a server or a database (or both) is what the name implies: a copy exists of the server/data enabling it to be switched to in the event of a failure in the primary system. There are two main ways of achieving this, and the required uptime of the system determines which is the most appropriate. The simplest one to achieve is a copy, taken at a specified time to the secondary system. The two systems are therefore only ever briefly synchronised and in the event of a system failure bringing the secondary system online means that the data will be at worst out-of-date by the time interval between copying. Replication of this form is usually overnight (thereby utilising less busy periods for systems) meaning that the copy is at worst nearly 24 hours out-of-date. If changes to the system have been logged, then these may be run against the secondary system before bringing it online, but this will lengthen the time taken to do so.

The most common use of this form of replication is in data repositories (covered in Section 13.3, Chapter 13), thereby removing some of the workload from the primary system and improving its reliability (and response time).

The second method of replication is synchronised: that is, both copies are exactly the same at all times. There are two primary ways of achieving this. The simplest is a data splitter: all changes sent to the primary system are also sent to the secondary (see Section 12.5, Chapter 12 for details on messaging). This is clearly the best method when the system receives changes via an interface (which may be from an input screen or from a medical device) as the primary system only has to process the incoming message and the work of replication is external to it (e.g. in an interface engine).[2] The second is where every change to the primary system is transmitted to the secondary for it to implement. This introduces a processing overhead on the primary system, especially if it has to also receive acknowledgement that the change has been applied to the secondary system. As this method introduces a messaging system, the system could instead be designed to use the first method described.

The most basic (and therefore most common) method of replication, though, is the backup. This is simply a copy of the system (or a part of it) taken at a specified time, usually onto removable media (for high volumes, usually tape). As such backups are generally out-of-hours and unattended, systems that exceed the capacity of the media are backed up in portions, a different portion each night.

DOI: 10.1201/9781003316244-14

Whilst a backup enables a quick restore of lost or corrupted data (and simplifies system rebuilds in the case of major failure), data errors are usually not so swiftly noticed and may therefore also exist on the backup. The **Grandfather-Father-Son** (GFS) backup rotation system was developed to reduce the effect of such errors. In this, three tapes are deployed: on day one, tape 1 is used. On day 2, tape 2 is used and on day 3, tape 3 is used. On day 4, tape 1 is re-used and so on, meaning that there are always three generations (hence the name) of backup. Most backup regimes are variants on the GFS scheme and may include a different tape for each day of the week, 52 tapes (e.g. every Wednesday) also rotated or 12 tapes (e.g. every first Wednesday). In this way data errors tracing back as far as a year may be corrected.

14.3   ARCHIVING

Despite the massive increases in storage capacity in recent years, medical imaging has also advanced and thus produces even larger data sets. It has been estimated that 80% of PACS images are never viewed again. However, as a reliable method for identifying those 80% has not yet been achieved, all images must be kept, but keeping them online (on expensive storage) is not a sensible option. Thus old images are generally archived onto removable media[3] (again, tape is usual) or onto a slower, less expensive system and the original data deleted to free up space. There are several algorithms for identifying data suitable for archiving, but the most common is based on age: not the age of the data, but the time since it was last accessed. In order to implement such a system, it is therefore imperative that each access updates the record, either in the database (for single items) or by the operating system (in the case of files).

14.4   RESILIENCE USING RAID

Probably the most common type of resilience, especially on a server, is **Redundant Array of Inexpensive Discs** (RAID).[4] There are several forms of RAID and we consider two (levels 1 and 5) here. RAID level 1 is a simple disc image, as per replication above – only in real time. The replication is handled by the RAID controller[5] which writes any information to both discs simultaneously. If one disc fails, then the other can be used to keep the system operational. The failed disc may then be replaced[6] and the RAID controller builds the new disc into a copy of the current primary over a period of time, depending on the amount of data held and the processing load.

Other forms of RAID do not replicate the data directly but spread it across several discs instead, adding in some error correction as well. The form of spreading (known as "striping") and the type of error correction are different for each level of RAID.

RAID 5 uses block-level striping and parity data, spread across all discs in the array. In all disc storage, the disc is divided up into a set of blocks, a block being the smallest unit of addressable storage. A file will therefore occupy at least one block, even if the file itself is only one byte in size. A read or write operation on a disc will read or write a set of blocks. In RAID 5, these blocks are spread across several discs, with parity data stored on another one (see Figure 14.1). Thus RAID 5 always requires a minimum of three discs to implement (as in Figure 14.1 but more discs can be used).

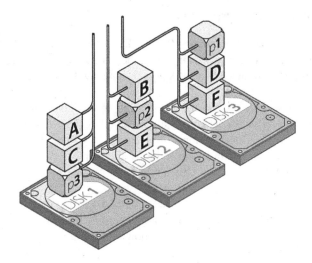

FIGURE 14.1 Diagram of a RAID 5 Setup – p1 contains the parity values from blocks A and B, p2 from C and D and p3 from E and F. p4 would be stored on DISK3 as the allocation rotates.

Parity, which we met earlier, is a computer science technique for reducing data corruption. It was originally designed for data transmission and consisted of adding an extra bit to the data.[7] This extra bit forced the sum of the bits to be odd (odd parity) or even (even parity). It was added at transmission and checked at reception (although this would only detect one error)[8]. More complex forms used more bits which enabled the data not just to be better checked but also to be corrected. Such codes are known as Hamming codes (see "Hamming Codes" in Section 12.2). The parity used within RAID 5 not only checks that the data is correct but also enables it to be re-built, should a disc fail and have to be replaced[9]. As a RAID controller can keep a system operational even when a disc has failed, so much so that users may not notice, it is therefore imperative to monitor such clusters as one failure may be easily fixed but two may be catastrophic.

In the figure, the distribution of the blocks and the parity can clearly be seen. Distributing the parity blocks distributes the load across all the discs, as this is where bottlenecks may appear (to read blocks B–C, for example, also requires two parity blocks to be read – in this case discs 1 and 3 only have one read operation to perform). RAID 5 has found favour as it is viewed as the best cost-effective option providing both good performance and good redundancy. As write operations can be slow, RAID 5 is a good choice for a database that is heavily read-oriented, such as image review or clinical look-up.

Traditionally computer systems and servers have stored operating systems and data on their own dedicated disc drives. With the data requirements expanding, it has now become more common to have separate large data stores using NAS or SAN technology. These differ in their network connectivity but both rely on RAID for resilience. NAS uses TCP/IP connections and SANs use Fibre Channel connections.

14.5   BUSINESS CONTINUITY

The other half of **Disaster Recovery** (DR) is **Business Continuity** (BC): while the scientists and engineers are recovering from the disaster and restoring service, how does the business continue (if at all)? A good BC plan will firstly describe downtime procedures.

From these downtime procedures it is possible to work out how long the service/ organisation can survive for. This then informs the DR plan. A service that can survive for a week needs little additional architecture. A service that can only tolerate 30 minutes requires redundant spares to be on-site. If it can only tolerate 5 minutes then hot-swaps are required. Less than a minute's tolerance requires a duplicate failover system to be running.

We will look at risk analysis later (see Section 16.4.1) and that should be the core of the DR and BC plans: what can go wrong (including infrastructure such as power and transport – human as well as data) and how much impact will it have.

NOTES

1 This chapter is reproduced from Ganney et al. 2022.
2 It is often postulated that giving the primary and secondary systems the same IP address will also achieve this. It is left as an exercise to determine whether or not this is a good idea.
3 There may be several "layers" of such storage, each slower to access than the previous, eventually reaching a removable media layer.
4 "I" is sometimes rendered "Independent".
5 A disc controller with additional functionality.
6 Often without halting the system – known as "hot swapping".
7 An alternative was to use one of the 8 bits in each byte for parity. Hence ASCII only uses 7 bits and simple integers often only have a range of 0 to 127.
8 For a 7-digit binary number, 3 bits are required to check for all possible errors – see "Hamming Codes" in Section 12.2, Chapter 12.
9 Therefore a failed disc can be ignored as the data on it can be computed in real time by the RAID controller.

REFERENCE

Ganney P, Maw P, White M, ed. Ganney R, 2022, *Modernising Scientific Careers the ICT Competencies*, 7th edition, Tenerife: ESL.

15 Software Engineering[1]

15.1 INTRODUCTION

Software engineering is the application of engineering to the design, development, implementation, testing and maintenance of software in a systematic method.

Typical formal definitions of software engineering are:

"the systematic application of scientific and technological knowledge, methods, and experience to the design, implementation, testing, and documentation of software;"
(Systems and software engineering 2010)

"the application of a systematic, disciplined, quantifiable approach to the development, operation, and maintenance of software;"
(IEEE Standard Glossary of Software Engineering Terminology 1990)

"an engineering discipline that is concerned with all aspects of software production;"
(Sommerville 2007)

"the establishment and use of sound engineering principles in order to economically obtain software that is reliable and works efficiently on real machines."
(Software Engineering. Information Processing 1972)

All of which say that it's the application of engineering principles to the development of software. The 1996 "Dictionary of Computing" (Dictionary of Computing, 1996) says that software engineering is the entire range of activities used to design and develop software, with some connotation of "good practice". It then lists a whole range of activities:

- User requirements elicitation
- Software requirements definition
- Architectural and detailed design
- Program specification
- Program development using some recognised approach such as structured programming
- Systematic testing techniques
- Program correctness proofs
- Software quality assurance
- Software project management
- Documentation
- Performance and timing analysis
- Development and use of software engineering environments.

DOI: 10.1201/9781003316244-15

Software engineering is expected to address the practical problems of software development, including those encountered with large or complex systems. It is thus a mix of formal and pragmatic techniques.

15.2   SOFTWARE

Software falls into one of three types:

1. Boot software – the stuff that gets the PC started.
2. **Operating system** (OS) – the stuff that does basic, but important things.
3. Applications – the stuff that does what you actually wanted a computer for in the first place.

The boot software is initiated when a computer starts up (either from power-on or a reset). It normally checks the hardware and then locates and loads an operating system (or, more likely, the loader for the OS). Dual (or larger) boot systems will normally pause in order to ask the user which operating system they wish to use. The boot software is part of the hardware and is stored in **Read Only Memory** (ROM)[2]. Therefore, one of the first things it will do is to load drivers for the storage media that holds the OS. It's the loader that asks you if you want to start Windows in safe mode, not the boot program.

15.2.1   OPERATING SYSTEMS

An operating system is a very complex piece of software that interfaces directly with the hardware upon which it is running. The first three introduced here are all multitasking operating systems which is why they are to be found on servers as well as end-user machines. The fourth has appeared more commonly in healthcare over the last decade, but predominantly as an end-user system.

15.2.1.1   Microsoft Windows

There are two families of this operating system: end-user (such as Windows XP, Windows 10 and Windows 95) and server (normally dated and often with a release number, e.g. 2008 R2). Both may also be referred to by their service pack status, e.g. XP SP3 indicates Windows XP where service packs 1-3 have been applied.

Windows is a graphical user interface, employing the WIMP (**Windows, Icons, Mouse, Pointer**[3]) paradigm pioneered by Xerox and brought into popularity by Apple and Atari. It is an event-driven operating system, in that events are generated (e.g. by a mouse click, a keyboard press, a timer or a USB device insertion) and these are offered by the operating system to the programs and processes that are currently running (including itself) for processing. For this reason, a program may not necessarily cancel just because a user has clicked on a button labelled "Cancel" – the event will merely sit in a queue, awaiting processing.

Windows has gained great popularity, partly because of its relative openness for developers (compared to, say, Apple) but mostly due to its common user interface: similarly displayed buttons and icons perform similar functions across programs (e.g. the floppy disc icon for saving – even though the program is unlikely to be saving to a floppy disc) and almost-universal keyboard shortcuts (e.g. Ctrl-C for "copy"). This makes the learning of new programs easier and more intuitive.

Windows is most likely to be found running desktop end-user machines, departmental servers and (in its embedded form) medical devices.

15.2.1.2 Unix

Unix is also a family of operating systems and may also run on end-user machines as well as on servers. However, it is on the latter that it is now most prevalent, due to its robust stability. Whilst there are graphical user interfaces, it is most commonly accessed through a command-line interface, a "shell" (such as BASH, Bourne or C) that accepts certain commands (usually programs in their own right rather than embedded into the OS) and logic flow. Originally developed for dumb-terminal access (where all the user has is a keyboard and VDU and the processing takes place on the server) it can therefore be accessed easily via a terminal emulator on any end-user machine (e.g. a Windows PC or an Android phone).

A key concept in Unix is that of the pipe. In this, the output of one program can be "piped" as the input to the next in a chain, thereby producing complex processing from a set of relatively simple commands.

For example, this is a simple pipeline command, creating a long list of files which is then piped into the *more* command:

```
ls -al | more
```

A slightly more complicated command which lists only the directories, as *grep* (with these parameters) only shows the records that begin with the letter *d*:

```
ls -al | grep '^d'
```

And a complex example listing all Apache log files for the month of August excluding 2004-6:

```
ls -al | grep Aug | grep -v '200[456]' | more
```

These examples (Alexander 2019) also introduce another key concept of Unix – that of regular expressions[4], a simple example of which might be "randomi[sz]e" in a text search, which would match both spellings of the word (i.e. "randomise" and "randomize").

Unix is generally seen as an expensive operating system, which is one of the reasons it is less likely to be found on end-user machines. Unix is also most likely to be found running departmental and critical enterprise servers.

15.2.1.3 Linux

Linux was developed as an open-source variant of Unix by Linus Torvalds (Linus Torvalds: A Very Brief and Completely Unauthorized Biography 2006 [online]). Its popularity rapidly increased because it is open-sourced and so developers are able to access the source code to the OS and produce new versions of the programs within it, thereby expanding functionality and correcting errors. The caveat is that this new version must also be freely available, with source code, to anyone who wishes to use it. There are therefore various different versions (referred to as "flavours") of Linux, such as Ubuntu and openSUSE.

As with Unix, Linux is mostly command-line, yet graphical user interfaces do exist and (as the cost factor does not exist) are more likely to be found on end-user machines. The low processing overhead (compared to Windows, for example) has led to the introduction of cheap yet powerful systems, such as the Raspberry Pi, which has seen Linux systems appearing in many areas of healthcare. Linux is most likely to be found running desktop end-user research machines, departmental servers and especially web servers.

15.2.1.4   iOS/macOS

iOS and macOS are OSs created and developed by Apple Inc. exclusively for its hardware. iOs is the mobile version (powering devices such as iPhone, iPad[5], and iPod) with macOS found on devices such as iMac and MacBook. They are very similar in terms of user experience, meaning that moving from one Apple device to another is very intuitive, once the first has been mastered (see the comment on Microsoft and standards, above).

Whereas macOS uses fairly traditional input devices such as a keyboard and a mouse, the iOS user interface is based upon direct manipulation, using multi-touch gestures such as swipe, tap, pinch, and reverse pinch which all have specific definitions within the context of the iOS OS and its multi-touch interface.

The tight development relationship between the hardware and software (compared to Linux which is designed to run on a vast range of hardware) means that both benefit from optimisations.

iOS is most likely to be found in mobile healthcare settings (such as clinical record taking) whereas macOS is more found in graphics-heavy settings, such as image processing. They are more likely to be end-user machines than servers, unless as part of an integrated system.

15.2.1.5   General

An operating system consists of a central part (often called the kernel) and a number of supplementary programs – this can especially be seen in UNIX-like OSs, where these supplementary programs can simply be replaced by a new executable and therefore the behaviour can be enhanced or corrected. An operating system is a very complex piece of software that interfaces directly with the hardware upon which it is running: all user software runs under the operating system and will call routines within the OS in order to achieve hardware effects, such as displaying information (on screen, printer or other display device), receiving input (from keyboard, mouse, graphics tablet, touch screen etc.) and reading from or writing to storage (hard discs, **Random Access Memory** (RAM) drives, tape systems or network-based storage), as illustrated in Figure 15.1.

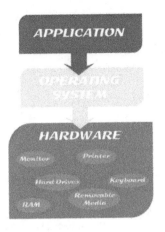

FIGURE 15.1 The operating system in a hierarchy.

It is possible to call OS routines directly from programs, for example to list all files with the extension.c in the current folder, using c:

```
struct _finddata_t c_file;
long hFile;

if ((hFile = _findfirst("*.c", &c_file)) == -1L)
printf("No *.c files in current directory");
else
{
    do
    {
        printf("%s\n", c_file.name);
    } while ( _findnext(hFile, &c_file) == 0 )
    _findclose(hFile);
}
```

15.2.1.6 Paradigms

There are two main paradigms to consider when looking at modern operating systems: event-driven and interrupt-driven.

An interrupt is something that takes place (key press, mouse click, input from I/O board) and interrupts whatever the OS is doing and sets it working on the new task straight away.

An event is something that places a new occurrence (key press, mouse click, input from I/O board) into a queue for processing when the current task in which the OS is engaged is complete.

Windows, in common with most modern operating systems, is an event-based, message-driven OS. At the heart of the OS is a mechanism known as a message pump. In code it looks as deceptively simple as this:

```
MSG msg;
while (GetMessage(&msg, NULL, 0, 0)) {
    TranslateMessage(&msg);
    DispatchMessage(&msg); // send to window proc
}
```

Every time a user takes an action that affects a window, such as resizing it, moving it or clicking the mouse in it, an event is generated. Every time an event is detected, the OS sends a message to the program so it can handle the event.

Every Windows-based program is based on events and messages and contains a main event loop that constantly and repeatedly checks to see whether any user events have taken place. If they have, then the program handles them. For example, a resize event will generate a message from the OS to ask each window whether or not the event is for them. If it is, then the program handles the event and the OS considers it done. If not, it keeps passing it on until a program does handle it or all open programs have been told about it.

Once a program receives a message that is for it, it will extract the event from the main OS queue and then process it. Often this involves passing the message though its own routines until a handler is found. Once the event (in this case a window resize) has been handled, then the program is ready to receive another event, checking its queue and with the OS. The application repeats this process until the user terminates the program.

There are three main categories of messages that a Windows program needs to handle (see Figure 15.2):

- WM_ prefixed messages (except for WM_COMMAND), handled by Windows and views[6].
- Control notifications including WM_COMMAND messages from edit controls such as listboxes indicating that the selection has changed
- Command messages such as WM_COMMAND notification messages from user-interface objects such as menus and toolbar buttons.

Table 15.1 shows the standard command route for a windows-based program.

The most common interrupt-driven OSs in use are "hard[7]" real-time systems, where specific events must take place at specific times. Examples of such are MTOS, Lynx and RTX which might be used in air traffic control, vehicle subsystems control or nuclear power plant control.

The main advantage of an interrupt-drive OS is in real time processing, such as signal acquisition. An interrupt-driven OS may not process all the signals in real time, but it won't miss any. An event-driven one will process all it sees, but may miss some.

In reality, OSs all incorporate some form of interrupt, which for event-driven ones is normally handled by a sub-system. The main form of such interrupts are I/O.

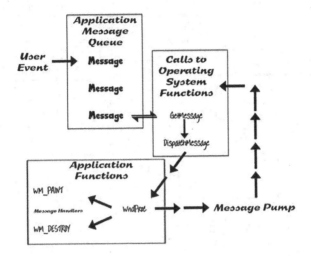

FIGURE 15.2 Messages and events in a windows-based program.

TABLE 15.1
Standard Command Route

Objects of this Type Receiving Commands	Order of Objects Offered the Command to Handle
Multiple Document Interface (MDI) frame window	Active document frame window
	This frame window
	Application
Document frame window	Active view
	This frame window
View	This view
	Document attached to the view
Document	This document
	Document template attached to the document
Dialog box	This dialog box
	Window that owns the dialog box
	Application

15.3 THE SOFTWARE LIFECYCLE

The lifecycle of a software system runs from the identification of a requirement until it is phased out, perhaps to be replaced by another system. It defines the stages of software development and the order in which these stages are executed. The basic building blocks of all software lifecycle models include:

1. Gathering and analysing user requirements
2. Software design
3. Coding
4. Testing
5. Installation & maintenance

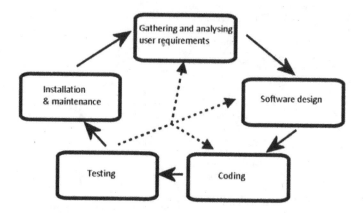

FIGURE 15.3 The stages in a software lifecycle.

Each phase produces deliverables required by the next stage in the lifecycle. These requirements are translated into design and code is produced driven by the design. Testing then verifies the deliverable of the coding stage against the requirements. This tested software is installed and maintained for its lifespan. Maintenance requests may involve additions or revisions to the requirements and the cycle then repeats, as shown in Figure 15.3. If testing reveals errors, then the process may return to any one of the three preceding stages. Release would normally come between the "Test" and "Maintain" stages.

To see how software progresses through its lifecycle, we will consider an example. A need has been identified for software for recording medical equipment maintenance. In the following sub-sections, we'll look at what's involved at each stage of this software's lifecycle.

15.3.1 REQUIREMENTS SPECIFICATION: GATHERING AND ANALYSING USER REQUIREMENTS

This phase of the lifecycle identifies the problem to be solved and maps out in detail what the user requires. The problem may be to automate a user task, to improve efficiency or productivity within a user group, to correct shortcomings of existing software, to control a device, etc. Requirements on software are usually a complex combination of requirements from a variety of users associated with a problem. Requirements specification may therefore also include systems analysis meaning that this stage may range from simply taking notes during a meeting to needing to interview and observe a wide range of users and processes.

For our example, requirements would include: a register of equipment to be maintained; a record of the service history and maintenance tasks undertaken (possibly together with parts and cost); management reports of frequency of breakdown; cost of purchase and maintenance; volume of equipment of the same type and a register of maintenance staff.

Various techniques are available for requirements gathering and analysis:

1. Observation of existing workflows, pathways and processes and any related documentation
2. Interviews with potential users – individually or in groups
3. Prototyping the concept, not necessarily as software but as mock screen-shots or storyboards
4. Use cases from different perspectives – user (at different levels), programmer, maintainer

A Use Case can help capture the requirements of a system in a simple diagram (see Figure 15.4 for an example). It consists of actors, the system, use-cases and relationships. Actors (the "stick men" in the diagram) are someone or something that uses the system, this could be a person, organisation, another system, or external device. Actors can be divided into primary and secondary, where primary actors are those which initiate the use of the system and secondary actors are reactionary in that the system requires assistance from secondary actors in order to achieve the primary actor's goal.

A central rectangle denotes the system being developed and the actors are placed outside of this. Use Cases are actions which accomplish a task, and are placed within the system, using ovals. They are presented in a logical order whenever possible[8].

Relationships complete the diagram. These denote what is done, not how it is done. Four types of relationships exist: association, include, extend and generalisation. The actors must have a relationship with at least one of the Use Cases[9].

- Associations are direct interactions and are shown with solid lines.
- Include relationships are where there is a dependency between a base Use Case and another Use Case – every time the base Use Case happens, the included Use Case also occurs. This relationship is depicted with a dotted line and an arrow pointing from the base Use Case to the included Use Case.
- Extend relationships are similar to include Use Cases, but the second Use Case only occurs occasionally, under certain circumstances, for example when it steps outside of the organisation's control. This relationship is either depicted with a dashed line from the extended Use Case to the base Use Case or, as in this diagram, with the notation <<extend>>.
- A generalisation relationship is where a parent Use Case has specialised (child) Use Cases. This relationship is shown with an elbow connector towards the base (parent) Use Case.

Requirements should be analysed for consistency and feasibility. It is essential for software requirements to be verifiable using available resources during the software acceptance stage. A change management process for requirements (which we'll come to under "Software Quality Assurance", section 16.1.2) will also be helpful depending on the expected lifespan of the software.The final step is to record these requirements including functional ones (e.g. what training data needs to be recorded),

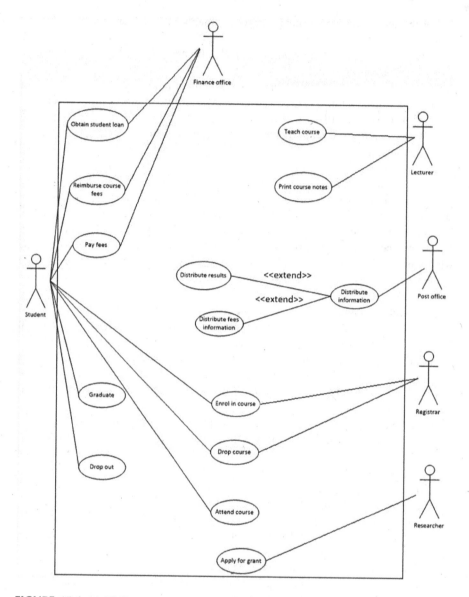

FIGURE 15.4 A UML use case for an education system. (Adapted from Agile Modeling 2022 [online].)

non-functional ones (e.g. how to ensure data confidentiality) and design aspects (what tools are required to build the software), etc. in a **Software Requirements Specification** (SRS). A prototype may also be built during this stage to help the user visualise the proposed software solution. This prototype might just consist of screen layouts – it doesn't have to actually do anything at this stage.

15.3.2 Software Design

Software design typically involves two levels of design – namely architectural and detailed design. The architectural design specifies the basic components of the software system such as the user interface, database, reporting module, etc. often using tools such as **Data Flow Diagrams** (DFD) (such as Figure 15.5) and **Entity-Relation Diagrams** (ERD) (such as Figure 15.12). Detailed design elaborates on each of these components in terms of tables and fields in the database, layout and data to be displayed on the graphical user interface, and often pseudo-code for any data analysis modules.

Common design strategies adopted include:

Procedural – Software is broken down into components based on functionality following a top-down approach and the process continues with each component until sufficient detail has been achieved. (We will do this later in this

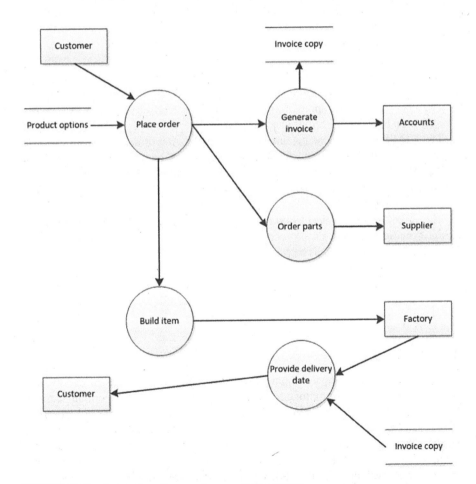

FIGURE 15.5 A data flow diagram for an ordering system.

chapter.) For our example this would mean different components for staff login, data management, reporting and so on. Data management can then be further sub-divided into data entry, data storage and so on.

Object-oriented – Software is described in terms of objects and their inter-actions with each other. This enables a component-based approach to be followed enabling modular deployment and improving reuse. In develop-ing object-oriented software, the design can be greatly assisted through the use of UML tools. The objects for our example would be staff member (which could be used as a base class for manager, technician etc.), item of equipment, maintenance record etc. Interactions are defined between these objects. For example, 'staff member' updates 'item of equipment', adding one or more 'maintenance record'.

Data-oriented – In this case, the input and output data structures are defined and functions and procedures are simply used as a tool to enable data trans-formation. For our example, data structures would be defined for an item of equipment together with detailed functions for its insertion, updating, verification and deletion and the software is then constructed around this basic framework.

A **Requirements Traceability Matrix** (RTM) (see Table 15.2) is used to map design components to requirements and can also be used to ensure that all the key requirements have been captured. The requirements traceability matrix report shows a correlation between the requirements and the tests.

The RTM links requirements throughout the validation process and its purpose is to ensure that all requirements defined for a system are tested in the test protocols. The RTM is thus a tool both for the validation team, to ensure that requirements are not lost during the validation project, and for auditors, to review the validation documentation.

The RTM is usually developed along with the initial list of requirements. As design specifications and test protocols are developed, the RTM is updated to include these. Requirements can thus be traced to the specific test step in the testing protocol in which they are tested and any omissions easily identified.

Risk analysis and management options are commonly carried out at this stage. Identifying potential problems or causes of failure at this stage may influence the development stage of the software.

15.3.3 CODING

This phase of the software development lifecycle converts the design into a complete software package. It brings together the hardware, software and communications elements for the system. It is often driven by the detailed design phase and must take into consideration practical issues in terms of resource availability, feasibility and technological constraints. Choice of development platform is often constrained by availability of skills, software and hardware. A compromise must be found between the resources available and the ability to meet software requirements. A good pro-grammer rarely blames the platform for problems with the software.

TABLE 15.2

A Requirements Traceability Matrix. In this Example, the Columns Show the Requirements and the Rows Show the Tests to be Performed. The Xs Show Which Requirements are Covered by Which Tests[10]

RTM / Test	Requirement #	1 A Register of Equipment to be Maintained	2 A Record of the Service History	3 Maintenance Tasks Undertaken	4 Frequency of Breakdown Report	5 Cost of Purchase and Maintenance Report	6 Volume of Equipment of the Same Type Report	7 Register of Maintenance Staff
# Covered		X	X	X	X	X	X	X
Data structures: field list	1	X	X	X				X
Reports: information returned	2				X	X	X	
User interface: data entry	3		X	X				X

If the required development platform is not already well understood, then this stage will also have to encompass staff development. As far as possible it is good practice for departments to select a family of tools and to use them for several projects so that local expertise can be efficiently developed. It is also at this point that any limitations in the design stage will become apparent.

Installing the required development environment, development of databases, writing programs, and refining them are some of the main activities at this stage. More time spent on detailed design can often cut development time, however technical stumbling blocks can sometimes cause delays. Software costing should take this into account at the project initiation stage.

For our example, a database will need to be constructed to hold the staff details, equipment details and historical records. User interfaces will be required for data management, along with the database routines that these invoke. Logic will be required to implement the different functional roles (technician, manager etc.). The backend should be selected taking into account the expected number of records, simultaneous multi-user access and software availability.

Software coding should adhere to established standards and be well documented. The basic principles of programming are simplicity, clarity and generality. The code should be kept simple, modular and easy to understand for both machines and humans. The written code should be generalised and reusable as far as possible and adaptable to changing requirements and scenarios. Automation and reduced manual intervention will minimise human errors.

Unit testing is often included in this phase of the software lifecycle as it is an integral part of software development. It is an iterative process during and at the end of development. This includes testing error handling, exception handling, memory usage, memory leaks, connections management, etc., for each of the modules independently.

15.3.4 Testing

Testing software is often viewed as an art rather than a science, due to a number of factors:

- It frequently takes longer and requires more effort than anticipated (it is often estimated that debugging takes more time than programming[11])
- Software will fail. If we are very lucky, when the first attempt is made to run it, it will crash spectacularly before any serious harm is done. In some cases problems will make it difficult or impossible to actually run the software, for example because the user interface does not function as expected. In a few cases, the results delivered by the system will be so gloriously wrong that anyone with sufficient familiarity with the domain in which the software is used will realise that an error has occurred. Finally, the scenario which should keep us awake at night: an error small enough (or worse, infrequent enough) to be missed by users causes a significant error in the diagnosis or treatment of a patient.

- Successful tests do not affect the software's quality. It is very tempting to believe that in the process of putting our software through numerous successful tests we have somehow improved it. Unfortunately this is not the case. Only when tests are failed (and corrections implemented) is the quality of the software improved[12].
- Testing will not identify all defects in software. Probably the biggest disillusionment which newcomers experience upon entering the world of software testing is that for any non-trivial project it is simply not possible to exhaustively test it in the strict sense of "exhaustive". A moment's reflection will reveal why. Let us consider a simple system in which all the data to be processed is available at the beginning of the software's execution. Suppose each data item (datum) can be represented by a variable able to take one of a range of values. Even if each datum can only take one of a relatively small number of values the number of possible combinations rises exponentially with the number of inputs. In many cases (and particularly in scientific computing) the number of possible values each datum may take is huge (think about the number of possible values a decimal value may take – which, depending upon implementation details, is typically 2^{32}). To exercise all possible input combinations of a simple program to add two such decimal values would require 2^{64} (= 1.8 x 10^{19}) separate tests. Assuming that the system is being executed on a reasonably well-specified PC which might perform 10^{11} floating point[13] operations per second, then just the execution of such a test (without even considering the time taken to evaluate it) would take 1.8 x 10^8 s (= 5.8 years). Most real-world problems (which will be rather more complex than summing two decimal values) are inherently beyond the reach of such a "brute force" approach. Adding in further complexity, consider the simple piece of 2-threaded code shown in Figure 15.6. On simple inspection, this code would be expected to produce a final value of x of between 10 and 20. However, it can produce values as low as 2 in 90 steps. (As an aside on complexity, this simple piece of code has in excess of 77,000 states) (Hobbs 2012). Testing therefore becomes a statistical activity in which it is recognised that the same code, with the same input conditions, may not yield the same result every time.

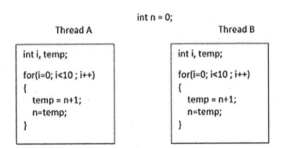

FIGURE 15.6   A simple two-threaded program.

- Testing will not identify all defects in a software product. A number of strategies have been developed in the hope that a carefully selected set of test cases will exercise the software sufficiently to expose any bugs which exist. Some of these are discussed further below, but inevitably bugs will remain in the software when released[14].

The "V" model of software development (see Figure 15.7) is often viewed as the most appropriate for medical device software. This model builds testing into the development process such that each of the stages of software design are linked to the creation and execution of a corresponding test. As each test level is completed, a test report is produced detailing the results achieved. Note that this scheme divides testing between validation (which demonstrates that what we have produced satisfies the user's requirements) and verification (which demonstrates that the output of each stage of the development process satisfies the requirements of the preceding stages).

Newcomers to the "V" model may find promoting test design above code construction somewhat disconcerting. Just remember that it should take no longer in this revised order and that it is a task which has to be completed anyway!

The model also determines when the tests are to be performed. Put simply, they are all performed at the earliest stage at which it makes sense to execute each test specification. For example, once individual units (e.g. classes, subroutines or functions) have been written they are each tested in isolation in the unit test stage. When these units are assembled into functional groups or modules then these are tested in the integration test stage, which reveals any issues in program organisation and the interfaces between the units. Having already tested each of the individual units we should not (hopefully) be distracted from the integration test by unit failures.

Software testing is an ongoing process along the development to maintenance path. There are 3 main levels of testing:

1. Unit testing: individual modules are tested against set criteria.
2. Integration testing: the relationships between modules are tested.
3. System testing: the workings of the software system as a whole are tested.

Test criteria should include functionality, usability, performance and adherence to standards. Test cases are usually generated during the design stage for each level of testing. These may be added to or modified along the pathway but should, at least, cover the basic criteria. Testing objectives also influence the set of test cases. For example, the test cases for acceptance testing might differ from those for a beta test or even a usability test.

There are many ways of testing software and these are very closely linked to validation and verification. The primary purpose of validation and verification (when applied to medical software) is safety. Functionality is a secondary purpose, although without functionality the code is pointless. The way to reconcile this is to consider the consequences of a failure of the code: a system that is 75% functional but 100% safe is still usable (albeit annoying) – if the figures are reversed, it is not.

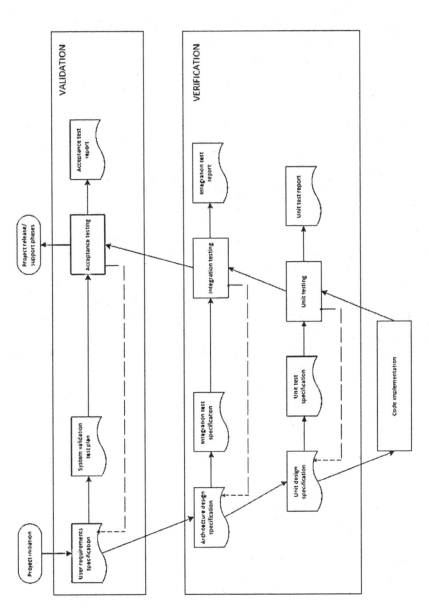

FIGURE 15.7 "V" Software Development Model. The dashed arrows indicate iteration in the design which may be required in the light of the experience gained by working through a set of tests. In some representations of this model a distinction is drawn between software units and software components (the latter may represent more than one unit) and this is depicted as an additional layer between architecture and unit specification. (Green et al. 2016).

In validating and verifying a system as safe, one starts from the premise that all software contains "bugs". These "bugs" may be classified as faults, errors or failures.

A fault is a mistake in the design or code, which may lead to an error (but equally may not), such as declaring an array to be the wrong size. An error is unspecified behaviour in execution, which may lead to a failure, such as messages starting with non-numeric codes being discarded as they evaluate to zero. A failure is the crossing of a safety threshold due to an uncontained error.

There are two main approaches to testing, often referred to as "black box" and "white box". Applying this to software testing, the "box" is the program, or module, that is to be tested.

In black box testing, the contents of the box are unknown[15]. Therefore, tests comprise of a known set of inputs and the predetermined output that this should provide. This is very useful when the software has been commissioned using an **Output-Based Specification** (OBS) or for end-user testing. It also removes any effect that may be caused by the application of the debugger environment itself[16].

In white-box testing (also known as clear box, glass box or transparent box testing, which may be a better descriptor of the process) the contents of the box are known and are exposed. In software terms, this may mean that the source code is available or even that the code is being tested in the development environment via single-stepping. It is therefore usually applied to structures or elements of a software system, rather than to its whole. It is also not unusual for a black box failure to be investigated using white box testing.

In generic terms, therefore, black box testing is functional testing whereas white box testing is structural or unit testing. Thus a large system comprising multiple components will often have each component white box tested and the overall system black box tested in order to test the integration and interfacing of the components. In upgrading a software system, it will be black-box testing that is undertaken as the code will not be available.

Testing should normally be undertaken by someone different to the software author. A draft BCS standard (2001) (British Computer Society Specialist Interest Group in Software Testing 2001 [online]) lists the following increasing degrees of independence:

a) the test cases are designed by the person(s) who writes the component under test;
b) the test cases are designed by another person(s);
c) the test cases are designed by a person(s) from a different section;
d) the test cases are designed by a person(s) from a different organisation;
e) the test cases are not chosen by a person.

There are multiple test case design techniques with corresponding test measurement techniques. The following non-exhaustive list is based on ISO 29119-4 "Software and systems engineering - Software testing - Part 4: Test techniques" (ISO/IEC/IEEE 29119-3:2013). Not all will be required or applicable for each project.

- Specification-based test design techniques (Black Box)
  - Equivalence partitioning
  - Classification Tree Method
  - Boundary-value analysis
  - Syntax testing
  - Decision table testing
  - Cause-effect graphing
  - State transition testing
  - Scenario testing
  - Random testing
- Structure-based test design techniques (White box)
  - Statement Testing
  - Branch Testing
  - Decision Testing
  - Modified Condition Decision Coverage (MCDC) Testing
  - Data Flow Testing
- Experience-based test design techniques
  - Error guessing

It is instructive to examine one of these, together with its corresponding measurement technique. The one we will select is Boundary Value Analysis. This takes the specification of the component's behaviour and collates a set of input and output values (both valid and invalid). These input and output values are then partitioned into a number of ordered sets with identifiable boundaries. This is done by grouping together the input and output values which are expected to be treated by the component in the same way: thus they are considered equivalent due to the equivalence of the component's behaviour. The boundaries of each partition are normally the values of the boundaries between partitions, but where partitions are disjoint the minimum and maximum values within the partition are used. The boundaries of both valid and invalid partitions are used.

The rationale behind this method of testing is the premise that the inputs and outputs of a component can be partitioned into classes that will be treated similarly by the component and, secondly, that developers are prone to making errors at the boundaries of these classes.

For example, a program to calculate factorials

This program would have the partitions (assuming integer input only):

- Partition a: $-\infty$ to 0 (not defined)
- Partition b: 1 to n (where n! is the largest integer the component can handle, so for a standard C integer with a maximum value of 2147483647, n would be 12)
- Partition c: n+1 to $+\infty$ (unable to be handled)

The boundary values are therefore 0, 1, n and n+1. The test cases that are used are three per boundary: the boundary values and ones an incremental distance to either side. Duplicates are then removed, giving a test set in our example of {-1, 0, 1, 2, n-1, n, n+1, n+2}. Each value produces a test case comprising the input value, the boundary tested and the expected outcome. Additional test cases may be designed to ensure invalid output values cannot be induced. Note that invalid as well as valid input values are used for testing.

It can clearly be seen that this technique is only applicable for black-box testing.

The corresponding measurement technique (Boundary Value Coverage) defines the coverage items as the boundaries of the partitions. Some partitions may not have an identified boundary, as in our example where Partition a has no lower bound and Partition c no upper bound. Coverage is calculated as follows:

$$\text{Boundary Value Coverage} = \frac{\text{number of distinct boundary values executed}}{\text{total number of boundary values}} \cdot 100\%$$

In our example, the coverage is 100% as all identified boundaries are exercised by at least one test case (although $+\infty$ and $-\infty$ were listed as the limits of partitions c and a, they are not boundaries as they indicate the partitions are unbounded). Lower levels of coverage would be achieved if all the boundaries we had identified were not all exercised, or could not be (for example, if 2147483647 was required to be tested, where 2147483648 is too large to be stored). If all the boundaries are not identified, then any coverage measure based on this incomplete set of boundaries would be misleading.[17]

15.3.4.1 Acceptance Testing

Acceptance testing is the stage at which the final product is assessed. In practice it can be divided into two components: system testing and user acceptance testing.

System testing demonstrates that the complete product fulfils each of the requirements set out in the user requirements specification. The system test specification describes a series of test cases based upon each of these requirements and defines the expected result. As this has been written prior to details of the code implementation being considered, it is regarded as "black box" testing. That is, no regard is taken of the inner logic or workings of the system.

Areas typically covered in the system test include:

- Correctness: The results of the computation as delivered by the system as a whole are those predicted for particular sets of valid input data.
- Resilience: Incorrect, out-of-range, or malformed input data are detected and flagged and the system subsequently behaves in a predictable manner.
- Performance: The software delivers the required result within an agreed time under specified conditions of load (e.g. typical number of users in a multi-user system).
- Maximum load: The system is able to successfully perform an agreed maximum number of concurrent operations and deliver results within a specified time.
- Endurance: The system continues to behave within its design constraints after extended periods (especially periods of heavy load).

- Security: access to information and the extent to which it may be edited or deleted is limited to those users specifically assigned such rights. Data stored within the system is protected against accidental or deliberate tampering.
- Environment dependencies: the requirements specification should include details of the environment(s) in which the system will be expected to operate. For example, it may require particular hardware (e.g. memory, CPU performance, hard disk availability) or operating system (e.g. a particular version of Windows). The system should flag to the user if these requirements are not met, and either shut down or degrade gracefully[18].

User Acceptance Testing differs from the other test regimes described above in that it is normally conducted either by the user or group for whom the software was developed (or a person acting on their behalf) independent of the development team. Passing the User Acceptance Test indicates that as far as this client is concerned the project's goals have been achieved. Typically this testing will be based around a series of tasks which a user will attempt to perform. Failure might indicate that the coding or algorithms implemented are faulty, however errors of this type should already have been detected in unit and integration tests. Therefore failures recorded during acceptance testing will frequently result from usability issues (such as confused layout/clarity of displayed data, poor design of the user interface, and inadequate support for visually impaired users).

15.3.5 INSTALLATION AND MAINTENANCE

Software maintenance is defined in the IEEE Standard for Software Maintenance[19] as "the modification of a software product after delivery to correct faults, to improve performance or other attributes, or to adapt the product to a modified environment". It lasts for the lifespan of the software and requires careful logging and tracking of change requests as per the guidelines for change management set out at the end of the requirements phase.

A maintenance request often goes through a lifecycle similar to software development. The request is analysed, its impact on the system as a whole determined and any required modifications are designed, coded, tested and finally implemented. Training and day-to-day support are also core components of the software maintenance phase. It is therefore essential for the maintainer to be able to understand the existing code[20]. Good documentation and clear and simple coding at the development stage will be most helpful at this point especially if the developer is not available or if there has been a long time gap since development.

There are many tools that provide help with the discipline of software development, such as, in Unix/Linux, SVN[21] and 'make', which codifies instructions for compiling and linking. Document management systems such as OSS wiki and Q-Pulse (Ideagen 2022 [online]) can help departments log their activities, including the queries that are raised during the complete software lifecycle.

15.4   SOFTWARE LIFECYCLE MODELS

The software lifecycle models help to manage the software development process from conception through to implementation within time and cost constraints. We consider four such here.

15.4.1   WATERFALL MODEL

The Waterfall methodology (also known as a linear-sequential model) forms the heart of all other development methodologies, so it is worth examining first. It is also very simple to understand and use and follows a structured sequential path from requirements to maintenance. It sets out milestones at each stage which must be accomplished before the next stage can begin.

The waterfall methodology follows a simple idea: each part of the process forms a complete unit and progress only passes to the next unit once its predecessor is complete. There is clearly a great alignment with the PRINCE 2 (PRINCE2 2022 [online]) project management technique (for "unit" read "stage" and it becomes very clear, so we will use this terminology from now on) which is probably why, despite being much maligned, Waterfall is still in very common use today. It is also very simple to understand (and therefore simple to explain to non-technical members of a project team or board – see Figure 15.8 for a graphical representation) and transfers easily into milestones and deliverables which can be measured.

The rigidity of the Waterfall model aids project management with well-defined milestones and deliverables. It does however restrict flexibility and does not provide much scope for user feedback until software development has been completed. It is only suitable for small-scale projects where user requirements are clearly defined and unlikely to change over the software lifespan.

The major flaw of a Waterfall model is that there is no way to return to an earlier stage. In other words, features that have been omitted from the design cannot be added in: the final product is exactly what was specified, whether or not that is now what is required. New features may be added post-implementation, but that is a project in itself which will follow the same lifecycle model.

15.4.2   INCREMENTAL MODEL/PROTOTYPING MODEL

The incremental model is an intuitive amendment to the waterfall model. In this model, there are multiple iterations of smaller cycles involving the key stages of formalised requirements, design, development, and testing. Each cycle produces a prototype of the software. Subsequent iterations thus improve or build on the previous prototype. A graphical representation of this is shown in Figure 15.9.

This approach is more suited to the small scale development often experienced by Physics and Engineering Departments but only really works if it is properly controlled. It is essential that both the disadvantages and advantages of this approach are understood before commencing a project using such a methodology (and especially before deciding against the Spiral method which we will cover in the next section).

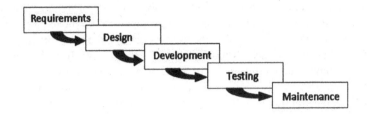

FIGURE 15.8 The waterfall model.

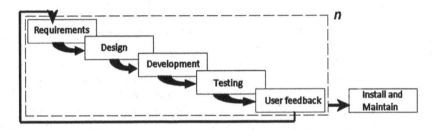

FIGURE 15.9 The incremental waterfall model.

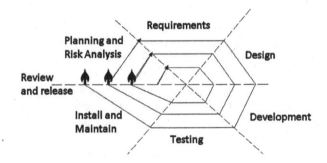

FIGURE 15.10 The spiral model.

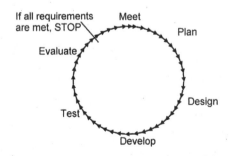

FIGURE 15.11 The agile model.

The obvious advantage that this methodology gives is that requirements that are difficult to completely define in advance of development commencing can progress through several prototypes, assisting the user in planning the final software's use and it can help to determine the full requirements for the system. It is also an effective method to demonstrate the feasibility of a certain approach for novel systems where it is not clear whether constraints can be met or whether algorithms can be developed to implement the requirements. Testing and code management is simpler with smaller cycles and errors can be detected early, allowing them to be corrected by re-design rather than by patches. As users have a tendency to change their minds in specifying requirements once they have seen the software, this method of developing systems allows team-development (see process models, section 15.5).

The major problems with incremental models are the "scope creep" that can often result from frequent user feedback incrementally increasing the requirements at each iteration and also the associated risk of getting stuck in a never-ending development loop with no final product being released. Another potential problem is that, as the requirements for the entire system are not gathered at the start of the project, the system architecture might be affected at later iterations: the initial design (especially the software infrastructure such as entities and classes) may not be the optimal one for the final product. This introduces the risk of a build-and-patch approach through development, leading to poor code design and increasing the complexity of testing.

15.4.3 Spiral Model

The Spiral Model was designed to include the best features from the Waterfall and Prototyping Models. It is similar to the incremental model, but at the end of each iteration (called a spiral in this model) a robust working version of the software is released for user evaluation. This release will not be complete in functionality, but – having completed testing – will be safe. Users are thus able to work with the software (albeit in a reduced capacity) early in the development cycle, which can alleviate some of the constraints around development time. A graphical representation is shown in Figure 15.10.

The development of each version of the system is carefully designed using the steps involved in the Waterfall Model. The first version is called the 'baseline spiral' and each subsequent spiral builds on the baseline spiral, each producing a new version with increased functionality. The theory is that the set of requirements is hierarchical in nature, with additional functionality building on each preceding release. The spiral model specifies risk analysis and management explicitly which helps to keep the software development process under control. This is a good model for systems where the entire problem is well defined from the start, such as modelling and simulation software, but not so much for database projects where most functions are essentially independent.

15.4.4 Agile Methodology

Web-based software and mobile applications have led to a requirement for quick releases and less focus on the traditional requirements→design→development process. The Agile methodology stresses collaboration over documentation and constant evolution over detailed design.

Agile software development is a set of frameworks and practices based on the Agile Manifesto[22] that provides guidance on how to adapt and respond quickly to requirement changes in a constantly evolving environment.

Popular Agile frameworks include Scrum and Kanban. An agile software development process always starts by defining the users and the vision for the problem to be solved. The problem is then further sub-divided into user stories which are prioritised and delivered in chunks at regular intervals called sprints. Each sprint delivers a releasable product increment. A graphical representation is shown in Figure 15.11.

The main difference between the agile methodology and the three previously described is a change in mindset. Unlike the waterfall model where a solution to the entire problem is considered at the outset, Agile aims to look at specific use cases or user stories[23] at a time and develop and test those with users. This provides the ability to change tracks quickly when requirements change.

The models discussed here (including the "V" model we met in software testing, Figure 15.7) are some of the basic models and individual software projects may sometimes combine techniques from different models to suit their specific needs. Medical Physics/Clinical Engineering projects often follow an iterative approach since the requirements are not always clearly defined and they may undergo frequent revisions. It is essential in such cases to maintain a robust software change control process (which leads nicely into "Software Quality Assurance", next chapter) and to document **Corrective And Preventative Actions** (CAPA), as no software is ever perfect, bug-free or finished, therefore.

So why use one (Adapted from CoFounders Lab 2016 [online])? A 2015 study by the Standish Group across 50,000 software projects discovered that:

- Only 29% of projects were completely successful
- 52% had major challenges
- 19% failed completely

And it was these reasons:

1. The wrong development partner/team
2. The wrong development model. (Agile vs. Waterfall)
3. Poor understanding of software development.
4. Unrealistic/risky goals for development.
5. Wrong expectations set by developer.

It has been argued (e.g. by the blog that furnished this information) that Agile is far superior to Waterfall for reasons such as:

- Speed-to-market: it is estimated that about 80% of all market leaders were first to market. Agile development philosophy supports the notion of early and regular releases, and 'perpetual beta'.

- Quality: A key principle of agile development is that testing is integrated throughout the lifecycle, enabling regular inspection of the working product as it develops.
- Business Engagement/Customer Satisfaction: Agile development principles encourage active 'user' involvement throughout the product's development and a very cooperative collaborative approach.
- Risk Management: Small incremental releases made visible to the product owner and product team through its development help to identify any issues early and make it easier to respond to change or to rollback to an earlier version.
- Cost Control: An Agile approach of fixed timescales and evolving require- ments enables a fixed budget so that the scope of the product and its features are variable, rather than the cost.
- Right Product: Agile development requirements emerge and evolve, and the ability to embrace change (with the appropriate trade-offs) enables the team to build the right product.
- More Enjoyable: debatable, especially for the project manager.

15.5 OVERVIEW OF PROCESS MODELS AND THEIR IMPORTANCE

This section has been adapted from Scacchi 2001.

In contrast to software lifecycle models, software process models often represent a networked sequence of activities, objects, transformations, and events that embody strategies for accomplishing software evolution. Such models can be used to develop more precise and formalised descriptions of software lifecycle activities. Their power emerges from their utilisation of a sufficiently rich notation, syntax, or semantics, often suitable for computational processing.

Software process networks can be viewed as representing multiple interconnected task chains. Task chains represent a non-linear sequence of actions that structure and transform available computational objects (resources) into intermediate or finished products.

Non-linearity implies that the sequence of actions may be non-deterministic, iter- ative, accommodate multiple/parallel alternatives, as well as partially ordered to account for incremental progress. Task actions in turn can be viewed as non-linear sequences of primitive actions which denote atomic units of computing work, such as a user's selection of a command or menu entry using a mouse or keyboard. These units of cooperative work between people and computers are sometimes referred to as "structured discourses of work", while task chains have become popularised under the name of "workflow".

Task chains can be employed to characterise either prescriptive or descriptive action sequences. Prescriptive task chains are idealised plans of what actions should be accomplished, and in what order. For example, a task chain for the activity of object-oriented software design might include the following task actions:

- Develop an informal narrative specification of the system.
- Identify the objects and their attributes.
- Identify the operations on the objects.

- Identify the interfaces between objects, attributes, or operations.
- Implement the operations.

Clearly, this sequence of actions could entail multiple iterations and non-procedural primitive action invocations in the course of incrementally progressing toward an object-oriented software design.

Task chains join or split into other task chains resulting in an overall production network or web. The production web represents the "organisational production system" that transforms raw computational, cognitive, and other organisational resources into assembled, integrated and usable software systems. The production lattice therefore structures how a software system is developed, used, and maintained. However, prescriptive task chains and actions cannot be formally guaranteed to anticipate all possible circumstances or idiosyncratic foul-ups that can emerge in the real world of software development. Thus, any software production web will in some way realise only an approximate or incomplete description of software development.

Articulation work is a kind of unanticipated task that is performed when a planned task chain is inadequate or breaks down. It is work that represents an open-ended non-deterministic sequence of actions taken to restore progress on the disarticulated task chain, or else to shift the flow of productive work onto some other task chain. Thus, descriptive task chains are employed to characterise the observed course of events and situations that emerge when people try to follow a planned task sequence.

Articulation work in the context of software evolution includes actions people take that entail either their accommodation to the contingent or anomalous behaviour of a software system, or negotiation with others who may be able to affect a system modification or otherwise alter current circumstances. This notion of articulation work has also been referred to as software process dynamism.

15.5.1 COMPARISON OF PROCESS MODELS

There are several process models, some of which we covered in the software life-cycle (especially the Spiral model). Others include:

15.5.1.1 Joint Application Development

Joint Application Development (JAD) is a technique for engaging a group or team of software developers, testers, customers, and prospective end-users in a collaborative requirements gathering and prototyping effort. JAD is quintessentially a technique for facilitating group interaction and collaboration. Consultants often employ JAD as do external software system vendors who have been engaged to build a custom software system for use in a particular organisational setting. The JAD process is based on four ideas:

1. People who actually work at a job have the best understanding of that job.
2. People who are trained in software development have the best understanding of the possibilities of that technology.
3. Software-based information systems and business processes rarely exist in isolation -they transcend the confines of any single system or office and

affect work in related departments. People working in these related areas have valuable insight on the role of a system within a larger community.

4. The best information systems are designed when all of these groups work together on a project as equal partners.

Following these ideas, it should be possible for JAD to cover the complete development lifecycle of a system. The JAD is usually a 3 to 6 month well-defined project, when systems can be constructed from commercially available software products that do not require extensive coding or complex systems integration. For large-scale projects, it is recommended that the project be organised as an incremental development effort, and that separate JAD's be used for each increment[24]. Given this formulation, it is possible to view open source software development projects that rely on group email discussions among globally distributed users and developers, together with internet-based synchronised version updates, as an informal variant of JAD.

15.5.1.2   Assembling Reusable Components

The basic approach of reusability is to configure and specialise pre-existing software components into viable application systems. Such source code components might already have associated specifications and designs associated with their implementations, as well as have been tested and certified. However, it is also clear that software domain models, system specifications, designs, test case suites, and other software abstractions may themselves be treated as reusable software development components. These components may have a greater potential for favourable impact on reuse and semi-automated system generation or composition. Therefore, assembling reusable software components is a strategy for decreasing software development effort in ways that are compatible with the traditional lifecycle models.

The basic dilemmas encountered with reusable software componentry include (a) acquiring, analysing and modelling a software application domain, (b) how to define an appropriate software part naming or classification scheme, (c) collecting or building reusable software components, (d) configuring or composing components into a viable application, and (e) maintaining and searching a components library. In turn, each of these dilemmas is mitigated or resolved in practice through the selection of software component granularity[25].

The granularity of the components varies greatly across different approaches. Most approaches attempt to utilise components similar to common (textbook) data structures with algorithms for their manipulation: small-grain components. However, the use/reuse of small-grain components in and of itself does not constitute a distinct approach to software development (otherwise all software would be written from scratch and nothing would be re-used). Other approaches attempt to utilise components which resemble functionally complete systems or subsystems[26], known as large-grain components. The use/reuse of large-grain components guided by an application domain analysis and subsequent mapping of attributed domain objects and operations onto interrelated components is an alternative approach to developing software systems.

There is, though, one downside to reusable components in that a component that works perfectly in one environment may not do so in another. The Therac-25 error (discussed in "Safety Cases" section 18.8.1) was caused by code that worked fine in the Therac-20, because the Therac-20 had hardware safety locks to prevent overdoses (which the Therac-25 did not).

There are many ways to utilise reusable software components in evolving software systems. However, studies suggest that their initial use during architectural or component design specification is a way to speed implementation. They might also be used for prototyping purposes if a suitable software prototyping technology is available.

15.5.1.3 Application Generation

Application generation is an approach to software development similar to reuse of parameterised, large-grain software source code components. Such components are configured and specialised to an application domain via a formalised specification language used as input to the application generator. Common examples provide standardised interfaces to database management system applications, and include generators for reports, graphics, user interfaces, and application-specific editors.

Application generators give rise to a model of software development whereby traditional software design activities are either all but eliminated, or reduced to a data base design problem. The software design activities are eliminated or reduced because the application generator embodies or provides a generic software design that should be compatible with the application domain. However, users of application generators are usually expected to provide input specifications and application maintenance services. These capabilities are possible since the generators can usually only produce software systems specific to a small number of similar application domains, and usually those that depend on a data base management system.

15.6   SYSTEMS DESIGN METHODS

Systems design is the process of defining the architecture, components, modules, interfaces, and data for a system to satisfy specified requirements. Systems design could be seen as the application of systems theory to product development. There is some overlap with the disciplines of systems analysis, systems architecture and systems engineering.

There are several systems design methodologies, of which possibly the most famous is **Structured Systems Analysis And Design Method** (SSADM). SSADM was produced for the Central Computer and Telecommunications Agency (now the Office of Government Commerce), a UK government office concerned with the use of technology in government, from 1980 onwards. It is essentially a Waterfall-based methodology.

Others include Joint Application Development which we covered earlier.

If the broader topic of product development "*blends the perspective of marketing, design, and manufacturing into a single approach to product development*" (Ulrich et al. 2019), then design is the act of taking the marketing information and creating the design of the product to be manufactured. Systems design is therefore the process of defining and developing systems to satisfy the specified requirements of the user.

Until the 1990s, systems design had a crucial and respected role in the data processing industry. At that time standardisation of hardware and software resulted in the ability to build modular systems. The increasing importance of software running on generic platforms has enhanced the discipline of software engineering.

Object-oriented analysis and design methods are becoming the most widely used methods for computer systems design. UML has become the standard language in object-oriented analysis and design. It is widely used for modelling software systems and is increasingly used for designing non-software systems and organisations.

The architectural design of a system hinges on the design of the systems architecture which describes the structure, behaviour, and other views of that system and analysis.

The logical design of a system pertains to an abstract representation of the data flows, inputs and outputs of the system. This is often conducted via modelling, using an over-abstract[27] (and sometimes graphical) model of the actual system. Logical design includes ERDs.

The ERD in Figure 15.12 contains six entities - supplier, customer, order, component, product and shipment. There are six relationships such as supplied by, ordered by etc. The "1" and "M" represents "one" and "many" in describing the relationship.

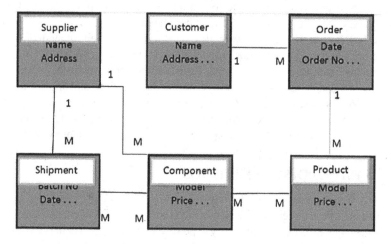

FIGURE 15.12　An entity relationship diagram example showing a simple database for the business order process.

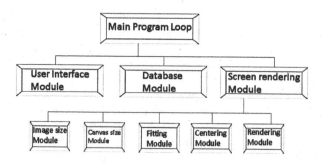

FIGURE 15.13　Top-down design example.

The physical design relates to the actual input and output processes of the system. This is explained in terms of how data is input into a system, how it is verified/ authenticated, how it is processed, and how it is displayed. In physical design, the following requirements about the system are decided:

- Input requirements,
- Output requirements,
- Storage requirements,
- Processing requirements,
- System control and backup or recovery requirements.

Put another way, the physical portion of systems design can generally be broken down into three sub-tasks:

- User Interface Design
- Data Design
- Process Design

User Interface Design is concerned with how users add information to the system and with how the system presents information back to them. Data Design is concerned with how the data is represented and stored within the system. Finally, Process Design is concerned with how data moves through the system, and with how and where it is validated, secured and/or transformed as it flows into, through and out of the system. At the end of the systems design phase, documentation describing the three sub-tasks is produced and made available for use in the next phase.

Physical design, in this context, does not refer to the tangible physical design of an information system such as the monitor resolution or keyboard placement or even the input device (mouse/trackpad etc.). Physical design in this context refers to the input and output processes such as a detailed design of a database structure processor and a control processor (i.e. usually software rather than hardware).

Two major approaches are used in system design: top-down and bottom-up.

Top-down is essentially problem decomposition: breaking the system down into components, each of which may be further subdivided until modules that can be constructed are defined, as shown in Figure 15.13

In comparison, bottom-up design works the other way around: it defines the modules and then builds them up into a complete system. This approach can be very helpful where libraries of code are to be utilised, especially those interfacing with hardware. New base-level functionality cannot be constructed, so the system has to be built from what is available. Bottom-up may also be a better way of handling constraints within a specification[28].

15.6.1   TOP-DOWN EXAMPLE

We start out by defining the solution at the highest level of functionality and breaking it down further and further into small routines that can be easily documented and coded. As an example, we'll look at reading in a set of numbers from a file, sorting them, and writing them back out again to another file.

First we need to look at how we're going to sort them and we'll use bubblesort, generally viewed as the simplest of the efficient methods. Bubblesort repeatedly works its way through a list, comparing items in pairs and swapping them if they're in the wrong order.

Structured programming is an iterative approach, in that it takes the output from one step as the input to the next. So in step 1 we identify the major modules. In our example they are: Read from a file; Sort (using bubblesort); Write to a file.

Step 2 is to break each major module down into smaller modules, so for the "read from file" module we may write:

- Ask the user for the name of the file to read from
- Open the file
- Read the numbers in the file
- Close the file

Then we repeat it again, so for "Read the numbers in the file" we break it down into:

```
While there are numbers to be read
    {
        Read the next number
    }
```

Note the use of the curly brackets to enclose a group of commands (even though there's only one in this example) that is controlled by the statement just before the open bracket ("While" in this case).

If we apply this technique to the bubblesort itself, we get:

```
For as many times as there are numbers
{
    For each pair of numbers
    {
        if they are in the wrong order
        {
            Swap them
        }
    }
}
```

Again note the use of the curly brackets to enclose blocks – in these examples there is only one statement in each block, but there could be multiples. Note also the "blocks within blocks" and the use of indentation to easily see where a block starts and ends – a useful way of spotting when a closing bracket has been omitted.

It should also be noted that this is a very good example of pseudocode – it has no language-dependent lines, saying what is to be done but not how to do it.

The basic principle of structured programming is therefore: If something is complex/difficult, break it down into easier pieces. This is often known as "Problem decomposition" and is common in most maths-based science.

NOTES

1 The sections on Operating Systems and Software Lifecyle are mostly reproduced from Ganney et al. 2022.
2 Originally bootstrap programs had to be keyed in by hand using switches on the front panel, in binary.
3 Some variants interpret "M" as "menus" and others interpret "P" as "pull-down menus"
4 Regular expressions pre-date Unix, but the syntax defined in Unix is now the most widely used.
5 In 2019 iOS for iPad became iPadOS and at this point the OSs diverged.
6 The bit of the program that paints the window and its data.
7 That is, a formal guarantee of always meeting the hard deadline is required.
8 Because some cases are accessed by multiple actors, the logical order may be correct for one but not for another.
9 Otherwise why are they in the diagram?
10 This fits very well with the "V" model of development, which we met in Figure 15.7.
11 "Everyone knows that debugging is twice as hard as writing a program in the first place. So if you're as clever as you can be when you write it, how will you ever debug it?" – "The Elements of Programming Style", Brian W. Kernighan and P. J. Plauger, 2nd edition, chapter 2.
12 We might however be ready to place higher confidence in the product after extensive successful testing.
13 A "floating point" number is one written as $nx10^m$. Floating point operations are numeric operations (add, divide etc.) on such numbers.
14 Estimates vary according to the software development methodology in use, but are typically in the range 1 – 25 errors per 1000 lines of code of delivered software.
15 But not in a Schrodinger sense.
16 It is extremely annoying to find that code runs perfectly under the debugger but not outside of it.
17 A fuller worked example can be found at http://www.testingstandards.co.uk/Component%20Testing.pdf.
18 Graceful degradation is where the system maintains limited functionality after suffering failure. The purpose is to maintain safety by preserving minimal functionality or shutting down in a controlled fashion.
19 Originally 1219-1998 but superseded by 14764:2006.
20 Which brings us to the art of comments, discussed in Section 16.1.5.
21 Apache Subversion.
22 https://agilemanifesto.org/.
23 User stories are an agile tool designed to help understand project requirements from a user's perspective, using simple, structured and non-technical phraseology; <role><feature><reason>. Against each story technical requirements, constraints and acceptance criteria are listed. For example, "As a member of staff, I want to create a user account so I can use the system" which yields the requirements for a table to hold the information and an interface to perform the function.
24 cf the spiral lifecycle model.
25 i.e., size, complexity, and functional capability.
26 e.g. user interface management system.

27 i.e. abstracted by more than one level.
28 C.f. "Assembling Reusable Components", section 15.5.1.2.

REFERENCES

Agile Modeling 2022, [online]. Available: http://www.agilemodeling.com/artifacts/useCaseDiagram.htm [Accessed 11/05/22].

Alexander A, 2019, Linux pipe command examples (command mashups) [online]. Available: http://alvinalexander.com/blog/post/linux-unix/linux-unix-command-mashups [Accessed 11/05/22].

British Computer Society Specialist Interest Group in Software Testing 2001, [online]. Available: http://www.testingstandards.co.uk/Component%20Testing.pdf [Accessed 11/05/22].

CoFounders Lab 2016 [online]. Available: https://cofounderslab.com/discuss/is-agile-really-that-good [Accessed 11/05/22].

Dictionary of Computing, 1996, 4th Edition. Oxford: Oxford University Press, p. 459.

Ganney P, Maw P, White M, ed. Ganney R, 2022, *Modernising Scientific Careers the ICT Competencies*, 7th edition, Tenerife: ESL.

Green A, Cosgriff P, Ganney P, Trouncer R and Willis D, 2016, The case for software testing: bugs have feelings too, *Scope* 25 (2) pp. 20–22.

Hobbs C, 2012, Build and Validate Safety in Medical Device Software, Medical Electronics Design [online]. Available: http://www.softwarecpr.com.php72-38.lan3-1.websitetestlink.com/wp-content/uploads/Free/Articles/Article-BuildandValidateSafetyinMedicalDeviceSoftware-010412.pdf [Accessed 22/02/19].

Ideagen 2022, [online]. Available: https://www.ideagen.com/products/q-pulse/ [Accessed 11/05/22].

IEEE Standard Glossary of Software Engineering Terminology, IEEE std 610.12-1990, 1990.

"ISO/IEC/IEEE International Standard – Systems and software engineering – Vocabulary," in ISO/IEC/IEEE 24765:2010(E), pp. 1–418, 15 Dec. 2010, doi: 10.1109/IEEESTD.2010.5733835.

ISO/IEC/IEEE 29119-3:2013 Software and systems engineering -- Software testing, 2013, [online] Available: https://www.iso.org/standard/56737.html [Accessed 11/05/22].

Linus Torvalds: A Very Brief and Completely Unauthorized Biography 2006, [online]. Available: http://www.linfo.org/linus.html [Accessed 11/05/22].

PRINCE2 2022, [online]. Available: https://www.prince2.com/ [Accessed 11/05/22].

Scacchi W, 2001, Process Models in Software Engineering, Institute for Software Research [online]. Available: http://www.ics.uci.edu/~wscacchi/Papers/SE-Encyc/Process-Models-SE-Encyc.pdf [Accessed 11/05/22].

Software Engineering. 1972, Information Processing. North-Holland Publishing Co. 71: 530–538.

Sommerville I, 2007 [1982]. *Software Engineering* (8th ed.) Harlow, England: Pearson Education. p. 7. ISBN 0-321-31379-8.

Ulrich K, Eppinger S and Yang MC, 2019, *Product Design & Development*, Maidenhead: McGraw-Hill.

16 Software Quality Assurance[1]

16.1 INTRODUCTION

The terms "**Software Quality Assurance**" (SQA) and "**Software Quality Control**" (SQC) are often mistakenly used interchangeably. The definition offered by the now defunct website[2] sqa.net is:

> Software Quality Assurance [is] the function of software quality that assures that the standards, processes and procedures are appropriate for the project and are correctly implemented.
>
> Software Quality Control [is] the function of software quality that checks that the project follows its standards processes, and procedures, and that the project produces the required internal and external (deliverable) products.

The two components can thus be seen as one (SQA) setting the standards that are to be followed with the other (SQC) ensuring that they have been. The process for SQC is one that should be specified as part of SQA so that the method for measurement and thus the criteria for compliance are known up-front. For example, the SQA may specify that ISO 14915 be used to define the multi-media user interfaces. SQC will therefore test to ensure that all multi-media user interfaces comply. It is thus clear that SQC will be undertaking a level of testing and it is not unusual for the full test suite to be part of the SQC process – testing not just the standards employed, but the functionality and safety of the software also.

16.1.1 ATTRIBUTES

It is common to break the software down into attributes that can be measured and in this there is a similarity with software testing (covered in Section 15.3.4). There are several definitions of software quality attributes: McCall (1977), Boehm (1978), Robert Grady's FURPS[3]+ (Grady and Caswell 1987) and Microsoft's Common Quality Attributes (Microsoft 2010 [online]).

Some such attributes are:

- Accuracy – the ability of the system to produce accurate results (and to what level of accuracy this is required).
- Availability – the proportion of time that the system is functional and working.
- Compatibility – the ability of the system to work with, for example, different input devices.
- Functionality – what the system is actually supposed to do.

DOI: 10.1201/9781003316244-16

- Manageability – the ease with which system administrators can manage the system, through tuning, debugging and monitoring.
- Performance – the responsiveness of a system to execute a required action with a set time frame.
- Security – the ability of a system to resist malicious interference.
- Supportability – the ability of the system to provide information to assist in rectifying a performance failure.
- Usability – how well the system meets the users' requirements by being intuitive and accessible. (Standards may seem limiting and anti-creative, but Microsoft's success is built on them, especially the common Windows user interface[4].)

Three further areas fall into the remit of SQA, all of which assume that the system (and especially the software at the heart of it) will not remain constant over time (it is often said that software is never finished, only implemented and passed to users): Configuration Management, Change Control and Documentation, which we will now examine.

16.1.2   CONFIGURATION MANAGEMENT AND CHANGE CONTROL

Software Configuration Management (SCM) is the tracking and controlling of changes in the software. It therefore requires a robust **Change Control** (CC) method. The TickIT guide (TickIT 2022 [online]) has two quality control elements covering this: "Maintain and Enhance" and "Support", listing the following control mechanisms:

- Change control approval points: authority to proceed with change
- Full release approval
- Regression tests
- System, integration and acceptance test
- Reviews and audits
- Configuration and change management
- Quality plans and projects plans

It can therefore be seen that changes to a live system should not be undertaken lightly. They require possibly more consideration than the initial system to ensure that a fix doesn't cause more problems[5]. A risk assessment, covering both the risk of making the change as well as the risk of not doing so, must be undertaken. Requiring authorisation ensures that SQA is undertaken. When compared with hazard logs (which we will meet in Safety Cases in Section 18.8) then the major addition is the "do nothing" risk.

16.1.3   DOCUMENTATION

Like the software, the documentation must be kept up to date. It must form an accurate reflection of the software in use and should thus undertake the same version control as the software (discussed in Section 16.2) at the same time. It is therefore clear that the system documentation (in several parts: user guide, programmers' guide, system administrators' guide, etc.) must have a clear structure to it. Without such structure, the documentation becomes difficult to maintain.

In order to enable programmers to work across multiple projects employing several languages, it is vital that the documentation has a consistent format to it so that information can be swiftly found and easily updated. In other words, the documentation also requires a standard, which should be part of the SQA process that it also enables.

There are three main types of documentation that we are concerned with in this book:

- User documentation, describing to the user how the system works and how to interact with it in order to achieve the desired results. There is a school of thought that this should be the first document written and should therefore form the specification. In a waterfall methodology this may work, but not in others. The structure of the user documentation should be derived from the Use Case (see "Software Engineering" Figure 15.4).
- Technical documentation, describing to technical staff how to maintain the code (including in-code documentation: see below), how to install it and what to do when something goes wrong. At UCLH we have a document called a "Panic File" for those occasions where, if it did not exist, that is what you would do. This technical file (which includes the technical design) is often required when seeking certification such as a CE mark. A technical file may be accessed in three possible circumstances and should be written with these in mind:
  - Routine maintenance
  - Emergency maintenance
  - Enhancement implementation
- Safety documentation (see "Safety Cases" Section 18.1).

Documentation doesn't have to be hard and can become second nature. For example:

The naming of variables may be prescribed by the language (e.g. in some forms of BASIC a variable ending in "$" is of type text; in FORTRAN variables whose names start with a letter between I and N are integers; in M (formerly MUMPS) variables prefixed with "^" indicate disc or permanent storage). The most common form of variable naming is what is known as "Hungarian notation" where the variable name (starting with a capital letter) is prefixed with one or more lower case letters that denote its type, e.g. nRank is numeric, sName is string (text) and bCapsLockOn is boolean.

16.1.4   HUNGARIAN NOTATION

Hungarian notation is named after Charles Simonyi, a Hungarian employee at Microsoft who developed it. Hungarian names are, unusually for Europe, rendered surname first. Thus Hungarian notation renders the variable name with the type ("n") first, followed by the unique name ("Rank") to make the full name of the variable ("nRank")[6]. This simple example also demonstrates its use as "nRank" may contain a positional index whereas "sRank" may contain a military title. As with many things in programming conventions, you only really start to realise how useful they are when you come back to a huge program which you wrote five years ago and now need to make some changes to. I have programs for which I spent ages building the infrastructure correctly and setting everything up correctly. When I need to make a change to them, it's easy: the code just flows and corrections and improvements are simple and effective. I also have others which I wrote in a hurry, so didn't build the right infrastructure and when I want to make changes to them I have to follow lots of code through to make sure that the changes I'm about to make aren't going to break something else. I therefore end up retro-fitting the infrastructure in order to better understand what I'm doing and improve my life in the future. The worst mistake you can make in programming is to assume that you will never need to change the code.

16.1.5   COMMENTS

All of the above aid the development of computer programs, but the common feature (sadly probably the least used) which greatly aids the maintenance of code is the comment. This is a piece of text which is ignored by the compiler/interpreter and plays no part whatsoever in the execution of the program. It does, however, carry information. There have been many attempts to codify the use of comments, from the simple "one comment every 10 lines of code" (resulting in useless comments such as "this is a comment" and failing to document more complex coding), to a more complex rigidly-defined comment block at the head of every subroutine describing variables used, the execution path and all revisions to the code since it was originally authored.

An example for a routine to calculate the number of days in a month might be, in C:

```
/* *********************************************
Routine to calculate number of days in a month
Written: Paul Ganney, 10/5/85
Edit[7]: Paul Ganney, 11/2/88 correction for leap years
Edit: Paul Ganney, 5/1/00 correction for 2000 not being a leap year

Inputs:
nMonth: month number (Jan=1, Dec=12)
nYear: year number as 4 digits

Outputs:
nDays: number of days in month

Execution: simple switch statement
*/
```

Note the use of /* and */ to denote the start and end of a comment block.

The most prevalent use of comments is probably as an aide-memoire to the programmer: a line of code that required the author to have to think carefully about when writing certainly requires a comment so that the logic behind it need not be recreated every time the code is altered. This though, is insufficient when multiple programmers are working on a system at the same time and a solid commenting methodology should form part of the design. There are two useful rules of thumb: "you cannot have too many comments" and "there are not enough comments yet". An example of such an aide-memoire (again in C) might be:

```
nCurrentItem=pDoc->NextPage(nCurrentItem-1)+1; // nCurrentItem-1 is the
one currently on screen
```

Note the different comment: it starts with // and ends at the end of the line.

One approach to writing code is to first write pseudocode, which describes the logic/action to be performed but in a human-readable form (see the bubblesort example in "Software Engineering" Section 15.6.1). Converting this into comments means that the logic is preserved when the pseudocode is converted into actual code.

And don't forget that the comments will only be read by humans – one of the joys of the lunar lander code being released was the sense of humour of the engineers, so do feel free to fill them with the occasional joke.

```
if(MakeLower(sInput)==MakeLower(sPassword))   //  CAPS  LOCK  –
Preventing Login Since 1980
```

16.2 VERSION CONTROL

As with any document, version control is vital in software development. There are two elements to this:

- Ensuring developers are all working on the latest version of the code.
- Ensuring users are all executing the latest version of the code.

There are multiple ways (and multiple products) to achieve this, so we will examine only one example of each.

Developers: A common code repository enables multi-programmer projects to be successfully developed. This repository allows all developers access to all the code for reading. When a developer wishes to work on a piece of code, this is "checked out" and no other developer may then access the code for alteration until it is "checked in" again. Dependency analysis is then utilised to alert all developers to other code modules which depend upon the altered code and may therefore require revision or re-validation.

Possibly the most popular code repository is git (or github) possibly because it is free open-source software distributed under the terms of the GNU General Public License version 2. Git is a distributed version-control system for tracking changes in source code during software development. While it was designed for coordinating work among programmers, it can actually be used to track changes in any set of files as it is not code-sensitive. Its goals include speed, data integrity and support for distributed, non-linear workflows. Git was originally created by Linus Torvalds[8] in 2005 for development of the Linux kernel. This is a snapshot-based repository in that a commit takes a snapshot of the code at that point, unlike other systems (such as Subversion) which record a set of file-based changes. To avoid inefficiency, git doesn't re-store unchanged files, merely a pointer to any such. Git therefore only ever adds data – a deletion still needs the old file to be stored so that deletion can be rolled back. Git has three main states that files can reside in: committed, modified and staged. Committed files are stored in the local database. Modified files have been changed but have not yet been committed to the database. Staged files are modified files marked to go into the next commit snapshot in this current version. The git workflow is therefore:

- Files in the working directory are modified.
- These files are staged, adding snapshots of them to the staging area.
- A commit is performed, which takes the files as they are in the staging area and stores that snapshot permanently to the Git directory.

Users: In a database project (or any software that accesses a database, regardless of whether that is its core functionality[9]) it is possible to ensure that all users are using the latest version of the software by hard-coding the version into the source code and checking it against the version in the database. If the database has a later version number, then the program can alert the user, advising them to upgrade or even halting their progress (depending on the nature of the upgrade).

16.3   SOFTWARE TOOLS AND AUTOMATION FOR TESTING

Although several hours of work may go into the design of each test case assembled throughout unit, integration and system testing, the actual execution of each test case may well only take a few moments. Manual verification that the correct results have been obtained (or otherwise) will frequently take longer than running the test itself. Running a comprehensive series of test cases "by hand" can be a lengthy and somewhat repetitive (and hence error-prone) task. The goal of test automation is to simplify the construction and execution of a series of test cases. The test cases are assembled into a single entity known as a test suite, and this is either run in its entirety by the automation software or particular test cases may be selected.

Test automation software is available for both "traditional" character-interface based and for **Graphical User Interface** (GUI) applications. Although not many applications are now written with character-based user interfaces, it is nevertheless very easy to execute simple test frames with packages such as the free open source DejaGnu (DejaGnu 2021 [online]). In the GUI domain, Selenium (Selenium 2022 [online]) is a popular free test framework for web applications distributed under the Apache 2.0 License.

Broadly, test automation software is configured as follows for each test case:

- The test environment is initialised to a known set of start conditions (thus any files accessed by the software under test can have predefined initial contents).
- Any initialisation parameters used by the software under test are set. This might include operating system environment variables or the argument values passed to a subroutine under test.
- Any interaction with the software to be performed during the test's execution is defined, e.g. via a script which emulates user input via mouse clicks or on-screen field updates.
- The expected output from the software under test for this test case is stored.

Once this has been done, the automation software executes the software under test, subjects it to the planned interactions and captures output generated by it. It will generally record performance statistics (the simplest of which might be the execution time). The output generated by the software under test is compared to the output expected and (if they match) a "pass" status is assigned to this test case. Test automation can be applied to each unit or to the complete system.

Once the test suite has been exhausted, the automation software builds a test report documenting the pass/fail status of each test case and may detail the output discrepancies of failed cases. The report will include additional information (e.g. name of the test suite in use and the time and date of execution) and a summary count of passed and failed test cases. In this way a test suite's execution is largely self-documenting.

It must be said that creating a test suite for test automation is a software development exercise in its own right. Test suites should be documented and kept under version control in just the same way as the software they are designed to test. This is not generally too arduous a task however because the tests will have been clearly defined in the test specification. Overall test automation is a timesaver, but don't expect to be able to beat the "50% rule" by adopting it[10].

The real value of test automation lies in the ease with which a complete set of tests (each of which might be quite complex) can be initiated. Retesting following any code change (this is known as regression testing) becomes a simple task, and since it is no harder to retest all the components of a system than it is to test just one unit this becomes a habit. Unexpected interactions resulting from such a change can be identified at the integration or system test levels. This benefit continues once the system has been released and is in the maintenance phase of its lifecycle. Following an upgrade of the software additional test cases can be added as required, but nevertheless kicking off a rerun of all the test suites associated with a project should be the work of a few moments. This helps to avoid the introduction of new bugs at the same time as existing bugs are fixed.

Finally, when a previously unidentified software fault is discovered by the users of your software, it is worth adding a test case to the test suite which can reproduce the problem. Some bugs have an uncanny ability to reappear even years later than they were thought to have been fixed.

There are more tools to undertake automated software testing than this text can cover, so we will examine the key attributes such a suite should have.

16.3.1   RECORD AND PLAYBACK

When automating, this is the first thing that most test professionals will do. They will record a simple script; look at the code and then play it back. This is very similar to recording a macro in Microsoft Excel. Eventually record and playback becomes less and less part of the automation process as it is usually more robust to use the built-in functions to directly test objects, databases and so on. However, this practice should be done as a minimum in the evaluation process because if the tool of choice cannot recognise the application's objects then the automation process will be a very tedious experience.

16.3.2   WEB TESTING

Web based functionality on most applications is now a part of everyday life. As such the test tool should provide good web based test functionality in addition to its client/ server functions.

Web testing can be riddled with problems if various considerations are not taken into account. A few examples are:

- Are there functions to tell when the page has finished loading?
- Can the test tool be instructed to wait until an image appears?
- Can the validity of links be determined?
- Can web based object functions (such as whether it is enabled, contains data, etc.) be tested?
- Are there facilities that allow programmatic searching for objects of a certain type on a web page or to locate a specific object?
- Can data be extracted from the web page itself? E.g. the title or a hidden form element.

With Client server testing the target customer is usually well defined so it is known what network operating system will be in use, the applications and so on but with the Internet it is very different. A person may be connecting from the USA or Africa; they may be visually impaired; they may use various browsers (and different versions of these which may not support the latest HTML standards); and the screen resolution on their computer may be different. They may speak different languages, may have fast connections or slow connections, may connect using MAC, Linux or Windows, etc. Therefore, the cost to set up a test environment is usually greater than for a client server test where the environment is fairly well defined.

16.3.3   DATABASE TESTS

Most applications will provide the facility to preserve data outside of itself. This is usually achieved by holding the data in a database. As such, checking what is in the backend database usually verifies the proper validation of tests carried out on the front end of an application. Because of the many databases available e.g. Oracle, DB2, SQLServer, Sybase, Informix, Ingres all of them support a universal query language known as SQL[11] and a protocol for communicating with these databases called **Open Database Connectivity** (ODBC) (**Java Database Connectivity** (JDBC) can be used on java environments).

16.3.4   DATA FUNCTIONS

As mentioned above, applications usually provide a facility for storing data offline. In order to test this, data needs to be created to input into the application. The chosen tool should allow the specification of the type of data required, automatically generate this data and interface with files, spreadsheets, etc. to create and extract data. This functionality is usually more important than the database tests as the databases will usually have their own interface for running queries. However applications do not usually provide facilities for bulk data input.

These functions are also very important as testing moves from the record/playback phase, to being data-driven and on to framework testing. Data-driven tests are tests that replace hard coded names, address, numbers, etc. with variables supplied from an external source – usually a **Comma Separated Variable** (CSV) file, spreadsheet or database. Frameworks are usually the ultimate goal in deploying automation test tools. Frameworks provide an interface to all the applications under test by exposing a suitable list of functions, databases, etc. This allows an inexperienced tester/user to run tests by just running the test framework with known commands/variables. A test framework has parallels to software frameworks where an encapsulation layer of software (framework) is developed around the applications, databases, etc. and exposes functions, classes, methods, etc. that are used to call the underlying applications, return data, input data and so on.

However to create a test framework may require a lot of skilled time, money and other resources.

16.3.5   OBJECT MAPPING

Hopefully most of the application has been implemented using standard objects supported by the test tool vendor, but there may be a few objects that are custom ones. These will need to be mapped in order to be able to test them.

Most custom objects will behave like a similar standard control and can therefore be mapped[12]. Some of the standard objects that are seen in everyday applications are:

- Pushbuttons
- Checkboxes
- Radio buttons
- List views
- Edit boxes
- Combo boxes

16.3.6 IMAGE TESTING

The tool should allow the mapping of painted controls[13] to standard controls but to do this the screen co-ordinates of the image have to be reliable. It may also provide **Optical Character Recognition** (OCR) and therefore be able to compare one image against another.

16.3.7 TEST/ERROR RECOVERY

This can be one of the most difficult areas to automate but if it is automated, it provides the foundation to produce a truly robust test suite. The test tool should provide facilities to handle the questions such as:

- If the application crashes during testing what can the tester do?
- If a function does not receive the correct information how can the tester handle this?
- If an error message is generated, how does the tester deal with that?
- If a website is accessed and a warning received, what does the tester do?
- If a database connection cannot be gained, how does the tester skip those tests?

16.3.8 OBJECT NAME MAP

As an application is tested using the test tool of choice it records actions against the objects that it interacts with. These objects are either identified through the co-ordinates on the screen or preferably via some unique object reference referred to as a tag, object ID, index, name, etc. The tool should provide services to uniquely identify each object it interacts with by various means. The least desirable method is by co-ordinates on the screen.

Once automation is established tens and thousands of scripts that reference these objects will have been developed. A mechanism that provides an easy update should the application being tested change is required.

All tools provide a search and replace facility but the best implementations are those that provide a central repository to store these object identities. The premise is that it is better to change the reference in one place rather than having to go through each of the scripts to replace it many times.

16.3.9 OBJECT IDENTITY TOOL

One of the primary means of identifying objects is via an ID Tool. This is a sort of spy that looks at the internals of the object providing details such as the object name, ID and similar. This allows reference to that object within a function call.

The tool should give details of some of the object's properties, especially those associated with uniquely identifying the object or window. The tool will usually provide the tester with a "point and ID" service where the mouse can point at the object and in another window all of that object's IDs and properties are displayed. A lot of the tools available allow searching of all the open applications in one pass and show the result in a tree that can be inspected when required.

16.3.10 EXTENSIBLE LANGUAGE

In this section we consider this question: if the standard test language does not support what is required then can a **Dynamic Link Library** (DLL) be created or can the language be extended in some way to do it? This is usually an advanced topic and is not usually encountered until the trained tester has been using the tool for at least 6 – 12 months. When this does arise the tool should support language extension – if via DLLs then the tester will need to have knowledge of a traditional development language e.g. C, C++ or VB.

Some tools provide this extension by enabling the creation of user defined functions, methods, classes, etc. but these are normally a mixture of the already supported data types, functions, etc. rather than extending the tool beyond its released functionality.

16.3.11 ENVIRONMENT SUPPORT

The number of environments that the tool supports out the box is important: whether it supports the latest Java release, which versions of Oracle, Powerbuilder, **Wireless Application Protocol** (WAP), etc. Most tools can interface to unsupported environments if the developers in that environment provide classes, DLLs etc. that expose some of the application's details but whether a developer has the time or the inclination to do this is another question.

Ultimately this is the most important part of automation. If the tool does not support the environment/application then it cannot be used and manual testing of the application will be required.

16.3.12 INTEGRATION

Finally, we consider how well the tool integrates with other tools. Ideally the tool should be able to be run from various test management suites, bugs should be raised directly from the tool and the information gathered from the test logs fed into it. It should integrate with products like Word, Excel and requirements management tools.

When managing large test projects with an automation team greater than five and testers totalling more than ten, the management aspect and the tools integration become more important. An example might be a major bank that wants to redesign its workflow management system to allow faster processing of customer queries. The anticipated requirements for the new workflow software might number in the thousands. To test these requirements 40,000 test cases may be identified; 20,000 of which can be automated. A test management tool is invaluable in such circumstances.

Integration also becomes very important when considering how to manage bugs that are raised as part of automation testing. Having separate systems that don't share data will require duplication of entry, which is clearly inefficient.

16.4   STANDARDS[14]

There are two major sets of standards that we are concerned with: regulatory, principally around medical devices, and developmental, around software. The most important European legislation designed to protect the patient is the **Medical Devices Regulations** (MDR), which we'll come to shortly, along with its predecessor the **Medical Devices Directive** (MDD) – of particular interest since Brexit.

NHS Digital produced a helpful document summarising the key data protection, data governance and medical device regulatory requirements involved in the development of decision supporting and decision-making software applications, or devices incorporating such technology, in the NHS and Adult Social Care. It's at https://digital. nhs.uk/services/clinical-safety/documentation. Whilst not intended as a comprehensive guide to all relevant legislation, it signposts detailed guidance and acknowledges ongoing work aimed at providing clarity and covering several issues not currently well catered for in the existing legislative framework.

16.4.1   IEC 601

IEC 60601-1 Medical electrical equipment – Part 1: General requirements for basic safety and essential performance, was first published in 1977. It is now in its third edition and became effective in Europe in June 2012. It is the main standard for electromedical equipment safety and is first in a family of standards, with 11 collateral standards[15] (numbered 60601-1-N) which define the requirements for certain aspects of safety and performance, e.g. Electromagnetic Compatibility (IEC 60601-1-2) and over 60 particular standards (numbered 60601-2-N) defining the standards for particular products, e.g. nerve and muscle stimulators (IEC 60601-2-10). Two important changes from the previous versions are the removal of the phrase "under medical supervision" from the definition of medical electrical equipment and the inclusion of the phrase "or compensation or alleviation of disease, injury or disability" (previously only diagnosis or monitoring) meaning that many devices previously excluded are now included in the standard's coverage. Possibly the largest change, though, is the requirement for manufacturers to have a formal risk management system that conforms to ISO 14971:2007 Medical devices – Application of risk management to medical devices, in place.

A risk management system includes within it two key concepts:

- Acceptable levels of risk
- Residual risk

Once acceptable levels of risk have been established, all residual risks (as documented in the hazard log – a part of the risk management file) can be measured against them. That way, risks can be demonstrably determined to be acceptable prior to manufacture and certainly prior to deployment. These also link nicely into the DCBs 0129 and 0160 described below.

Risk management responses can be a mixture of four main actions:

- Transfer – where the current score is higher than an acceptable target score the decision reached may be to transfer the consequence of a risk to another owner, e.g. purchase an insurance policy so that financial loss is covered.

- Tolerate – where the current score is within an acceptable limit the decision reached may be to tolerate or accept the risk with no further action required. In this case the controls must be monitored and the risk reviewed regularly.
- Treat – where the current score is higher than an acceptable score the decision reached may be to treat the risk. A SMART action plan will be developed, and when an action is complete the current score should be reviewed. The risk should be reviewed regularly.
- Terminate – where the current score is higher than an acceptable score and there is no option to transfer, tolerate or treat the risk, the decision reached may be to NOT proceed with the activity. In this case the only option is to choose to terminate the risk.

Furthermore, there are two compulsory **Data Coordination Board** (DCB) standards that are mandatory under the Health and Social Care Act 2012. DCB0129[16] describes the risk management processes required to minimise risks to patient safety with respect to the manufacture of health software products either as new systems or as changes to existing systems. DCB0160[17] is similar, describing the deployment of such products. Thus many healthcare sites need to comply with DCB0129 and all with DCB0160. A fully populated hazard log may be used to demonstrate compliance, and compliance with DCB0160 will encompass compliance with IEC 80001-1.

IEC 60601-1 covers all aspects of the medical device, including classifying the parts that connect (directly or indirectly) to patients. Of interest to this section of this book, though, is its applicability to computer-based medical devices. It is almost certain that a standard PC power supply will not meet the regulation so this must either be replaced, or an isolation transformer must be put in line. As it is unlikely that any computer-based medical device will exist in isolation, the connections to other items of equipment must also be examined. One commonly overlooked connection is to a network – this can be isolated either via transformer or optically. However, some on-board network adaptors are unable to supply the power required to drive such an isolator which may not pose a problem to a desktop machine (as a separate adaptor can be added), but would to a laptop, netbook or tablet device (especially if it is operating on battery power). Wi-Fi may provide the best electrical isolation but brings with it other problems leading to a risk balancing exercise.

IEC 60601-1 applies to medical devices and for a definition as to what constitutes such a device, we must turn to the Medical Devices Directive.

16.4.2 THE MEDICAL DEVICES DIRECTIVE

The **Medical Devices Directive** (MDD) of 1993 (less well known as Directive 93/42/EEC) contained within it a definition of a medical device. There have been several additional Directives (2003, 2005 and 2007[18]) which have modified the original Directive. Generally these modifications have been to increase the scope of the Directive to cover more types of device. At the time of writing, the most recent technical revision is Directive 2007/47/EC.

This Directive defines a 'medical device' as meaning

> any instrument, apparatus, appliance, software, material or other article, whether used alone or in combination, including the software necessary for its proper application intended by the manufacturer to be used for human beings for the purpose of:
>
> - diagnosis, prevention, monitoring, treatment or alleviation of disease,
> - diagnosis, monitoring, treatment, alleviation of or compensation for an injury or handicap,
> - investigation, replacement or modification of the anatomy or of a physiological process,
> - control of conception,
>
> and which does not achieve its principal intended action in or on the human body by pharmacological, immunological or metabolic means, but which may be assisted in its function by such means.

(EUR-Lex 2007 [online]) (This is part 1.2a of the Directive as mentioned in the flowchart in Figure 16.1).

It is worth noting that accessories (defined as "*an article which whilst not being a device is intended specifically by its manufacturer to be used together with a device to enable it to be used in accordance with the use of the device intended by the manufacturer of the device*" (EUR-Lex 2007 [online])) are treated as Medical Devices in their own right and must be classified, examined and regulated as though they were independent.

The most recent revision, which added the word "software" to the definition of a medical device means that software alone may be defined as a medical device and not just when it is incorporated within hardware defined as a medical device. This is made clear by item 6 of this revision, which states:

> It is necessary to clarify that software in its own right, when specifically intended by the manufacturer to be used for one or more of the medical purposes set out in the definition of a medical device, is a medical device. Software for general purposes when used in a healthcare setting is not a medical device.
>
> (EUR-Lex 2007 [online])

The ability to run software that is a medical device on a computer that is not originally designed as a medical device thus redesignates the hardware as a medical device and must be evaluated and controlled accordingly. The non-mandatory guidance document "Qualification and Classification of stand alone software" provides decision flowcharts and definitions in order to assist the determination as to whether stand-alone software is a medical device and, if so, the class to which it belongs. Without repeating the full content here, it is worth noting that it is recognised that software may consist of multiple modules, some of which are medical devices and some of which are not[19].

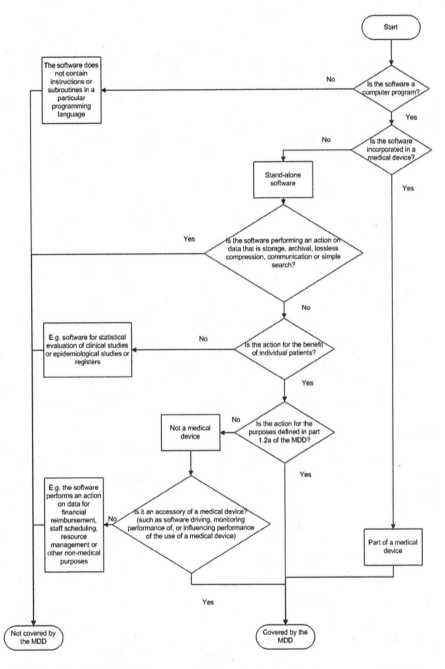

FIGURE 16.1 A decision diagram to assist the qualification of software as a medical device. (UK Government 2021 [online]).

Item 20 of the revision states:

> Taking account of the growing importance of software in the field of medical
> devices, be it as stand alone or as software incorporated in a device, validation of
> software in accordance with the state of the art should be an essential requirement.
>
> (EUR-Lex 2007 [online])

and is covered in Section 15.3.4, Chapter 15.

A medical device may be classified as Class I (including Is & Im), Class IIa, IIb
and III, with Class III covering the highest risk products[20].

Classification of a medical device will depend upon a series of factors,
including:

- how long the device is intended to be in continuous use
- whether or not the device is invasive or surgically invasive,
- whether the device is implantable or active
- whether or not the device contains a substance, which in its own right is
  considered to be a medicinal substance and has action ancillary to that of
  the device.

(Wellkang Tech Consulting 2009 [online])

The classification rules are set out in Annex IX of the directive which includes definitions of the terminology used in the classification rules[21].

16.4.3 THE MEDICAL DEVICES REGULATIONS

One of the cornerstones of the EU is the free movement of goods. This means that
a product that is allowed on the market in one of the Member States is also allowed
on the markets of other Member States. This concept is enabled by three conditions
within the 2016 version of the "Blue Guide" on the implementation of EU products,
which must be met:

1. Essential requirements for the products involved must be defined;
2. Methods must be established to describe how product compliance with the
   requirements is addressed;
3. Mechanisms to supervise and control the actions of all Economic Operators
   and others involved in the manufacturing and distribution of the products
   must be created.

The predecessors of the MDR – the **Active Implantable Medical Devices Directive**
(AIMDD) 90/385/EEC and the MDD – achieved this. They defined Essential
Requirements, introduced harmonised standards helping to demonstrate conformity
to the Essential Requirements, defined conformity assessment procedures and organ-
ised market surveillance functions by **Competent Authorities** (CAs) and **Notified
Bodies** (NBs).

The Directives had some inherent weaknesses, though. Changes in technology and medical science demanded additional legislation and the interpretation of the Directives was not consistent across all national governments. Directive 2007/47/EC modified the MDD and AIMDD in an attempt to address these concerns, but this amendment did not achieve all goals. The scandal involving defective breast implants manufactured by **Poly Implant Prosthesis** (PIP) in France demonstrated additional weaknesses in the system.

The major difference between a regulation and a directive is that a directive has to be taken and put into member state law through each different legislative system (thus giving scope for different interpretations in different member states), whereas a regulation becomes law as written in all member states. There is no room for differences of interpretation or application unless explicitly stated in the regulation that aspects are left up to member states.

The MDR came into force in 2017. Member states had until 2020 in which to implement it, meaning that most requirements did not fully apply until 26th May 2020 for Medical Devices and 26th May 2022 for In Vitro Diagnostic Medical Devices[22].

> It lays down rules concerning the placing on the market, making available on the market or putting into service of medical devices for human use and accessories for such devices in the Union. This Regulation also applies to clinical investigations concerning such medical devices and accessories conducted in the Union.
> (Official Journal of the European Union L117 Volume 60, 5 May 2017 [online])

The MDR has 175[23] pages and is vastly more "legal" than the MDD, which took more of a good will approach. It will thus result in far more work for the regulators.

The MDD focused on the path to CE marking, whereas the MDR promotes a lifecycle approach (in this it is similar to the US' **Food and Drug Administration** (FDA) and many other international standards).

Previous guidance documents (MEDDEVs) have been incorporated into the regulation[24]. This has made the optional guidance mandatory. Clinical data and evaluations will have greater emphasis and equivalence[25] will be more rigorously interpreted, increasing the challenge for demonstrating performance of clinical safety for medical devices. Article 1 of the MDR[26] brings products without an intended medical purpose that are listed in Annex XVI into the scope of the MDR[27]. Medical devices, accessories and the products in Annex XVI are referred to as "devices". In the definition of accessories, no exception is made for products without a medical purpose that will be considered medical devices and therefore their accessories will also fall within the scope of the MDR.

Annex XVI may have new groups of products added during the lifetime of the MDR.

Article 2 lists 71 definitions, compared to the MDD's 14. Amongst these, **In Vitro Diagnostics** (IVD) are covered and accessories are now defined to "assist" as well as "enable" a medical device's usage. Likewise, "label" is defined[28], as is "risk"[29].

"Standalone software" is no longer mentioned.

> Software may have a medical purpose, in which case it falls within the scope of the MDR. Annex VIII, Classification Rules now refers to "software that drives a device or influences the use of a device" versus software that is "independent of any other device."
>
> (Loh and Boumans 2017)

The major definitions[30] (from a computing viewpoint) are:

(1) 'medical device' means any instrument, apparatus, appliance, software, implant, reagent, material or other article intended by the manufacturer to be used, alone or in combination, for human beings for one or more of the following specific medical purposes:
   - diagnosis, prevention, monitoring, prediction, prognosis[31], treatment or alleviation of disease,
   - diagnosis, monitoring, treatment, alleviation of, or compensation for, an injury or disability,
   - investigation, replacement or modification of the anatomy or of a physiological or pathological process or state,
   - providing information by means of in vitro examination of specimens derived from the human body, including organ, blood and tissue donations,
   and which does not achieve its principal intended action by pharmacological, immunological or metabolic means, in or on the human body, but which may be assisted in its function by such means.
   The following products shall also be deemed to be medical devices:
   - devices for the control or support of conception;
   - products specifically intended for the cleaning, disinfection or sterilisation of devices as referred to in Article 1(4) and of those referred to in the first paragraph of this point.
(2) 'accessory for a medical device' means an article which, whilst not being itself a medical device, is intended by its manufacturer to be used together with one or several particular medical device(s) to specifically enable the medical device(s) to be used in accordance with its/their intended purpose(s) or to specifically and directly assist the medical functionality of the medical device(s) in terms of its/their intended purpose(s);
(3) 'custom-made device' means any device specifically made in accordance with a written prescription of any person authorised by national law by virtue of that person's professional qualifications which gives, under that person's responsibility, specific design characteristics and is intended for the sole use of a particular patient exclusively to meet their individual conditions and needs.
   However, mass-produced devices which need to be adapted to meet the specific requirements of any professional user and devices which are mass-produced by means of industrial manufacturing processes in accordance with the written prescriptions of any authorised person shall not be considered to be custom-made devices;
(4) 'active device' means any device, the operation of which depends on a source of energy other than that generated by the human body for that purpose, or by gravity, and which acts by changing the density of or converting that energy. Devices intended

to transmit energy, substances or other elements between an active device and the patient, without any significant change, shall not be deemed to be active devices.
Software shall also be deemed to be an active device[32];

...

(12) 'intended purpose' means the use for which a device is intended according to the data supplied by the manufacturer on the label, in the instructions for use or in promotional or sales materials or statements and as specified by the manufacturer in the clinical evaluation;

...

(25) 'compatibility' is the ability of a device, including software, when used together with one or more other devices in accordance with its intended purpose, to:
  (a) perform without losing or compromising the ability to perform as intended, and/or
  (b) integrate and/or operate without the need for modification or adaption of any part of the combined devices, and/or
  (c) be used together without conflict/interference or adverse reaction.

(26) 'interoperability' is the ability of two or more devices, including software, from the same manufacturer or from different manufacturers, to:
  (d) exchange information and use the information that has been exchanged for the correct execution of a specified function without changing the content of the data, and/or
  (e) communicate with each other, and/or
  (f) work together as intended.

(29) 'putting into service' means the stage at which a device, other than an investigational device, has been made available to the final user as being ready for use on the Union market for the first time for its intended purpose;

(30) 'manufacturer' means a natural or legal person who manufactures or fully refurbishes a device or has a device designed, manufactured or fully refurbished, and markets that device under its name or trademark;

(31) 'fully refurbishing', for the purposes of the definition of manufacturer, means the complete rebuilding of a device already placed on the market or put into service, or the making of a new device from used devices, to bring it into conformity with this Regulation, combined with the assignment of a new lifetime to the refurbished device;

(Official Journal of the European Union L117 Volume 60, 5 May 2017 [online])

Additionally, clause 19 from the preamble is worth mentioning[33]:

It is necessary to clarify that software in its own right, when specifically intended by the manufacturer to be used for one or more of the medical purposes set out in the definition of a medical device, qualifies as a medical device, while software for general purposes, even when used in a healthcare setting, or software intended for life-style and well-being purposes is not a medical device. The qualification of software, either as a device or an accessory, is independent of the software's location or the type of interconnection between the software and a device.

(Official Journal of the European Union L117 Volume 60, 5 May 2017 [online])

Chapter II – "Making Available on the Market and Putting into Service of Devices, Obligations of Economic Operators, Reprocessing, CE Marking, Free Movement" provides substantive definitions and responsibilities but also delineates between the responsibilities of the **Authorized Representative** (AR), the distributor and the importer. Interestingly, "Distance sales" are regulated, meaning that devices sold to EU citizens through the Internet have to comply with the MDR[34]. Thus manufacturers of such devices must appoint ARs if they are not based in Europe.

Paragraph 4 of this chapter states: *"Devices that are manufactured and used within health institutions shall be considered as having been put into service."* (Official Journal of the European Union L117 Volume 60, 5 May 2017 [online]) but is tempered by Paragraph 5, which gives conditions (all of which must be met) under which the requirements of the regulation do not apply. These are:

- The device may not be transferred to another legal entity.[35]
- An appropriate[36] **Quality Management System** (QMS) must be in place.
- A documented justification that the target patient group's specific needs cannot be met[37] by an equivalent[38] device that is already on the market.
- Information must be available on the use of the device to the competent authority, which includes the justification of the manufacturing, modification and use.
- A publicly available declaration must exist[39], which includes details of the manufacturing institution, details necessary to identify the device and that the GSPR has been met (or a justification as to why they haven't).
- Documentation should exist describing the manufacturing facility and process, the design and performance data and intended purpose, so that the competent authority can determine that the GSPR have been met.
- Experience gained from the clinical use of the device must be reviewed and all necessary corrective actions must be taken.
- Member states can restrict the manufacture and use of these devices and can inspect the activities of the institution.
- This paragraph shall not apply to devices that are manufactured on an industrial scale[40].

Article 10 also specifies requirements for custom-made devices, such that full technical documentation is not required but documentation in accordance with Section 2 of Annex XIII is. Custom-made and investigational devices are also exempt from the requirement to draw up an EU declaration of conformity. Investigational (but not custom-made) devices are exempt from the requirement to maintain a QMS[41]. Article 5.5 (described above) sets out the conditions for what the MHRA refer to as the **Health Institution Exemption** (HIE). If this light-touch regulation is not met, then the regulation applies in full[42].

Chapter II introduces the person responsible[43] for regulatory compliance within the manufacturer or AR (Article 15). This employee should be highly educated and experienced.

Annex 1, the **General Safety and Performance Requirements** (GSPR) explains the "reduction of risk as far as possible" as reducing risk "without adversely affecting the risk benefit ratio." It also inserts the statement *"taking into account the generally acknowledged state of the art"* which will assist non-medical products treated as medical devices and products for which there are no sufficient standards. The GSPR apply even if the HIE does.

The new GSPR checklist has more than 220 items to review (Loh and Boumans 2017) and includes the requirement for the manufacturer to use a risk management system.

Chapter 2, "requirements regarding design and manufacture", has added several sections, of which the most pertinent relating to software are:

- Software in devices and software that are devices in and of themselves;
- Risks concerning medical devices for lay persons.
- Possible negative interactions between medical device software and other IT are to be considered (see 80001-1:2010), as are the environments in which mobile computing platforms are used (sections 14, 14.2, 14.5, 15.1 and 17.3).

Chapter 3 contains the requirement for the label to state that the product is a medical device, leading to speculation (Loh and Boumans 2017) that this may lead to the introduction of a "MD" symbol equivalent to the current "IVD" one.

There are, as has been noted in the footnotes, some ambiguities and areas requiring clarification. One such is whether a new version of software is a new device, as is how software relates to the definitions on parts and components. Overall, though, the concepts and requirements of the MDR are recognised best practice enshrined in regulation.

16.4.3.1 Scripts

There has been some confusion and debate regarding the subject of scripts (i.e. pieces of code that run on a medical device). One view is that they are covered by the device's CE mark as the device is intended to host scripts and therefore is operating as intended. This does not make the manufacturer liable for any errors in programming, however, as the author is still responsible for the safe application of the device (and therefore the script). Certainly, scripts that are actually recorded macros (designed to be re-played, even if edited slightly) are not MDs in their own right, they are simply the normal operation of the MD in question.

The alternative view is that scripts form software modules and therefore are independent. The MDD defined software as being able to be broken down into modules where each one correlates with an application of the software, some having a medical purpose and some not (MEDDEV 2.1/6 [online]). While the first view is probably the most accurate (certainly in the majority of cases where scripts mainly aggregate data), a risk analysis may reveal a requirement to treat the script as independent, thus leading to the second view.

16.4.3.2 Brexit

There has been confusion as to how much (if at all) of the MDR will pass into UK law. At the time of writing, Great Britain will follow MDD based regulations until 2023 and Northern Ireland implemented the MDR from May 2021. The UK MDR2002 was updated in Jan 2021[44]. One of the main changes was the introduction of the **UKCA**[45] mark, which must be placed on medical devices placed on the GB market (note there are different rules for Northern Ireland, which has a **UKNI** mark).

> A UKCA mark is a logo that is placed on medical devices to show they conform to the requirements in the UK MDR 2002. It shows that the device is fit for its intended purpose stated and meets legislation relating to safety.
>
> (MHRA (1) 2020 [online])

It is worth noting that an explanatory memorandum to the UK MDR2002 says

> Any devices that are in conformity with EU legislation (MDD, AIMDD, IVDD, MDR, IVDR) can continue to be placed on the market in GB until 30 June 2023. This is to provide manufacturers with time to adjust to future GB regulations that will be consulted on and published at a later date.
>
> (The Medical Devices (Amendment etc.)
> (EU Exit) Regulations 2020 [online])

This has given businesses longer to apply UKCA marking (originally January 2022 – although current guidance says they must be ready to apply it by January 2023 (Businesses given more time to apply new product safety marking 2021 [online])).

There is great similarity between the UK MDR2002 and the MDD – the MHRA's advice contains flowcharts on conformity with the instruction to read "CE marking" as "UKCA marking" and any reference to Competent Authority as "MHRA".

The UK MDR 2002 encapsulates these directives:

- Directive 90/385/EEC on active implantable medical devices (EU AIMDD)
- Directive 93/42/EEC on medical devices (EU MDD)
- Directive 98/79/EC on in vitro diagnostic medical devices (EU IVDD)

This means that since 1 January 2021, the Great Britain route to market and UKCA marking requirements are still based on the requirements derived from current EU legislation.

Regardless, the following will be true:

All medical devices, active implantable medical devices, IVDs and custom-made devices will need to be registered with the MHRA prior to being placed on the UK market. As this is an extension to the existing registration requirements, there will be a grace period to allow time for compliance with the new registration process:

- 4 months: Class III medical devices, Class IIb implantable medical devices, Active implantable medical devices, IVD List A

- 8 months: Class IIb non-implantable medical devices, Class IIa medical devices, IVD List B, Self-test IVDs
- 12 months: Class I medical devices, Self-certified IVDs, Class A IVDs

The MDR regulations will still be adopted within the UK[46]. However, the UK will be treated as being a "3rd country" – that is, a country outside of the EU27. For European medical devices imported to the UK, UK-based importers and distributors will no longer be treated as European distributors (i.e. almost as if they were within the same country as the manufacturer). Instead, European medical device manufacturers will need an AR established within the UK to:

- ensure & maintain conformity
- communicate with the Secretary of State upon request
- provide PMS (Post Market Surveillance)
- inform the manufacturers about complaints & reports
- terminate the legal relationship with the manufacturer in the event the manufacturer acts contrary to its obligations.

The Cumberledge report[47] recommended that UK legislation be at least as stringent as the EU MDR, so the pragmatic advice would seem to be to follow the EU MDR.

The UK has introduced a new **Medical Device Information System** (MDIS) with which all MDs manufactured in the UK (or have a UK Responsible Person) have to be registered. Initially only Class 1, custom-made and IVD devices were required to be registered but all others were to be done by September 2021. The MDIS is accessible to the public, although hospitals and suppliers are more likely to use it.

16.4.4   CE Marking

Directive 2007/47/EC embraces the concept of software as a medical device and therefore clearly links it to a discussion of CE[48] marking, which concludes this section. The key points are that CE marking on a product:

- is a manufacturer's declaration that the product complies with the essential requirements of the relevant European health, safety and environmental protection legislation.
- indicates to governmental officials that the product may be legally placed on the market in their country.
- ensures the free movement of the product within the **European Free Trade Association** (EFTA) & **European Union** (EU) single market (total 27 countries) and
- permits the withdrawal of the non-conforming products by customs and enforcement/vigilance authorities.

CE marking did not originally encompass Medical Devices, but they were brought into the scope of the general directive by a series of subsequent directives from 2000 onwards. It is worth noting that a device must comply with all relevant directives (i.e. all the ones that apply to it).

EU directives often use a series of questions in order to classify the level of risk and then refer to a chart called "Conformity Assessment Procedures". This chart includes all of the acceptable options available to a manufacturer to certify their product and affix the CE mark.

Products with minimal risk can be self-certified, where the manufacturer prepares a "*Declaration of Conformity*" and affixes the CE mark to their own product. Products with greater risk are usually (depending on the directive) independently certified, which must be done by a "*Notified Body*".

A Notified Body is an organisation that has been nominated by a member government and has been notified by the European Commission. They serve as independent test labs and perform the steps required by directives. Manufacturers are not required to use notified bodies in their own country, but may use any within the EU.

Custom made devices, devices undergoing clinical investigation and in-vitro medical devices for clinical investigation do not currently require CE marks but must be marked 'exclusively for clinical investigation'.

The MHRA is the UK body that provides advice and guidance on matters concerning the relationship between the MDD and the CE requirements and their website (MHRA (2) 2022 [online]) contains helpful documents.

Of great interest to Medical Physics and Clinical Engineering departments is the issue of custom-made devices. The UKCA mark that replaces the CE mark in Great Britain must not be placed on a medical device if it is a custom-made device – although it must still meet the requirements in the UK MDR 2002 and the type of device should be labelled clearly. There is no requirement to have a third party certify conformance with the requirements but a statement declaring compliance is required[49].

Following Brexit, CE marking will continue to be recognised in Great Britain until 30 June 2023 and certificates issued by EU-recognised Notified Bodies will continue to be valid for the Great Britain market until 30 June 2023.

16.4.5 OTHER STANDARDS

Three international standards offer valuable contributions to those working to provide and support Medical Devices:

- IEC 62304-2015: Medical Device Software Lifecycle – Software Lifecycle processes. This offers a risk based approach and makes reference to the use of **Software of Unknown Provenance** (SOUP).
- IEC/ISO 90003-2004 Guidelines for the application of ISO 9000-2000 to computer software, which offers similar concepts to those embraced in the TickIT scheme (TickIT 2022 [online]).
- ISO 13485-2003 Medical Devices – Quality Management Systems – Requirements for Regulatory Purposes. This deals with the production and management of Medical Devices in a manner that parallels ISO 9000.

ISO 13485-2003 has several sections, of which section 4 describes the system itself. The standard is based on the Plan-Do-Check-Act cycle and whilst mainly concerned with documentation and the control of those documents the purpose is to ensure the device is safe and fit for purpose. Section 7 ("Product Realization") describes requirements for planning, product requirements review, design, purchasing, creating the product or service and controlling the equipment used to monitor and measure the product or service.

Of more recent interest is IEC 80001-1:2021[50]. This standard, titled "Application of risk management for IT-networks incorporating medical devices. Safety, effectiveness and security in the implementation and use of connected medical devices or connected health software" contains many definitions. The main definition for consideration here is that of the *"Medical IT Network"*, which was originally defined as *"an IT-NETWORK that incorporates at least one MEDICAL DEVICE"*, but now incorporates health software and health IT systems. An IT-NETWORK is defined as *"a system or systems composed of communicating nodes and transmission links to provide physically linked or wireless transmission between two or more specified communication nodes"* and is adapted from IEC 61907:2009, definition 3.1.1. The MEDICAL DEVICE definition is from the MDD. A hospital that connects even one medical device into its standard network (or, indeed, loads medical device software onto a non-medical device so connected or incorporates it into other systems) has thereby created a medical IT-Network. The bounds of this network are that of the responsible organisation[51] but do bring different responsibilities into play, as detailed in the standard. In particular the role of the medical IT-network risk manager is specified.

This family of standards could stimulate cross disciplinary teams in hospitals, involving IT departments, informatics departments and clinical departments to establish quality systems for clinical computing. This would include assurance of finance, planning of procurement and upgrades and the monitoring of adequate support arrangements. Medical Physicists and Clinical Engineers would have an important role to play in these groups, particularly with regard to day to day running and relationships with device suppliers and the maintenance of a medical device asset register.

16.4.6 PROCESS STANDARDS

Software should be developed according to a logical sequence of events, referred to as a methodology, and quality management[52] standards seek to impose a structure and discipline on this process. In terms of the software development lifecycle (see Section 15.3, Chapter 15), the standards tend to have a bias towards 'front-end' aspects (i.e. requirements specification and design) as many large-scale commercial software projects have failed due to poor communication between the users and the developer. In the medical field, about 25% of all medical device recalls are due to "software failure".

Requirements analysis issues are generally less of a problem for most in-house development projects in medical physics, since the user and developer are either the same person, or the user is a medical colleague. Nonetheless, the frameworks

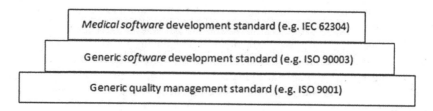

FIGURE 16.2 The standards pyramid showing a foundation general quality management system at the bottom, with more specific standards layered on top.

described by these standards are undoubtedly useful as they prescribe a well-recognised order in which things should be done, and what should be documented and tested.

Standards are best applied in a 'pyramidal fashion', with a generic 'foundation standard' at the bottom and more sector specific (so-called vertical) standards on top, as shown in Figure 16.2.

The most obvious foundation standard is the generic quality systems standard ISO 9001: 2015, which has been implemented in UK medical physics/clinical engineering departments, making it easier to implement the more specific standards concerned with software development in general, and medical software development in particular. ISO 9003: 2014. Software engineering – Guidelines for the application of ISO 9001:2008 to computer software is one such specific standard.

Process standards relevant to the in-house production of medical software include:

- ISO/IEC 62366-1:2015. Medical Devices – Part 1: Application of Usability Engineering to Medical Devices
- ISO 14971: 2012. Application of risk management to medical devices
- IEC 62304:2006/A1:2015[53]. Medical device software – lifecycle processes.
- ISO 13485: 2016. Medical devices – quality management systems – requirements for regulatory purposes.
- IEC 60601:2006/A12:2014. Safety requirements for programmable electronic medical systems.
- IEC 61508: 2000. Functional Safety of Electrical/Electronic/Programmable Electronic Safety-related Systems.
- IEC 80001-1: 2010. Application of risk management for IT networks incorporating medical devices: Part 1: Roles, responsibilities and activities.
- ISO 15189: 2012. Medical laboratories – Requirements for quality and competence.
- IEC 12207: 2017. Systems and software engineering – software life cycle processes.
- IEC 82304-1: 2016. Health software – Part 1: General requirements for product safety.

The standards attracting the most attention within medical physics/clinical engineering will now be briefly discussed.

16.4.6.1   ISO/IEC 62366-1: 2015 Medical Devices – Part 1: Application of Usability Engineering to Medical Devices

ISO/IEC 60601 originally contained a standard on usability (60601-1-6), which was replaced in 2007 by ISO/IEC 62366. This international standard was aimed at reducing the risk of medical errors due to poor interface design, accounting for over a third of all adverse incidents involving medical devices. It also applies to any documentation that may accompany the device, and the training of users. IEC 62366 has been accepted by the FDA as a means of satisfying the quality systems/design control requirement to ensure that the needs of the user are appropriately defined, verified and validated and can also be used to demonstrate compliance with new EU regulations on (electronic) **instructions for use** (IFU).

16.4.6.2   ISO 14971:2012 Application of Risk Management to Medical Devices

The 1992 EC General MDD includes a requirement (somewhat hidden in the annexes[54]) that the medical device manufacturer must carry out a risk analysis as part of the production process. This led, in stages, to the development of this standard. The relationship of ISO 14971 to the medical device directives is described in the updated appendices. The standard may be seen as fulfilling the risk management aspects of the wider quality assurance standard ISO 13485. Finally, the IEC have produced an associated Technical Report (IEC/TR 80002-1:2009. Guidance on the application of ISO 14971 to medical device software) for medical software producers.

16.4.6.3   IEC 62304:2006/A1:2015 Medical Device Software – Lifecycle Processes

IEC 62304 is effectively a medical sector derivative of IEC 61508. It emphasises the importance of software engineering techniques in the software development process. It has a risk-based approach similar to that previously described in IEC 60601-1:2006+A12:2014.

As with all risk assessments carried out in a clinical setting, it is the device manufacturer's responsibility to determine the safety classification of the software, based purely on the degree of personal injury that a software malfunction could potentially cause. This 3 level classification then determines, in advance, the rigour and formality required for the software production process and the subsequent maintenance.

IEC 62304 is a 'harmonised standard', so medical device manufacturers adopting it will be deemed to have satisfied at least some of the relevant essential requirements contained in the **(General) Medical Devices Directive (93/42/EEC)** (GMDD) and its amendment M5 (2007/47/EC) relating to software development. Although its use is voluntary, if IEC 62304 is not applied, the manufacturer has to provide other objective evidence to demonstrate that the software is compliant with the GMDD. IEC 62304 is regarded as current state-of-the-art and is used by the Notified Bodies as a 'frame of reference' for assessment.

By design, IEC 62304 does not cover software validation, usability issues or final release, so other standards or procedures need to be followed to demonstrate compliance with the relevant essential requirements. The standard also goes some way to clarify the situation regarding incorporation of general purpose or **commercial off-the-shelf** (COTS) software, which it collectively refers to as SOUP.

IEC 62304 does describe the interlinked processes that a QMS should include. The QMS should document these processes:

- Software Planning
- Requirements analysis
- Architectural and detailed design
- Unit implementation and verification
- Integration and integration testing
- System testing
- Software release
- Maintenance Plan
- Risk management
- Change Control
- Post market surveillance

IEC 62304 is being used in a few UK medical physics departments as the basis for a quality standard to support the **manufacture of in-house medical equipment** (MIME), but is yet to find use in Radiotherapy or Nuclear Medicine departments.

Note: A general revision (Edition 2) of IEC 62304 (containing a new section on legacy software) was in preparation and was expected to be published in 2020. However, the draft was rejected in May 2021 and the project closed meaning that IEC 62304:2006/A1:2015 remains valid for some years to come.

16.4.6.4   ISO 13485: 2016 Medical Devices – Quality Management Systems – Requirements for Regulatory Purposes

The main purpose of this standard is to facilitate harmonised medical device regulatory requirements for quality management systems. As a result, it includes some particular requirements for medical devices and excludes some general requirements of ISO 9001 that are not essential for regulatory compliance. ISO 13485 has a stated relationship to IEC 62304 (in the latter's appendix) and also makes reference to ISO 14971. It has recently been used within a Medical Physics Department as a basis for standards-driven software development in physiological measurement, with particular emphasis on verification and validation.

16.4.7   CODING STANDARDS

Quality management standards implicitly assume that adoption of a well-recognised development process leads to the production of usable, safe and reliable products. As a result, the over dependence on process standards is often questioned. Although overall design is clearly important, scientists and engineers are naturally drawn toward technical (e.g. coding) standards and product-specific 'drop testing' techniques.

The key is to have a departmental policy for in-house software development, which forces consideration of the key issues, so it is interesting to note that in a recent survey of Canadian Radiotherapy Physicists less than 20% of departments had a written policy for production of in-house software (Salomons and Kelly 2015).

Coding standards documents tend to be programming language specific, but the same basic principles apply. Clearly, well written code will make future maintenance a simpler task, and the author should always assume that someone else will be fulfilling this role. Although spreadsheets do not necessarily involve coding per se, there is a well-established need to follow design guidelines when developing them.

16.4.8 STANDARDS AND GUIDELINES ISSUED BY PROFESSIONAL BODIES

Some professional bodies have issued guidelines and standards that are relevant to in house software development in areas of medical physics. For example, the **British Nuclear Medicine Society** (BNMS) has published scientific and technical standards to accompany its voluntary organisation audit scheme. These contain a requirement that there should be a policy regarding software quality assurance, and that records of software quality assurance should be available for commercial software and in-house developed software. To date, there is no similar guidance for Radiotherapy, an area that could have a severe impact on patients if software or spreadsheet use is not properly controlled.

Previous **Institute of Physics and Engineering in Medicine** (IPEM) reports and guidelines have attempted to address in house development issues in nuclear medicine and Radiotherapy, but the take up of recommendations has been patchy. Useful guidelines on software validation and spreadsheet development have also been published by the UK **National Physical Laboratory** (NPL).

Of particular interest is the recent IPEM document "Best-practice guidance for the in-house manufacture of medical devices and non-medical devices, including software in both cases, for use within the same health institution" (Best-practice guidance for the in-house manufacture of medical devices and non-medical devices, including software in both cases, for use within the same health institution 2021 [online]) which summarises the post-Brexit legislative landscape and provides some advice.

16.5   MARKET

The final area we will consider is that of taking a development to market.

The proposed changes to the MDD/MDR have muddied the definition of "market" somewhat and some proposals even went as far as including a patient-specific prescribed device as being "placed on the market". The attendant problems this would bring to areas such as Rehabilitation Engineering led to calls for this proposal to be re-defined, which it was. If not, every single device would have needed to be CE marked (see the note on HIE in section 16.4.3).

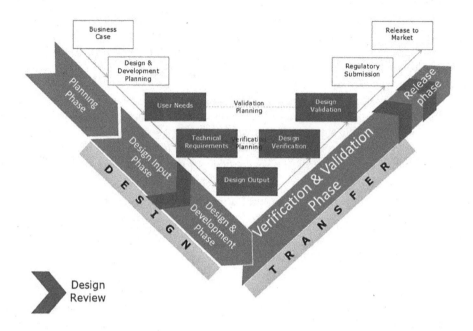

FIGURE 16.3 Design controls as required by the regulations.

All medical devices that are "placed on the market" (however that is defined) need to be CE marked (as discussed earlier). At the heart of the accreditation process is a quality system, which defines the controls placed upon the design process.

The benefit of many of the controls can only really be seen later in the process, as Figure 16.3 suitably illustrates. Note the two phases of the project: design and transfer. Note also the similarity with the V model of software development seen in "Software Engineering" Figure 15.7.

NOTES

1 The sections Introduction, MDD, MDR, CE marking, other standards, documentation and version control are reproduced from Ganney et al. 2022.

2 It existed in 2020 but is no more. The definitions are oft-quoted, with one source attributing them to NASA (although they're not on NASA's website either). They're good definitions though.

3 Functionality, Usability, Reliability, Performance, Supportability. The "+" in the model refers to additional categories that are often thought of as constraints (such as design, implementation, interface and physical).

4 Left-click to select, right-click for a context menu, commonality of icons so "save" always looks the same, etc.

5 An IT director once told me that his aim for his support staff was that they fixed more problems than they caused. Sadly, this was aspirational at the time.

6 There are those that dislike Hungarian notation, feeling that it can obscure code and is only alerting you to errors that the compiler should pick up anyway. However, in non-declarative languages such as Python, the compiler will not provide this assistance and so

Hungarian Notation is to be preferred. See https://www.joelonsoftware.com/2005/05/11/ making-wrong-code-look-wrong/ for an interesting discussion.

7 The edit that should have appeared in the previous code examples.

8 Torvalds said: "I'm an egotistical bastard, and I name all my projects after myself. First 'Linux', now 'git'." (PC World. 14 July 2012).

9 It is worth remembering that a password list is a simple database.

10 50% of your time is debugging.

11 Although there are variants on the syntax, so this also needs to be taken into consideration.

12 Those that cannot will require tests to be specifically written for them.

13 Controls which are an image in a specific part of the screen.

14 The introduction and section on IEC601 are unique to this book.

15 Of which 1, 5 and 7 are discontinued.

16 Clinical Risk Management: Its Application in the Manufacture of Health IT Systems.

17 Clinical Risk Management: Its Application in the Deployment and Use of Health IT Systems.

18 Although the year of adoption is not the same as the year of implementation.

19 Indeed, an individual algorithm may be a medical device.

20 Put simply, Class I is low risk, IIa low-to-medium, IIb medium-to-high and III high risk.

21 There are some flowcharts to assist with classification on the cited website.

22 The Implementation date for Medical Devices was postponed to May 26, 2021, due to COVID-19.

23 Although the first 15 pages are known as Recitals; so Recital (19) is relevant to software and Recital (30) is relevant to in-house manufacture. They 'set the scene', but are not requirements. In Standards language, they are "informative" and not "normative".

24 Possibly not all, but certainly guidance on authorized representation, clinical evaluation, vigilance, and post-market clinical follow-up.

25 Previously references to studies done with other devices.

26 Articles are definitions – Annexes describe application.

27 Such as contact lenses, substances or items created for facial or dermatological use by means of injection or another type of introduction (excluding tattoos) and high intensity electromagnetic radiation equipment for use on the body, as used in depilation and tattoo removal treatments. The Regulation considers that these types of products with an aesthetic rather than a medical purpose may be similar to medical devices in terms of functionality and risk profile, and as such must be considered medical devices.

28 The physical label on the device or package.

29 As per ISO EN 14971:2012 – Risk classification is in Annex 8.

30 There is now only one definition of a medical device whereas previously there were three.

31 Prediction and prognosis are significant additions from the MDD's definition.

32 Previously passive – this may thus increase the risk rating of existing software.

33 As it means that "Lifestyle" software (activity trackers etc.) are not classified as medical devices.

34 It is unclear how this will be controlled.

35 Not defined, but clarification has been issued that the NHS is not a legal entity so responsibility falls to individual Trusts.

36 There is no definition as to what might be appropriate.

37 This may be because an appropriate level of performance cannot be met.

38 What constitutes "equivalent" is ambiguous as it is based on the patient need, which is not very well defined. Effectively this clause insists that a device be CE marked if an equivalent CE-marked device already exists.

39 A local website may be sufficient.

40 Not defined.

41 Meaning, of course, that all other in house manufacture devices must be controlled, usually by a QMS.

42 For an excellent discussion of this see "MDR – the Health Institution Exemption and MHRA draft guidance" by Justin McCarthy in Scope 27:3.

43 "*at least one*" although small enterprises need not have such an employee but must "*have such person permanently and continuously at their disposal*".

44 The latest guidance is at https://www.gov.uk/guidance/regulating-medical-devices-in-the-uk.

45 UK Conformity Assessed.

46 Imported devices will have been manufactured to them and exporters will need to comply with them.

47 The **Independent Medicines and Medical Devices Safety** (IMMDS) Review was commissioned in February 2018 and the report, "First do no harm" was published in July 2020.

48 "CE" is an abbreviation of the French phrase "Conformité Européene" ("European Conformity"). Whilst the original term was "EC Mark", it was officially replaced by "CE Marking" in the Directive 93/68/EEC in 1993 which is now used in all EU official documents.

49 There does not appear to be a requirement to file this statement with anyone, but including it in the QMS would be very sensible.

50 The previous edition, 80001-1:2010 is now withdrawn.

51 Therefore a connection to the Internet does not render a network a medical IT-network.

52 i.e. process.

53 The nomenclature means that standard is dated 2006, with an amendment (the only one, in this case) in 2015.

54 The MDR is much clearer in this requirement.

REFERENCES

Best-practice guidance for the in-house manufacture of medical devices and non-medical devices, including software in both cases, for use within the same health institution, IPEM 2021 [online]. Available: https://www.ipem.ac.uk/media/31gnvr05/ihmu-best-practice-guidance-final.pdf [Accessed 16/05/22].

Businesses given more time to apply new product safety marking, UK Government 2021, [online]. Available: https://www.gov.uk/government/news/businesses-given-more-time-to-apply-new-product-safety-marking [Accessed 16/05/22].

DejaGnu 2021, [online]. Available: http://www.gnu.org/software/dejagnu/ [Accessed 16/05/22].

EUR-Lex 2007, [online]. Available: https://eur-lex.europa.eu/legal-content/en/ALL/?uri= CELEX%3A32007L0047 [Accessed 19/05/22].

Ganney P, Maw P, White M, ed. Ganney R, 2022, *Modernising Scientific Careers The ICT Competencies*, 7th edition, Tenerife: ESL.

Grady R, Caswell D, 1987, Software Metrics: Establishing a Company-wide Program, Hoboken: Prentice Hall. p. 159.

Loh E and Boumans R, 2017, Understanding Europe's New Medical Devices Regulation, Emergo [online]. Available: https://www.emergogroup.com/resources/articles/ whitepaper-understanding-europes-medical-devices-regulation [Accessed 16/05/22].

MEDDEV 2.1/6: MEDICAL DEVICES: Guidance document – Qualification and Classification of standalone software, European Commission n.d., [online]. Available: http://www. meddev.info/_documents/2_1_6_ol_en.pdf [Accessed 16/05/22].

MHRA 2020, (1) [online]. Available: https://www.gov.uk/guidance/medical-devices-conformity-assessment-and-the-ukca-mark [Accessed 16/05/22].

MHRA 2022, (2) [online]. Available: https://www.gov.uk/government/organisations/medicines-and-healthcare-products-regulatory-agency [Accessed 16/05/22].

Microsoft 2010, [online]. Available: https://msdn.microsoft.com/en-us/library/ee658094.aspx [Accessed 16/05/22].

Official Journal of the European Union L117 Volume 60, 5 May 2017 [online]. Available: http://eur-lex.europa.eu/legal-content/EN/TXT/?uri=OJ:L:2017:117:TOC [Accessed 16/05/22].

Salomons GJ and Kelly D, 2015, A survey of Canadian medical physicists: software quality assurance of in-house software, *Journal of Applied Clinical Medical Physics*, volume 16, number 1, 2015.

Selenium 2022, [online]. Available: http://www.seleniumhq.org/ [Accessed 16/05/22].

The Medical Devices (Amendment etc.) (EU Exit) Regulations 2020, UK Government [online]. Available: https://www.legislation.gov.uk/ukdsi/2020/9780348213805/memorandum/contents [Accessed 16/05/22].

TickIT 2022, [online]. Available: http://www.tickitplus.org/ [Accessed 16/05/22].

UK Government 2021, [online]. Available: https://assets.publishing.service.gov.uk/government/uploads/system/uploads/attachment_data/file/999908/Software_flow_chart_Ed_1-08b-IVD.pdf [Accessed 16/05/22].

Wellkang Tech Consulting 2009, [online]. Available: http://www.ce-marking.org/Guidelines-for-Classification-of-Medical-Devices.html [Accessed 16/05/22].

17 Project Management[1]

17.1 INTRODUCTION

Project Management (PM) is the process and activity of planning, organising, motivating and controlling resources to achieve specific goals. A project is a temporary endeavour designed to produce a unique product, service or result with a defined beginning and end (usually time-constrained and often constrained by funding or deliverables), undertaken to meet unique goals and objectives, typically to bring about beneficial change or added value. The temporary nature of projects stands in contrast with business as usual (or operations), which are repetitive, permanent or semi-permanent functional activities to produce products or services. In practice, the management of these two systems is often quite different, and as such requires the development of distinct technical skills and management strategies.

The primary challenge of PM is to achieve all of the project goals and objectives while honouring the preconceived constraints. The primary constraints are scope, time, quality and budget. The secondary – and more ambitious – challenge is to optimise the allocation of necessary inputs and integrate them to meet pre-defined objectives.

PM allows you to work out WHAT problem you need to solve, whereas software development methods tell you HOW to build software. It is too often assumed that building software is the solution.

Writing a piece of software is not a project. Writing a piece of software to solve a specific problem, putting it into practice and keeping it running; that's a project[2].

Project management also helps to create a "controlled development environment", which (along with quality management) is important for activities that can have risk implications such as implementing novel algorithms, automating activities and situations involving patient safety.

There are many project management methodologies around, but the NHS has focused largely on **PRojects IN Controlled Environments** (PRINCE) 2, which was developed by the UK Government although it is used widely in the private sector. It attempts to ensure:

- An organised and controlled start, i.e. organise and plan things properly before leaping in
- An organised and controlled middle, i.e. when the project has started, make sure it continues to be organised and controlled
- An organised and controlled end, i.e. when you've got what you want and the project has finished, tidy up the loose ends

A PRINCE 2 project progresses through many stages (you may spot a similarity with software lifecycle – especially Waterfall – here) in order to meet its goal. Management of a PRINCE 2-controlled project is done by two bodies: a **Project Board** (PB) and a

DOI: 10.1201/9781003316244-17

Project Team (PT). The PB steers the project and is responsible for the resourcing of it, the PT delivers the project. A **Project Manager** (PM) is responsible for the day-to-day running of the project and sits on both bodies, reporting to the PB and managing the PT. The PM will create regular status reports for the PB, describing progress made, obstacles to be overcome and the use of resources to date (especially finance), often using a simple traffic-light system (or **Red-Amber-Green** (RAG) status) to provide an at-a-glance overview of the various components. Obstacles that cannot be overcome or other variations to the project plan are managed as exceptions and the PM will present an exception report to the PB for approval. A PM would normally be a certified PRINCE 2 Practitioner.

17.2 STARTING OFF

To initiate a project, a **Project Initiation Document** (PID) is required. This defines the scope of the project: the timescales, the purpose, the constraints and the benefits. It may also contain the maintenance methodology and eventual decommissioning plan. It should be written and approved by the project sponsors[3] before work commences. That does not mean that exploratory work cannot be undertaken in advance, including prototyping, but a PID should be in place prior to the detailed work on the project commencing.

A PID provides a useful tool for both management and for those working on the project, clarifying what is needed as well as identifying possible issues before they arise. It needs to identify the requirements of the software, the risks and benefits as well as the staff, skills and time needed. The resulting paperwork may be as little as one page of A4.

Software development in a Medical Physics/Clinical Engineering environment needs to sit within a project framework and, as the problems are usually novel, there may be implications for service activities and there could also be significant safety issues involved. PM should prompt some introspection to decide whether the available in-house skills and resources are adequate to meet the demands of the project. The PID forms the basis of the management decision to actually undertake the project and put resources behind it (as well as the agreement from clinical sponsors that this is indeed what is being asked for). It may be that the project never progresses beyond this stage because defining the project shows it is too risky, too expensive, or just isn't needed. The formal assessment exercise is important as it is obviously better not to start a project than have to abandon it half way through.

Identifying what is needed is not always straightforward. Even if the authors are actively involved with the clinical application (and may actually use the software themselves) their approach may differ to other users and the workflow may look very different from their perspective. For example, automatic radiation dose film analysis software may speed up the process for some, but if it can't be integrated with second checks then other users may find it too difficult to use, the management may not support it or if film stops being used then the software will be a waste of time. Project Definition requires understanding of the clinical application and workflows, users, and even possible future changes. People who have "domain knowledge[4]" are needed to help develop the user requirements. Seeking the opinions of users who will be affected by the new software can significantly reduce the risks and will help to engender a sense of "ownership" encouraging acceptance of the changes to be introduced.

A PID does not contain a full design specification – it is a high-level description of the stages required, of which detailed design is one.

A Risk Analysis should be undertaken at this stage – not of putting the software into practice (that is a later stage) but of simply undertaking the project in the first place. This requires consideration of such issues as:

- What would happen if the software was not written?
- If the software is not written, would other related projects fail?
- Is there something available commercially or in other use elsewhere that can do (most of) the job?

An initial Project Plan can then be devised (which should form part of the PID – it is often on this that the decision whether or not to proceed is made). Estimates of the actual resources are made here:

- How long (approximately), in person-days, will the project take?
- How long will it take to put the infrastructure in place?
- Are any parts of the development dependent on other tasks or resources being put in place?
- What contingencies have been made if things do not go to plan?
- What documentation will be generated throughout the project, who will produce it, when will this be done, and how will it be reviewed (including QMS requirements)?
- When will the expected quality measures (e.g. code reviews) be performed and by which groups of people?
- When will the various levels of testing be performed and by whom?
- When will the system (or its constituent parts) be shown to the user?
- Which other parts of the hospital need to be involved? (MP/CE is no longer an island)

Finally, the criteria that will be used to determine the success of the project are determined and recorded in the PID, such as:

- Safety
- Availability/uptime
- Correctness
- Economy
- Optimality
- Timeliness
- Deliverability
- The expected benefits of the project (realising which may be a complete plan in itself)

17.3   KEEPING IT GOING – MANAGING THE PROJECT

The PID contains an overview Project Plan (see Figure 17.1, Figure 17.2 and Fig 17.3 for examples), often in the form of a GANTT chart[5]. A full plan is now required (for simple projects, the PID may have contained it). This plan outlines the way in which the project will progress, shows dependencies when multiple elements are being progressed simultaneously and gives the times at which each stage is expected to complete. It is worth noting the constraints being recorded in the GANTT chart in Figure 17.3, as these can greatly assist scheduling.

A PB consists usually of senior figures, as the PB controls the allocation of resources: it is their role to ensure the release of these, especially clinical staff time. Three major roles on the PB are therefore:

- The customer – the person who the project is for (who generally chairs the PB).
- The senior user – not necessarily someone who will use the completed project[7], but represents staff who will and is able to control this resource (e.g. staff time for training).
- The senior technical officer – for the implementation of a commercial product, this may be someone from the supplier side. Equally it may be a shared role, if it requires technical resources from the organisation.

PRINCE 2 can seem very pedantic and focused more on paperwork than problem solving. However, it is undoubtedly a useful tool when undertaking a complex, multi-stranded project. Simpler projects are probably better suited to a simpler method, although some elements[8] are useful in all projects.

An often overlooked but very useful feature of project management is the sign-off. A PRINCE 2 project cannot progress from one stage to the next unless the PB completes the sign-off (agree as completed). In the TickIT[9] methodology, the role of sign-off is given purely to the customer, who in small-scale developments is the specifier, funder and end-user.

Experience shows that two of the main benefits that project management techniques can provide are:

- Sign-off: this is evidence that what has been produced is what was specified.
- The creation of a benefits expectation plan, which is vital for testing.

17.4   STOPPING (THE HARD BIT)

At the end of a project, a "lessons learned" document is produced, detailing those aspects that, with hindsight, could have been improved. This is used to inform future projects and their management.

The benefits outlined in the PID are re-examined at the end of the project to produce a Benefits Realisation Report and it is by this that the success (or otherwise) of the project is judged. If all has been successful, then the project's temporary existence is over and the software is passed into BAU for ongoing support.

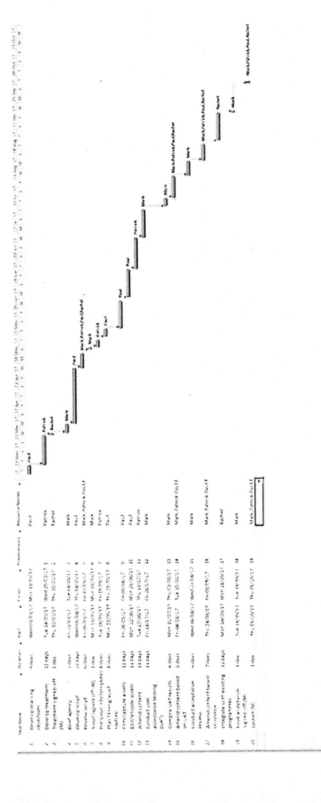

FIGURE 17.1 An example project plan, from MS Project.

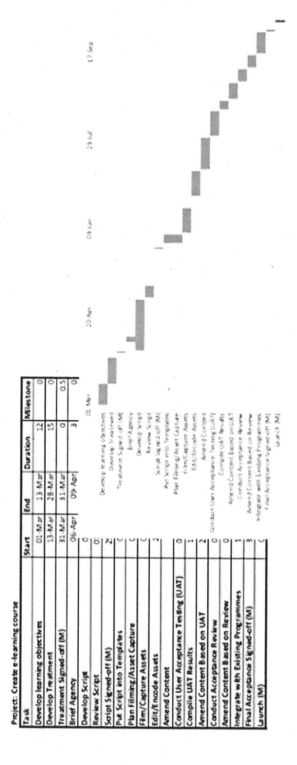

FIGURE 17.2   A simpler GANTT chart created in Excel.[6]

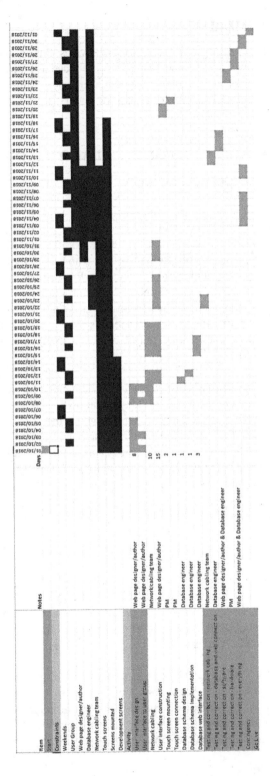

FIGURE 17.3 A GANTT chart showing constraints.

17.5   RISK MANAGEMENT

Risk management is central to developing medical devices. It is a requirement of the MDD that benefits outweigh risks and that risks are reduced as far as possible.

A documented, structured process for risk management will improve the design, helping to identify potential failures early and to understand patient risks, and is covered in detail in chapter 18 on Safety Cases.

17.6   TEAM MANAGEMENT (PERSONNEL AND TECHNICAL)

You only have to have a cursory working knowledge of "The Apprentice" to know how getting the wrong people in the wrong jobs backfires spectacularly. There is far more to getting a good team together and making them work efficiently as a team than coming up with a great team name.

Merely assigning people to tasks doesn't create a project team. Any good project manager knows that a team is a collection of people who are committed to common goals and who depend on one another to do their jobs. Project teams consist of members who can and must make a valuable and unique contribution to the project.

As soon as project team members are identified (and the PM may not have any choice in this), steps to define and establish the team's identity as well as its operating practices need to be taken. The following elements, if the team understands and accepts them, will do this:

- Goals: What the team as a whole and members individually hope to accomplish
- Roles: Each member's areas of specialty, position on the team, authority, and assignments (it is important to recognise who is the expert in a particular field and that all accept that person will have the final say in that area. Likewise, establishing the areas where no-one has expert opinion and therefore debate is going to be the only way a solution will be found is key)
- Processes: The techniques that team members will use to perform their project tasks
- Relationships: The attitudes and behaviours of team members toward one another

As soon as people join the team, they should review the approved project plan (as this will most likely be in place prior to the team being formed – the plan determines the team, not the other way round) to reinforce the project's goals, clarify the work planned, confirm the feasibility of time and resource estimates, and identify any potential problems. Meeting as a group to discuss people's thoughts and reactions, after they've reviewed the plan, is key:

- Team members who contributed to the proposal can remind themselves of the project's background and purpose, their planned roles, and the work to be done. They can also identify situations and circumstances that may have changed since the proposal was prepared and then review and reassess project risks and risk-management plans.

- New team members can understand the project's background and purpose, find out about their roles and assignments, raise concerns about time frames and budgets, and identify issues that may affect the project's success.

Team members commit to the project when they believe their participation can help them achieve worthwhile professional and personal goals. Team members can be helped to develop and buy into a shared sense of the project goals by doing the following:

- Discuss the reasons for the project, its supporters, and the impact of its results.
- Clarify how the results may benefit your organisation's clients. (If it doesn't then why are you doing it?)
- Emphasise how the results may support your organisation's growth and viability.
- Explore how the results may impact each team member's job.

Most team members for any reasonably sized project will have been seconded to the role. It's therefore useful to encourage people to think about how their participation may help them achieve personal goals, such as acquiring new skills and knowledge, meeting new people, increasing their visibility in the organisation, and enhancing their opportunities for job advancement (i.e. why they volunteered to do it). Obviously, projects aren't only about helping team members achieve personal benefits. However, when team members can realise personal benefits while performing valued services for the organisation, the members' motivation and commitment to project success will be greater.

17.7 PROJECT PLANNING (RESOURCE AND TECHNICAL)

Estimating the required resource for a project is not a simple task. Whilst one person may not be able to get the network diagram designed in a day, assigning two people to the task will not necessarily mean it can be done any quicker as so much of design is thinking. However, the time taken to test cable runs can be greatly shortened by placing one person at each end. And (as hinted at in the introduction) what happens when you are lacking a skill? Do you buy it in, train someone up, or just hope that Wikipedia is accurate[10]?

In his seminal work "The Mythical Man-Month", Fred Brooks asserts that:

- More programming projects have gone awry for lack of calendar time than all other causes combined.
- All programmers are optimists: "All will go well"
- The man-month[11] is a fallacious and dangerous myth, for it implies that men and months are interchangeable.
- Brooks's Law: Adding manpower to a late software project makes it later. (Brooks 1995)

An important part of resource planning is availability: therefore each resource in a GANTT chart should have his or her own timeline. In some project planning software this is entered into a database and is hidden from the main view, but it is there regardless. Thus annual leave can be accounted for and times when no leave can be granted can be planned and communicated in advance of project inception. Other resources that may require such planning are room availability (e.g. running cables during a planned refurbishment or ensuring work is done prior to a planned deep clean).

The GANTT chart will also indicate the required level of resource required: if a task is on the critical path[12] (which we'll look at in Chapter 19) then additional resource may be useful – but note the examples given earlier regarding thinking time.

The precursor to the GANTT chart is worth noting here: the project activity list. This lists all the activities that must be done in order to complete the project, together with its dependencies (resources, other activities that must be completed beforehand etc.)

A good resource plan consists of a schedule that is as detailed as possible for the information known, and the types of resources needed for each task. A good resource plan will have a single task owner on each task.

Two columns in a project activity list are 'duration' and 'resource type'. Duration refers to the timeframe in which the task will be performed. Resource type is the skill set required to accomplish the task. In order to assign tasks to individuals, it is necessary to know both of these factors. Before assigning individuals to tasks, it is recommended to associate a task with a resource type. Then the expected duration of that task based on the resource chosen can be entered. This provides the ability to analyse a project schedule, assuming that there are no resource constraints on an individual's availability.

Duration is the expected timeframe needed to complete the task while taking into consideration the skill level and general availability of the resource. Duration should account for reality. If the activity 'identification of users to test the system' is expected to take two weeks, but historically, employees are only available 70% of the time due to general meetings, public holidays, leave, etc., then planning for a duration of three weeks would be more reasonable.

The first step is to produce a detailed list of all the individual resources needed to complete the project. Firstly each of the major resource groups (e.g. labour, equipment and materials) are listed, followed by individual components of each group.

- Labour: identify all the roles responsible for or involved with the completion of any activity specified in the Project Plan. Remember to include any external or contract staff that will be brought in for specific tasks.
- Equipment: identify all the equipment which will be needed to complete the project, e.g.: office equipment (PCs, photocopiers, mobile phones[13] etc.), telecommunications equipment (cabling, switches etc.) and machinery (heavy and light machinery).
- Materials: consumables (e.g.: photocopier paper, stationery, ink cartridges) are often needed to complete project activities. Other materials (e.g. wood, steel and concrete) may be needed to produce physical deliverables. Draw up a detailed list of all the materials required to complete the project. This should be as accurate as possible, since it will be used to produce the Resource Schedule[14] and Expense Schedule.

17.7.1 Quantifying the Resource Requirements: Labour

Table 17.1 can be used to list all the roles required to undertake the project. The number of people required to fulfil each role can then be identified and the responsibilities and skills needed to undertake each role successfully described. Additionally the timeframe during which the role will exist can be specified.

Equipment: Table 17.2 can be used to list each item of equipment required to complete the project. The amount of each item needed can be quantified and the purpose and specifications of each item described. Additionally, timeframes for which the equipment will be required can be specified.

Material: Table 17.3 can be used to list each item of material required to complete the project. The amount of each item needed can be quantified and the timeframe during which the materials will be required can be specified.

TABLE 17.1
Resource Requirements: Labour

Role	Number	Responsibilities	Skills	Start Date	End Date
List each project role.	Identify the number of people required for each role.	Summarise the responsibilities for each role.	Summarise the skills required to fulfil each role.	Enter the start date for the role.	Enter the end date for the role.

TABLE 17.2
Resource Requirements: Equipment

Item	Amount	Purpose	Specification	Start Date	End Date
List each item of equipment required.	Identify the amount of each item of equipment required.	Describe the purpose of each item.	Describe the specifications of each item.	List the date by which the equipment is needed.	List the date when the equipment can be released.

TABLE 17.3
Resource Requirements: Material

Item	Amount	Start Date	End Date
List each item of material.	Quantify the amount of each item of material needed.	List the date by which the material item is needed.	List the date upon which the use of the material ends.

TABLE 17.4
A Resource Schedule

Resource	Jan	Feb	Mar	Apr	May	Jun	Jul	Aug	Sep	Oct	Nov	Dec	Total
Labour													
Labour Type													
Equipment													
Equipment Type													
Materials													
Material Type													

17.7.2 Constructing a Resource Schedule

All the information required to build a detailed resource schedule has now been collected. The next step is to list these resources with the amount (i.e.: value) of each resource required during the periods it will be needed, in the resource schedule shown in Table 17.4.

A detailed Resource Schedule enables a Project Manager to identify the total quantity of each type of resource needed on a daily, weekly and monthly basis.

Any assumptions made during this resource planning process should be listed, e.g.:

- The scope of the project will not change.
- Identified resources will be available upon request.
- Approved funding will be available upon request.

Any constraints identified during this resource planning process should also be listed, e.g.:

- The project team must create all of the physical deliverables within the allocated budget.
- All work must be undertaken within normal working hours.

17.8 EDUCATION AND TRAINING

There are three aspects to this: team training, user training and expert training. None are as quick as you'd like them to be and therefore require proper planning.

Team training fills gaps in the knowledge of the team that cannot be bought in, either because it isn't available in the required timeframe or is too expensive for the budget. As the team members have been scheduled on the GANTT chart, this is relatively easy to accommodate.

User training should appear on the GANTT chart as a specific activity. There will be two phases of this: initial users who will test the system and accept it; and all end-users. If cascade training is utilised, then the availability of the initial users must also be included in the plan.

Expert training identifies the expertise required after the project has been completed, in order to keep it running.

17.9   COST ESTIMATION

Whilst it is often easy to calculate the cost of delivering a project once it is scheduled[15], that is only one part of the equation and often overlooks the true cost of ownership. A much better method is **Total Cost of Ownership** (TCO). As an example, buying a car

Total Cost of Ownership Calculator

Gray cells are calculated for you. You do not need to enter anything into them.

Cost per year	CRT monitor	LCD monitor	LCD monitor savings
Purchase cost	17.59	33.08	(21.50)
Cost of power	5.33	2.53	2.81
Cost of space	1,184.38	312.54	871.84
Visual task time	$6,260.83	$5,378.63	$882.21
Total	**$7,450.55**	**$5,693.70**	**$1,756.85**

Assumptions		Unit of measurement
Product 1 [CRT monitor]		
Purchase price	150	Dollars
Power consumption	100	Watts per hour
Standby power	8	Watts
Width	16	Inches
Depth	18	Inches
Life expectancy	20000	Hours
Product 2 [LCD monitor]		
Purchase price	500	Dollars
Power consumption	47	Watts per hour
Standby power	5	Watts
Width	19	Inches
Depth	4	Inches
Life expectancy	30000	Hours

Indirect costs		
Cost		Unit of measurement
Normal annual office hours	2,345	Hours per year
Average employment cost	14.71	Dollars per hour
Time spent using computer	33%	Percentage of working week
Time spent visual searching	15%	Percentage of computer use
Speed increase in visual tasks	25%	Percent by which speed is increased
Reading time	40%	Percentage of computer use
Speed increase in reading	10%	Percent by which speed is increased
Cost of electricity	6.63	Cents per 100 KW per hour
Average office cost	592.19	Dollars per sq. ft per year
% left on with power saving	10%	Percentage of computers
% left on without power saving	5%	Percentage of computers

FIGURE 17.4   Excel TCO calculator. (Excel Templates 2022 [online]).

is the easy bit. Keeping this car on the road is the expensive part: no-one would budget to buy a car and not factor into the equation the tax, insurance and MoT[16].

Classic business models built upon the failure of consumers to account for TCO are the razor and blade model, or (more recently) the printer and ink model. Both of these rely on low-profit initial purchases (the razor or printer) and higher-profit consumables (the blade or the ink) which are repeatedly purchased.

Gartner defines TCO as

> a comprehensive assessment of information technology (IT) or other costs across enterprise boundaries over time. For IT, TCO includes hardware and software acquisition, management and support, communications, end-user expenses and the opportunity cost of downtime, training and other productivity losses.

> (Gartner 2022 [online])

17.9.1 Tactical versus Strategic Purchasing Decisions

Strategic purchasing is the process of

> planning, implementing, evaluating, and controlling strategic and operating purchasing decisions for directing all activities of the purchasing function toward opportunities consistent with the firm's capabilities to achieve its long-term goals.

> (Carr and Smeltzer 1997)

Tactical purchasing is typically a subset of activities and processes within the strategic purchasing approach. It focuses on operational purchasing requirements based on information from a limited environmental scan (Lysons and Farrington 2016).

Essentially, strategic actions are long term: they may take a while to design, implement and achieve. Tactical actions are short term. Ideally they should fit within the overall strategy, if there is one.

A simple example would be this: a strategic purchase is one that evaluates every network card on the market and decides which one will best serve the hospital over several years, especially when TCO is taken into account. A tactical purchase says I need one now and runs up to PC World to get it.

NOTES

1 The opening sections of this chapter up to and including "stopping", together with the final section on Cost Estimation are reproduced from Ganney et al. 2022.
2 Actually it's not as "keeping it running" should be **Business As Usual** (BAU) but for the purposes of this chapter, we'll pretend it is.
3 Or PB if PRINCE 2 is being used as the controlling methodology.
4 Experts in the way things work and could work.
5 A GANTT chart is a form of bar chart, devised by Henry Gantt in the 1910s, illustrating a project schedule. The bars describe activities: the horizontal position and length are times/durations and linkages show dependencies.

6 The basic instructions can be found at https://www.ablebits.com/office-addins-blog/make-gantt-chart-excel/

7 It is generally recognised that it is best if this person will not use the software, but is not always possible.

8 Such as benefits realisation, resource control and breaking down into stages – including the GANTT.

9 A DTI-developed project management and certification methodology especially for software.

10 It's safest to assume it's not.

11 Originally published in 1975, so the language is dated (and sexist) but the point is still valid.

12 The path (and thereby the activities) that must complete on time for the entire project to complete on time – any delay in one activity on this critical path will give an attendant delay in the project completion.

13 Some would say that mobile phones are telecoms – it doesn't matter as long as all equipment is considered.

14 A resource schedule is sometimes called a resource matrix, due to the grid layout.

15 You know how many days it will take and how much those people cost, for example.

16 Actually, a lot do. But they shouldn't.

REFERENCES

Brooks Jr F P, 1995, *The Mythical Man-Month, 20th* Anniversary edition, Boston: Addison-Wesley.

Carr A S, Smeltzer L R, 1997, An empirically based operational definition of strategic purchasing, *European Journal of Purchasing and Supply Management* 3 (4), 199–207.

Excel Templates 2022, [online]. Available: http://exceltemplates.net/calculator/total-cost-of-ownership-calculator/ [Accessed 19/05/22].

Ganney P, Maw P, White M, ed. Ganney R, 2022, *Modernising Scientific Careers The ICT Competencies*, 7th edition, Tenerife: ESL.

Gartner 2022, [online]. Available: http://www.gartner.com/it-glossary/total-cost-of-ownership-tco/ [Accessed 19/05/22].

Lysons K and Farrington B, 2016, *Procurement and Supply Chain Management*, London: Pearson Education.

18 Safety Cases[1]

18.1 INTRODUCTION

Risk analysis has many associated techniques. Many of these are reactive (in that they produce analysis after events) whereas the safety case is a proactive technique that aims to quantify the safety of a product (and in the context of this book, software) prior to installation and often prior to inception of the project.

18.2 THE PURPOSE OF A SAFETY CASE

A clinical safety case report is a version controlled document, produced to support a specific product/system/activity/project phase/gate. The report presents a summary of the key components of the safety case relating to the safety of the product up to this point of its lifecycle with references to supporting information. It forms part of the risk management records and may be made available to an assessor or regulatory authority. It can be produced at any point in the software lifecycle prior to release, although "design" or "development" are the best options for initial construction, with "testing" providing the evidence for the case.

The safety case report is a structured argument, supported by a body of evidence, intended to provide a compelling, comprehensible and valid case that a system or part of a system is acceptably safe in the given context.

This report is a means of communication with all stakeholders including suppliers, customers, end users, top management and regulators, as appropriate. The content is a summary of relevant knowledge which has been acquired during development or use and which relates to the safety of the product and its use. The report identifies the methods and techniques employed to derive that knowledge and the criteria employed to justify the acceptance of the residual risks.

Prior to safety cases being introduced to the healthcare sector, the method of determining if a system was safe was through the measurement of harm via falls data, never events, etc. This relied on outcome measures with little understanding of how the outcome came about, as well as being reactive rather than proactive due to the use of historical data only. A safety case, however, is a proactive technique as the risks are identified and assessed prior to the system being implemented. It is a risk-based argument containing the supporting evidence of the system operating at a specified safety level (Maxwell and Marciano 2018).

It is not unusual for a safety case to be constructed while the project is still in progress. In fact, the safety case report can be an important piece of evidence in getting approval to "go live".

A hazard workshop therefore needs to identify hazards that may appear (e.g. "lack of time to complete login module causes security breach") and mitigations that have not yet been applied (this is especially true of acceptance testing) as well as ones already in place.

DOI: 10.1201/9781003316244-18

A safety case report is a living document that gets updated as the project progresses; therefore, it can point to testing that is planned to be done, and, once done, update the safety case appropriately (e.g. "testing demonstrated no errors in dose calculation, but did reveal a potential security flaw in the login procedure").

18.3   THE STRUCTURE OF A SAFETY CASE

18.3.1   CLAIMS

A safety case is essentially a claim that the product is safe for its intended use. It is a structured argument in support of this claim. Elements that a safety case should cover are[2]:

- reliability and availability
- security
- functional correctness
- time response
- maintainability and modifiability
- usability
- fail-safety
- accuracy

18.3.2   EVIDENCE

A major part of the evidence to support the argument resides in the Hazard Analysis, (see Section 18.8, Chapter 18), which includes risk assessment and records of risk reductions planned and achieved. It may take the form of a **Goal Structuring Notation** (GSN) diagram which summarises the argument (see Section 18.3.4, Chapter 18). Elements that may be used as evidence include:

- design
- development documentation
- simulation experience
- previous field experience

18.3.3   ARGUMENT

The entirety of the safety case is a structured argument to demonstrate that the item in question is safe for use as described. It effectively consists of justifications for each element of the case, i.e. "this is true because…" and may consist of three types of argument:

- deterministic
- probabilistic
- qualitative

A deterministic system is a system in which no randomness is involved in the development of future states of the system. A deterministic model/argument will thus always produce the same output from a given starting condition or initial state.

The behaviour of a probabilistic system cannot be predicted exactly, but the probability of certain behaviours is known. This will involve the use of statistical tests and confidence intervals to make the argument.

Qualitative arguments deal with descriptions and data that can be observed but not measured, e.g. colours, textures, smells, tastes, appearance, beauty, etc.

18.3.4 Inference

One way of writing a safety case and one which graphically describes the inferences within the report is to use a system known as GSN. GSN is a graphical notation for presenting the structure of engineering arguments. The approach may be used to present any situation where one wishes to make a claim and where the support for that claim will be based upon evidence and argument. This would include situations such as legal cases, but here we are most interested in the risk-based safety case.

GSN has gained some acceptance within the functional safety discipline as one of the primary means of depicting how particular safety claims or goals have been made and justified through citing or arguing the deductions from particular evidence. There are other tools and techniques, such as the **Claims Argument Evidence** (CAE), the

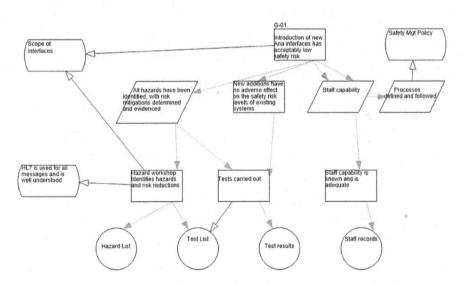

FIGURE 18.1 A GSN diagram from my First Safety Case. Note that I failed to reference anything besides the primary goal.

Adelard Safety Case Editor (ASCE) and even mind mapping diagrams that may be used in a similar manner to GSN to present engineering arguments.

A GSN diagram, like any other form of diagram, merely assists with the demonstration or clarification of how the set of evidence items may be combined together and argued to demonstrate the top claim. Such visual approaches:

a) Provide a greater chance of identifying gaps in the evidence.
b) Provide a clear summary to all interested parties of how the case has been constructed to meet the top level Goal.
c) Can be converted into a web-based format with the use of hyperlinks between all elements of evidence, context and justification as appropriate.

The term "goal structuring" is used to describe the process of linking the elements of the GSN together in a network. The resulting structure provides a diagrammatic method that helps to clarify how the goal is broken down into sub-goals and then these sub-goals are further broken down until the point where sub-goals can be linked directly to supporting evidence. This method facilitates an understanding of the argument strategies adopted, including where the argument has adopted the use of quantified and qualitative approaches.

18.3.5 THE GSN DIAGRAM

The GSN diagram helps to show how the set of evidence items may be combined together and argued to demonstrate the top claim (e.g. that the system is reliable enough to operate in a particular operating environment). In order to do this, the GSN shows how the goal and sub-goals are supported by specific claims and how these claims either lead to lower level sub-goals or ultimately to how the claims are supported by evidence.

Elements within the GSN diagram might be as shown in Figure 18.2. Note that the amount of text within the diagram is limited, but as each item is referenced, additional text may appear in the hazard log.

The structured argument is built top down, decomposing the top-level goal into lower level, more detailed goals (much like we saw with problem decomposition – see Section 15.6.1, Chapter 15) Good practice is to use a simple dome-shaped diagram to provide an overview of the overall argument and required evidence. The diagram is dome shaped as it builds out and down from the top-level goal and broadens as each successive layer is developed. The nodes can be colour coded to show the status of the evidence and ownership of the evidence defined within the evidence or solution element or at a higher level, e.g. sub-goals. The particular notation adopted and supporting evidence available will depend on the specific project or programme.

Looking again at the GSN presented in Figure 18.1, it can be read as "each statement is true because..." and following the arrows down.

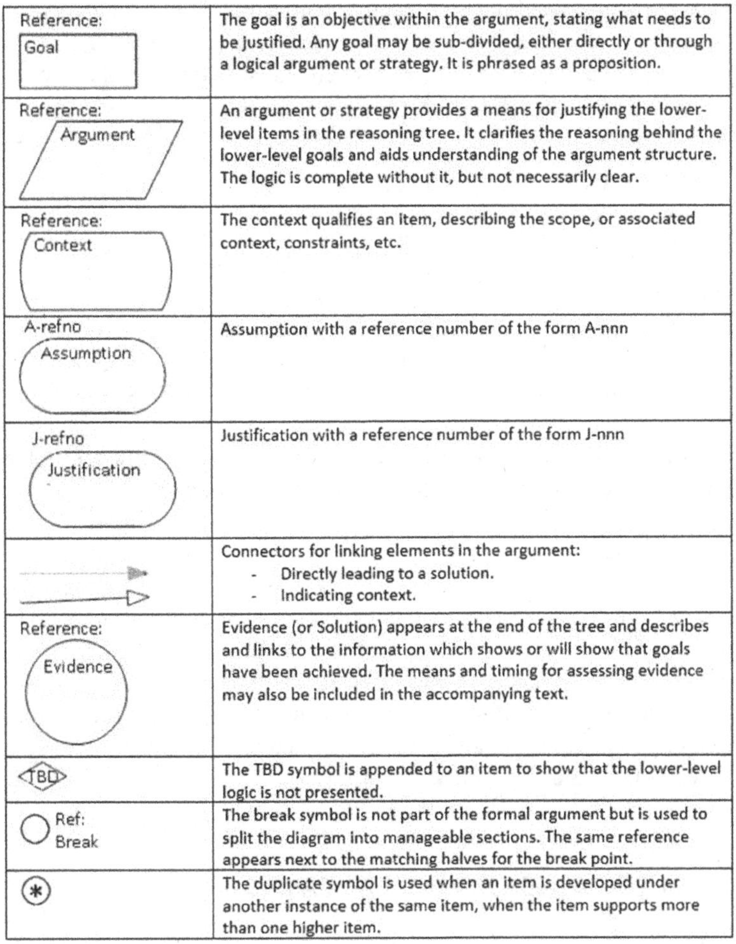

Reference: Goal	The goal is an objective within the argument, stating what needs to be justified. Any goal may be sub-divided, either directly or through a logical argument or strategy. It is phrased as a proposition.
Reference: Argument	An argument or strategy provides a means for justifying the lower-level items in the reasoning tree. It clarifies the reasoning behind the lower-level goals and aids understanding of the argument structure. The logic is complete without it, but not necessarily clear.
Reference: Context	The context qualifies an item, describing the scope, or associated context, constraints, etc.
A-refno Assumption	Assumption with a reference number of the form A-nnn
J-refno Justification	Justification with a reference number of the form J-nnn
(connectors)	Connectors for linking elements in the argument: - Directly leading to a solution. - Indicating context.
Reference: Evidence	Evidence (or Solution) appears at the end of the tree and describes and links to the information which shows or will show that goals have been achieved. The means and timing for assessing evidence may also be included in the accompanying text.
TBD	The TBD symbol is appended to an item to show that the lower-level logic is not presented.
Ref: Break	The break symbol is not part of the formal argument but is used to split the diagram into manageable sections. The same reference appears next to the matching halves for the break point.
(*)	The duplicate symbol is used when an item is developed under another instance of the same item, when the item supports more than one higher item.

FIGURE 18.2 GSN notation.

18.4 IMPLEMENTATION OF A SAFETY CASE

Safety management is integrated into project processes and activities, with the allocation of roles and responsibilities within those processes:

- Project planning and hazard analysis
- Suppliers' contribution to safety management
- Development and test activities, design control and risk mitigation
- Release management and safety justification
- Problem and change management
- Safety review/audit

Hazard analyses are carried out for system components as they are developed, then risk levels are allocated. Next, risk reduction measures are identified. As part of the release process, the achievement of risk reductions is reviewed and agreed as a criterion for go-live.

18.5   DESIGN FOR ASSESSMENT

The key point of a safety case is evidence: if you cannot provide that evidence, then you cannot use that particular argument. It's similar to research – if you can't show why a statement is true, then you cannot include it.

It is therefore imperative that the project lifecycle outputs documentation which forms evidence for the safety case. Did a colleague test it? Get an email with the results and get it signed. Was the design on a whiteboard? Take a photo. In the case of a safety argument, too much is never enough. The worst thing[3] that can happen to your code is that it provides a simple cheap solution to a well-known problem, but no-one can use it because you can't prove that it's safe enough to use.

18.6   THE SAFETY CASE LIFECYCLE

A safety case is not a finished document, but an ongoing one. Unforeseen issues are added to the hazard analysis as they appear and the mitigation that provides resolution (or acceptability) enhances the overall safety argument.

Whenever any aspect of the safety management process for the product or system is reviewed or revisited[4], the results will be reflected in an updated safety case report or in supplementary data (e.g. change records).

The lifecycle may thus be described as shown in Figure 18.3.

18.7   THE CONTENTS OF A SAFETY CASE

A safety case report includes the following:

- Introduction and product identification including:
  - general background
  - unique identification details including version
  - product description with summary of intended operational environments
  - any critical constraints
- Description of general and safety management arrangements
- Overview of hazard and risk assessment processes including risk evaluation and acceptance criteria covering:
  - identification of the conceptual hazard and risk methodology used
  - justification of risk acceptance criteria
  - justification of residual risk criteria
- Identification and justification of any residual risks:
  - list of residual risks that have been identified and any related operational constraints and limitations that are applicable to the ongoing maintenance of the identified residual risk at the stated level

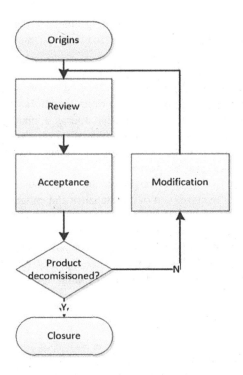

FIGURE 18.3 The safety case lifecycle.

- Overall safety justification:
  - summary of the outcomes of the assessment procedures applied
- Lifetime management arrangements including:
  - performance monitoring arrangements
  - incident/adverse event response arrangements
  - lifetime support arrangements

At the heart of risk analysis and safety cases lies the hazard log, so we will examine this now.

18.8   HAZARD LOG

A hazard log is a record-keeping tool applied for tracking all hazard analysis, risk assessment and risk-reduction activities for the whole-of-life of a safety-related system. Important aspects of its disposition and use are as follows:

- It is the single source of record for all risk management activities. Where necessary it cross-references other documents to demonstrate how hazards have been resolved.

- It is the primary mechanism for providing objective evidence of the risk-management process and assurance of the effective management of hazards and accidents.
- It is continuously updated throughout the project lifecycle (including decommissioning and disposal).
- It is first created as an outcome of the preliminary hazard analysis process.
- It is typically maintained by a systems assurance manager throughout a project's development and assessment phases.
- It is formally closed at the completion of a contracted development and handed over to the system's owner.
- It is preserved into a system's operation and maintenance phases to support functional safety management of maintenance and enhancement activities.

Chambers.com.au notes: "A hazard log is deemed to be closed out when the safety risks associated with all hazards identified have been reduced to a level that is acceptable to the system owner." (Chambers & Associates Pty Ltd 2022 [online]) (NB reduced, not eliminated) Figure 18.4 shows a sample hazard from a radiotherapy treatment system.

18.8.1 The Therac-25 Incident

In the medical physics field, one of the worst accidents in radiotherapy involved the Therac-25 computer-controlled radiotherapy machine, which between June 1985 and January 1987 massively overdosed 6 patients leading to 4 fatalities[5]. Professor Nancy Leveson, a leading American expert in system and software safety, identified the following as causal factors leading to the failure of the software (Leveson and Turner 1993):

- Overconfidence in the software. It was found that the Therac-25 engineers had too much confidence in the software even though almost all responsibility for safety rested on it[6].

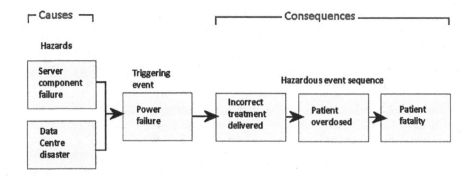

FIGURE 18.4   A sample hazard.

- Reliability and safety are not the same. The software was highly reliable – thousands of patients were successfully treated, but reliability is not the same as safety.
- Lack of defensive design. The software did not contain self-checks, error detection or error-handling which would have prevented dangerous machine states.
- Unrealistic risk assessments. The risk assessments which were conducted were wholly unrealistic and not based on any measurable fact.
- Inadequate software engineering practices. Software specifications and documentation were an afterthought. No quality assurance processes were in place. Software was overly complex when a simple solution would have sufficed. No formal analysis or external testing was performed.

A hazard log is simply a record, normally in tabular form, containing the information shown in Table 18.1[7].

There are many ways of enumerating risk in order to produce the risk rating. For software, a very good one is outlined in Scope 28:2 (Whitbourn et al. 2019) which includes enumeration methods such as "the greater the number of users the greater the chance of the software being used in an inappropriate way".

A hazard log is normally produced during a hazard workshop. This workshop is normally chaired by a suitably qualified and experienced safety expert who has also been tasked with writing (or signing off) the safety case.

A hazard workshop should contain people closely associated with the issue or system being discussed. For a clinical software system, this will include users of the

TABLE 18.1
Hazard Record Structure

Field	Description
Hazard number	A unique identifier so that this hazard can be referred to in the safety case report
Issue/risk description	A short description of the risk or issue. This may reference a longer description if necessary
Cause	The cause of the risk, e.g. poor manufacture, faulty design
Consequences	The consequence of the risk or issue manifesting itself
Risk Rating	A risk score composed of three parts: the Severity (1-5 where 5 is the worst) the Likelihood (1-5 where 5 is the most likely) the Score (the product of the other two numbers)
Mitigations	A description of actions that either reduce the severity (e.g. battery power instead of mains) or likelihood (e.g. staff training)
Post-mitigation Risk Rating	A risk score composed of three parts: the Severity (1-5 where 5 is the worst) the Likelihood (1-5 where 5 is the most likely) the Score (the product of the other two numbers)
Evidence	A description of the evidence that the mitigation has been successfully applied
GSN reference	A unique reference number to an element in the GSN diagram (i.e. it is unique to the GSN – it may occur repeatedly in this table)

system as well as people from the clinical environment in which it is to be used (who may, of course, be the same group).

The main questions to address are:

- What might go wrong? (NB "might" not "has" – the workshop should think widely)
- How likely is this to happen?
- What are the effects if it does?
- What can be done to reduce the likelihood and/or effect?
- What is the likelihood after this mitigation?
- What are the effects after this mitigation?

From these answers the hazard log and then safety case can be constructed.

NOTES

1   The sections on purpose and hazard logs are reproduced from Ganney et al. 2022.
2   These are analogous to the software quality attributes from Section 16.1.1, Chapter 16.
3   Maybe not the worst, but certainly one of them.
4   For example, on the release of an update or patch to the product, or when the product usage or interfaces are extended.
5   For a good description of the error (and plenty of others) see Parker M, 2019, *Humble Pi: A Comedy of Maths Errors*, London: Allen Lane.
6   While the final responsibility was the engineers', they had more confidence in the software performing this function than they should have – they had incorrectly delegated too much responsibility to it.
7   This hazard record format is fairly typical, for example it is the same structure as used by the UK **Civil Aviation Authority** (CAA), which can itself be downloaded from the Internet as a template document (CAA 2013).

REFERENCES

CAA, 2013 [online]. Available: https://view.officeapps.live.com/op/view.aspx?src=https%3A%2F%2Fwww.caa.co.uk%2Fmedia%2Fvjbbzzk2%2Fhazard-log-template-v3.docx [Accessed 23/05/22].

Chambers & Associates Pty Ltd 2022, [online]. Available: http://www.chambers.com.au/glossary/hazard_log.php [Accessed 23/05/22].

Ganney P, Maw P, White M, ed. Ganney R, 2022, *Modernising Scientific Careers The ICT Competencies*, 7th edition, Tenerife: ESL.

Leveson N and Turner CS, 1993, An Investigation of the Therac-25 Accidents, *IEEE Computer*, Vol. 26, No. 7, July 1993, pp. 18–41. [online]. Available: http://www.cse.msu.edu/~cse470/Public/Handouts/Therac/Therac_1.html [Accessed 23/05/22].

Maxwell and Marciano, 2018, What 'safety cases' mean for healthcare, *Health Service Journal* [online]. Available: https://www.hsj.co.uk/topics/technology-and-innovation/what-safety-cases-mean-for-healthcare/5053463.article [Accessed 23/05/22].

Whitbourn J, Boddy I, Simpson A, Kirby J, Farley R and Bird L, 2019, Winds of change in software regulations, *Scope* volume 28 issue 2.

19 Critical Path Analysis

19.1 INTRODUCTION

Critical Path Analysis (CPA) is an important tool in project scheduling. In a multi-threaded project, where there are many paths through the activities and stages from inception to completion and where each activity is dependent on the completion of one or more preceding activities, CPA determines the path (and thereby the activities) that must complete on time for the entire project to complete on time – any delay in one activity on this critical path will give an attendant delay in the project completion.

CPA may also be known as Network Analysis or PERT (**Program Evaluation and Review Technique**). There are three stages:

Planning
Analysing and Scheduling
Controlling

19.2 PLANNING STAGE

There are four elements to this stage:

Identify the component activities
Determine the logical order (dependence)
Draw the network
Estimate the duration of the activities

Activities (operations consuming time and possibly other resources) are represented by arrows in which the direction is important but the length and angle are not.

Events are defined as states of the project after the completion of all preceding activities but before the start of any succeeding activity. They are represented by circles with four divisions, as shown in Figure 19.1.

By convention, time flows from left to right and lower numbered events feed into higher numbered events ($E_i \rightarrow E_j$, $i < j$), as shown in Figure 19.2.

Dummy activities are used in order to ensure that all activities have unique references. An example is shown in Figure 19.3.

They are also used in cases where, for example, A and B are parallel and both C and D are dependent on A with D dependent on B alone, as can be seen in Figure 19.4.

A network:

FIGURE 19.1　An event in CPA.

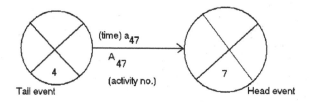

FIGURE 19.2　Two CPA events showing the flow of time with appropriate labelling.

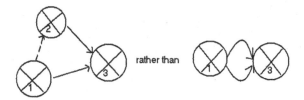

FIGURE 19.3　A dummy event.

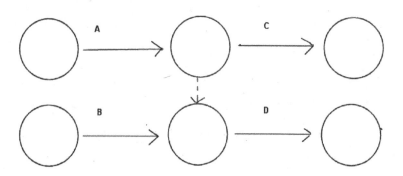

FIGURE 19.4　A dummy event for parallel processes.

- Has a unique first point
- Has a unique end point
- Every point except the last has at least one arrow (activity) starting from it
- Every point except the first has at least one arrow (activity) ending at it
- There are no loops or circuits

19.3 ANALYSIS STAGE

Analysis consists of three steps: a forward pass, a backward pass and a float calculation. From this the critical path can be determined.

19.3.1 THE FORWARD PASS

This determines the earliest start and finish times for each activity. It is the earliest arrival time for all paths to this event, i.e. the maximum – this shows the earliest completion time for the project. An example is shown in Figure 19.5.

The earliest arrival time for E_9 is 28 as this is the sum of the earliest arrival for E_4 and the activity A_{49} and no alternative paths exist.

The earliest arrival time for E_7 is 14 as this is the maximum of (the sum of the earliest arrival time for E_4 and the activity $A_{47} = 14$) and (the sum of the earliest arrival time for E_5 and the activity $A_{57} = 13$).

Thus, starting from time zero as the earliest arrival time for the first point the earliest arrival times for all events can be computed.

19.3.2 THE BACKWARD PASS

This determines the latest start and finish times for each activity which will still allow the project to complete on time. It is the latest start time for all paths to this event, i.e. the minimum – this shows the latest start time for the project. An example is shown in Figure 19.6.

The latest start time for E_5 is 4 as this is the difference between the latest start time for E_7 and the activity A_{57} and no alternative paths exist.

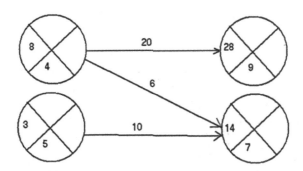

FIGURE 19.5 The forward pass.

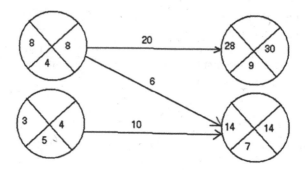

FIGURE 19.6 The backward pass.

The latest start time for E_4 is 8 as this is the minimum of (the difference between the latest start time for E_9 and the activity $A_{49} = 10$) and (the difference between the latest start time for E_7 and the activity $A_{47} = 8$).

The backward pass begins by transferring the earliest arrival time for the final event into its latest start time. The latest start times for all events can be computed. Note that the latest start time for the starting event may not be zero, but usually will be.

19.3.3　FLOAT

A float represents under-utilised resources and/or flexibility, both of which need to be identified for good project management. There are three types of float that may be calculated, of which only the Total Float is of interest here. This is calculated for each activity as

$$\text{Total float} = \text{Latest end} - \text{Earliest start} - \text{Time for path}^{\,1} \qquad (19.1)$$

An example is shown in Figure 19.7.

The total float for A_{49} is 30-8-20 = 2
The total float for A_{47} is 14-8-6 = 0
The total float for A_{57} is 14-3-10 = 1

The Critical Path may now be identified: it is that path from start to end with the minimum (usually zero) total float, in this example A_{47}.

19.4　SCHEDULING

Having analysed the network by drawing it and determining the critical path and total floats, then (if this is satisfactory) a complete schedule of the activity start and finish times can be drawn up, together with dates for ordering materials, etc.

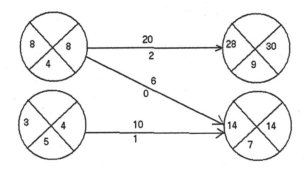

FIGURE 19.7 The total float (shown below the arrow).

19.5 CONTROL STAGE

The schedule is a forecast of activity finish times, etc. which will not be adhered to in practice. Regular reviews are necessary, and CPA may need to be undertaken again during the project's progress.

NOTE

1 i.e. use the outside set of numbers from the events as the difference between the two outer numbers gives the maximum time available for the activity, subtracting the time the activity takes gives the float.

Appendix

LIST OF ABBREVIATIONS

1NF	First Normal Form
2NF	Second Normal Form
3.5NF	Boyce-Codd Normal Form
3NF	Third Normal Form
4NF	Fourth Normal Form
5NF	Fifth Normal Form
ACID	Atomicity, Consistency, Isolation, Durability
AES	Advanced Encryption Standard
AI	Artificial Intelligence
AIM	AOL Instant Messenger
AIMDD	Active Implantable Medical Devices Directive
ANSI	American National Standards Institute
AP	Access Provider
API	Application Programming Interface
AR	Authorised Representative
ASCE	Adelard Safety Case Editor
ASCII	American Standard Code for Information Interchange
ASP	Active Server Pages
BaaS	Backup as a Service
BAU	Business as Usual
BC	Business Continuity
BCNF	Boyce-Codd Normal Form
BLOB	Binary Large OBject
BNMS	British Nuclear Medicine Society
bps	bits per second
CA	Competent Authority
CAE	Claims Argument Evidence
CAPA	Corrective and Preventative Actions
CC	Change Control
CC	Concurrency Control
CCS	Crown Commercial Service
CDA	Clinical Document Architecture
CDS	Clinical Decision Support
CE	Conformité Européene
CGI	Common Gateway Interface
CIDR	Classless Inter-Domain Routing
CIS	Clinical Information Systems
CORBA	Common Object Request Broker Architecture
COTS	Commercial Off-The-Shelf

CPA	Critical Path Analysis
CPU	Central Processing Unit
CR/LF	Carriage Return/Line Feed
CRISP-DM	Cross Industry Standard Process for Data Mining
CSMA	Carrier Sense Multiple Access
CSP	Cloud Service Provider
CSS	Cascading Style Sheets
CSV	Comma Separated Variable
CSV	Comma Separated Variable
CT	Computed Tomography
DaaS	Data as a Service
DBA	Database Administrator
DCB	Data Coordination Board
DCL	Data Control Language
DDL	Data Definition Language
DFD	Data Flow Diagram
DHCP	Dynamic Host Configuration Protocol
DICOM	Digital Imaging and Communications in Medicine
DICOM-RT	DICOM for Radiotherapy
DLL	Dynamic Link Library
DLP	Data Loss Prevention
DMBS	Database Management System
DML	Data Manipulation Language
DNF	Disjunctive Normal Form
DNS	Domain Name Service
DOM	Document Object Model
DPA	Data Protection Act
DPC	Data Protection Commissioner
DR	Disaster Recovery
DTD	Document Type Definition
DTI	Diffusion Tensor Imaging
DUAL	Diffusing Update Algorithm
DV	Distance-Vector
EBCDIC	Extended Binary Coded Decimal Interchange Code
EF	Entrance Facilities
EFTA	European Free Trade Association
EHRS	Electronic Health Record System
EIA	Electronic Industries Alliance
EIGRP	Enhanced Interior Gateway Routing Protocol
EPR	Electronic Patient Record
ER	Equipment Room
ER	Essential Requirement
ERD	Entity-Relationship Diagram
EU	European Union
EU GDPR	The European Union General Data Protection Regulation
FC	Fully Connected

FDA	Food and Drug Administration
FDDI	Fibre Distributed Data Interface
FHIR	Fast Healthcare Interoperability Resources
FIFO	First In First Out
FTP	File Transfer Protocol
FURPS	Functionality, Usability, Reliability, Performance, Supportability
GDPR	General Data Protection Regulation
GFS	Grandfather-Father-Son
GIS	Geographic Information Systems
GMDD	(General) Medical Devices Directive
GNU	GNU Not Unix
GSN	Goal Structuring Notation
GSPR	General Safety and Performance Requirements
GUI	Graphical User Interface
HATEOAS	Hypermedia as the Engine of Application State
HDD	Hard Disc Drive
HDF	HL7 Development Framework
HIE	Health Institution Exemption
HL7	Health Language 7
HTML	HyperText Markup Language
HTTP	Hypertext Transfer Protocol
HTTPS	Hypertext Transfer Protocol Secure
IaaS	Infrastructure as a Service
IC	Integrated Circuit
ICMP	Internet Control Message Protocol
ICO	Information Commissioner's Office
IDL	Interface Definition Language
IE	Information Entity
IFU	Instructions for Use
IG	Information Governance
IGP	Interior Gateway Protocol
IHMU	In House Manufacture and Use
IIOP	Internet Inter-ORB Protocol
IMAP	Internet Message Access Protocol
IMMDS	Independent Medicines and Medical Devices Safety
IOD	Information Object Definition
IP	Internet Protocol
IPEM	Institute of Physics and Engineering in Medicine
IR	Infra-Red
IRC	Internet Relay Chat
IS-IS	Intermediate System to Intermediate System
ISP	Internet Service Provider
IT	Information Technology
IVD	In Vitro Diagnostics
JAD	Joint Application Development
JDBC	Java Database Connectivity

JSON	JavaScript Object Notation
JSP	Java Server Pages
KDD	Knowledge Discovery in Databases
KVP	Peak kiloVoltage
LAN	Local Area Network
LF	Line Feed
LIFO	Last In First Out
LPS	Left-Posterior-Superior
LSR	Link State Routing
MAC	Media Access Control
MDA	Main Distribution Area
MDD	Medical Devices Directive
MDDS	Medical Device Data System
MDI	Multiple Document Interface
MDIS	Medical Device Information System
MDR	Medical Devices Regulations
MDX	MultiDimensional eXpressions
MEAN	Mongo, Express, Angular, Node
MES	Medical Electrical System
MHRA	Medicines and Healthcare products Regulatory Authority
MIME	Manufacture of In-house Medical Equipment
MIME	Multipurpose Internet Mail Extensions
MLC	Multi-leaf Collimator
MSCUI	Microsoft Common User Interface
MTU	Maximum Transmission Unit
MU	Monitor Units
MUTOA	Multiuser Telecommunications Outlet Assembly
MVD	Multi-Valued Dependency
NAT	Network Address Translation
NB	Notified Body
NNTP	Network News Transfer Protocol
NoSQL	Non SQL or Not Only SQL
NPL	National Physical Laboratory
NTP	Network Time Protocol
OBS	Output-Based Specification
OCR	Optical Character Recognition
ODBC	Open Database Connectivity
OLAP	Online Analytical Processing
OLTP	Online Transaction Processing
OMG	Object Management Group
ORB	Object Request Broker
OS	Operating System
OSI	Open Systems Interconnection
OSPF	Open Shortest Path First
OWASP	Open Web Application Security Project
PaaS	Platform as a Service

PACS	Picture Archiving and Communications System
PAS	Patient Administration System
PAT	Portable Appliance Test
PB	Project Board
PBX	Private Branch eXchange
PCA	Principal Component Analysis
PERT	Program Evaluation and Review Technique
PGP	Pretty Good Privacy
PID	Project Initiation Document
PIP	Poly Implant Prosthesis
PM	Project Management
PM	Project Manager
POE	Power over Ethernet
POP	Post Office Protocol
PPM	Planned Preventative Maintenance
PRINCE	PRojects IN Controlled Environments
PT	Project Team
PTV	Planning Target Volume
QA	Quality Assurance
QMS	Quality Management System
QoS	Quality-of-service
RAG	Red-Amber-Green
RAID	Redundant Array of Inexpensive/Independent Discs
RAM	Random Access Memory
RDBMS	Relational Database Management System
REST	Representational State Transfer
RFID	Radio Frequency Identification
RIM	Reference Information Model
RIP	Routing Information Protocol
ROM	Read Only Memory
RPC	Remote Procedure Call
RTM	Requirements Traceability Matrix
SaaS	Software as a Service
SAN	Storage Area Network
SCM	Software Configuration Management
SGML	Standard Generalized Markup Language
SIP	Session Initiation Protocol
SMTP	Simple Mail Transfer Protocol
SNMP	Simple Network Management Protocol
SOA	Service Oriented Architecture
SOAP	Simple Object Access Protocol
SONET	Synchronous Optical Network
SOUP	Software of Unknown Provenance
SQA	Software Quality Assurance
SQC	Software Quality Control
SQL	Structured Query Language

SRS	Software Requirements Specification
SSADM	Structured Systems Analysis and Design Method
SSH	Secure Shell
SSI	Server Side Includes
SSL	Secure Socket Layer
SVN	Apache Subversion
TCL	Transaction Control Language
TCO	Total Cost of Ownership
TCP	Transmission Control Protocol
TCP/IP	Transmission Control Protocol/Internet Protocol
TCP/UDP	Transmission Control Protocol/User Datagram Protocol
TE	Telecommunications Enclosure
TIA	Telecommunications Industry Association
TN	Transaction Management
TR	Telecommunications Room
UDDI	Universal Description, Discovery, and Integration
UDP	User Datagram Protocol
UID	Unique Identifier
UKCA	United Kingdom Conformity Assessed
UKNI	United Kingdom Conformity Assessed (Northern Ireland)
UML	Universal Modelling Language
URI	Uniform Resource Identifier
URL	Uniform Resource Locator
UTP	Unshielded Twisted Pair
vCPU	virtual CPU
VLAN	Virtual Local Area Network
VM	Value Multiplicity
VM	Virtual Machine
VOIP	Voice over IP
VR	Value Representation
W3C	World Wide Web Consortium
WAN	Wide Area Network
WAP	Wireless Application Protocol
WIMP	Windows, Icons, Mouse, Pointer although some variants interpret "M" as "menus" and "P" as "pull-down menus"
WSDL	Web Services Description Language
XDS	Cross-enterprise Document Sharing
XHTML	Extensible HyperText Markup Language
XML	Extensible Markup Language
YAML	YAML Ain't Markup Language

Index